Zeshui Xu

Linguistic Decision Making

Theory and Methods

Zeshui Xu

Linguistic Decision Making

Theory and Methods

With 8 figures

Science Press
Beijing

Springer

Author
Zeshui Xu
Institute of Sciences, PLA University of
Science and Technology Nanjing
Room 2609, Building 5, No.2, Biaoying, Yudao Street
Nanjing 210007, Jiangsu, P.R.China
Tel.: +86 25 84483382
E-mail: xuzeshui@263.net

ISBN 978-7-03-033105-2
Science Press Beijing

ISBN 978-3-642-29439-6 e-ISBN 978-3-642-29440-2
Springer Heidelberg New York Dordrecht London

Library of Congress Control Number: 2012935013

Printed on acid-free paper

Springer is part of Springer Science+Business Media (www.springer.com)

Preface

The complexity and uncertainty of objective thing and the fuzziness of human thought result in decision making with linguistic information in the fields of society, economy, medicine, management and military affairs, etc., such as personnel evaluation, military system performance evaluation, online auctions, supply chain management, venture capital and medical diagnostics.

In real-life, there are many situations, such as evaluating university faculty for tenure and promotion and evaluating the performance of different kinds of stocks and bonds, in which the information cannot be assessed precisely in numerical values but may be in linguistic variables. That is, variables whose values are not numbers but words or sentences in a natural or artificial language. For example, when evaluating the "comfort" or "design" of a car, linguistic labels like "good", "fair" and "poor" are usually used, and evaluating a car's speed, linguistic labels like "very fast", "fast" and "slow" can be used. Therefore, how to make a scientific decision with linguistic information is an interesting and important research topic that has been attracting more and more attention in recent years.

To date, a lot of methods have been proposed for dealing with linguistic information. These methods are mainly as follows:

(1) The methods based on extension principle, which make operations on the fuzzy numbers that support the semantics of the linguistic labels.

(2) The methods based on symbols, which make computations on the indexes of the linguistic terms.

Both the above methods develop some approximation processes to express the results in the initial expression domains, which produce the consequent loss of information and hence the lack of precision.

(3) The methods based on fuzzy linguistic representation model, which represent the linguistic information with a pair of values called 2-tuple, composed by a linguistic term and a value of the symbolic translation. Together with the model, the methods also give some computational techniques to deal with the 2-tuples without loss of information. But the model needs some transformation between a counting of information and the linguistic 2-tuple by a function in the aggregation process, and thus, the model is somewhat cumbersome in representation.

(4) The methods which compute with words directly.

Compared with the methods (1)$\sim$(3), the methods (4) can not only avoid losing any linguistic information, but also are straightforward and very convenient in calculation, and thus, are more practical in actual applications.

In recent years, the author has made an in-depth and systematical research on the methods (4) and their applications. The research results mainly include linguistic evaluation scales, linguistic aggregation operators, uncertain linguistic aggregation operators, dynamic linguistic aggregation operators, the priority theory and methods of linguistic preference relations, uncertain linguistic preference relations and incomplete linguistic preference relations, interactive approach to linguistic multi-attribute decision making, linguistic multi-attribute group decision making methods, dynamic linguistic multi-attribute decision making methods, uncertain linguistic multi-attribute decision making methods, and their applications in solving a variety of practical problems, such as the partner selection of supply chain management, personnel appraisal, investment decision making, military system efficiency dynamic evaluation, venture capital project evaluation, and enterprise technology innovation capacity evaluation, etc. This book will give a thorough and comprehensive introduction to these results, which mainly consists of the following parts:

The preface to this book gives a brief background introduction to the current study on the theory and methods of linguistic decision making, and summarizes the main contents and structure.

Chapter 1 mainly introduces the basis of linguistic decision making–Linguistic evaluation scales. Linguistic evaluation scales are classified into two types: additive linguistic evaluation scales and multiplicative linguistic evaluation scales. The unbalanced additive linguistic label sets and unbalanced multiplicative linguistic label sets are highlighted.

Chapter 2 introduces the aggregation techniques for linguistic information. A comprehensive survey of the existing main linguistic aggregation operators is provided.

Chapter 3 mainly introduces the concepts of linguistic preference relations, uncertain linguistic preference relations, incomplete linguistic preference relations, consistent linguistic preference relations, acceptable linguistic preference relations, and their desirable properties. Then the decision making approaches based on these linguistic preference relations are also overviewed.

Chapter 4 mainly introduces the approaches to linguistic multi-attribute decision making. Based on a variety of linguistic aggregation operators, such as the linguistic weighted averaging operator, dynamic linguistic weighted averaging operator, linguistic weighted geometric operator and dynamic linguistic weighted geometric operator, etc., a series of methods and models for multi-attribute decision making under linguistic environments are established, including the maximizing deviation procedure, ideal point-based model, goal programming model, interactive decision making approach, and multi-period multi-attribute decision making method, etc. Furthermore, most of these results are extended to accommodate multi-attribute decision making under uncertain linguistic environments.

This book is suitable for the engineers, technicians and researchers in the fields of fuzzy mathematics, operations research, information science, management science and systems engineering, etc. It can also be used as a textbook for the senior undergraduate and graduate students in the relevant professional institutions of higher learning.

Zeshui Xu
Nanjing
December, 2011

Contents

Some Commonly Used Symbols

S_i	Linguistic evaluation scale
$\bar{S}_i$	Extended linguistic evaluation scale
s_α, s_β	Linguistic labels
$\tilde{s}_i$	Uncertain linguistic variable
$\boldsymbol{w}, \boldsymbol{\omega}, \boldsymbol{\zeta}, \boldsymbol{\upsilon}, \boldsymbol{\xi}$	Weight vectors
x_i, e_k, G_k	Alternative, decision maker, attribute
$X, E, G, \Omega, \Theta, H, \tilde{S}_i$	Sets
d, ρ	Distance, similarity degree
$\langle u_i, s_{\alpha_i}\rangle, \langle u_i, \tilde{s}_i\rangle$	2-tuples
$\langle v_i, u_i, s_{\alpha_i}\rangle, \langle v_i, u_i, \tilde{s}_i\rangle$	3-tuples
$\boldsymbol{A}, \boldsymbol{A}_k, \boldsymbol{B}, \boldsymbol{B}_k, \tilde{\boldsymbol{A}}, \tilde{\boldsymbol{A}}_k, \tilde{\boldsymbol{B}}$	Matrices
$\tilde{\boldsymbol{B}}_k, \boldsymbol{R}, \tilde{\boldsymbol{R}}, \boldsymbol{R}_k, \tilde{\boldsymbol{R}}_k, \boldsymbol{P}$	Matrices
$a_{ij}, a_{ij}^{(k)}, b_{ij}, b_{ij}^{(k)}, r_{ij}, \tilde{r}_{ij}, r_{ij}^{(k)}, \tilde{r}_{ij}^{(k)}, p_{ij}$	Elements
c_j	Close degree
$\alpha_0, \tau_0, \eta_0, \beta_0$	Thresholds
ε_j^+	Upper deviation variable
ε_j^-	Lower deviation variable

Chapter 1
Linguistic Evaluation Scales

The complexity and uncertainty of objective thing and the fuzziness of human thought result in decision making with linguistic information in a wide variety of practical problems, such as personnel evaluation, military system performance evaluation, online auctions, supply chain management, venture capital, and medical diagnostics. In such problems, a realistic approach may be to use linguistic assessments instead of numerical values by means of linguistic variables, that is, variables whose values are not numbers but words or sentences in a natural or artificial language (Fan and Wang, 2003; 2004; Fan et al., 2002; Herrera and Herrera-Viedma, 2003; 2000a; 2000b; 1997; Herrera and Martínez, 2001a; 2001b; 2000a; 2000b; Herrera and Verdegay, 1993; Herrera et al., 2005; 2003; 2001a; 2001b; 2000; 1997; 1996a; 1996b; 1995; Herrera-Viedma, 2001; Herrera-Viedma and Peis, 2003; Herrera-Viedma et al., 2005; 2004; 2003; Wang and Chuu, 2004; Xu, 2010; 2009a; 2009b; 2009c; 2008; 2007a; 2007b; 2007c; 2007d; 2007e; 2006a; 2006b; 2006c; 2006d; 2006e; 2006f; 2006g; 2006h; 2006i; 2005a; 2005b; 2005c; 2005d; 2004a; 2004b; 2004c; 2004d; 2004e; 2004f; Xu and Da, 2003; 2002; Zadeh and Kacprzyk, 1999a; 1999b). This may arise for different reasons (Chen and Hwang, 1992): ① the information may be unquantifiable due to its nature; and ② the precise quantitative information may not be stated because either it is unavailable or the cost of its computation is too high and an "approximate value" may be tolerated. For example, when evaluating the "comfort" or "design" of a car, linguistic labels like "good", "fair" and "poor" are usually used, and evaluating a car's speed, linguistic labels like "very fast", "fast" and "slow" can be used (Bordogna et al., 1997; Levrat et al., 1997).

Considering that a proper linguistic evaluation scale should be predefined when a decision maker needs to provide his/her preferences over an object with linguistic labels (Carlsson and Fullér, 2000; Herrera et al., 1996a; Torra, 1996; Xu, 2004a; 2004b; 2004c; 2004d; 2004e; 2004f; Yager, 1995; 1992), in the following sections, we shall introduce some common linguistic evaluation scales:

1.1 Additive Linguistic Evaluation Scales

Yager (1995; 1992) defined an ordinal scale as

$$L = \{L_i | i = 1, 2, \cdots, m\} \tag{1.1}$$

such that $L_i > L_j$ if $i > j$.

Later, Cordón et al. (2002), Herrera et al. (2001a; 2000; 1996a), Herrera-Viedma et al. (2004; 2003), Martínez et al. (2005) and Torra (2001) introduced a finite and totally ordered discrete additive linguistic evaluation scale (linguistic label set):

$$S_1 = \{s_\alpha | \alpha = 0,\ 1,\ \cdots,\ \tau\} \tag{1.2}$$

where s_α represents a possible value for a linguistic label. In particular, s_0 and s_τ denote the lower and upper limits of linguistic labels used by a decision maker in practical applications, τ is a positive integer. The cardinality value of S_1 is odd such as 7 and 9, the limit of cardinality is 11 or not more than 13, it must be small enough so as not to impose useless precision on the decision makers, and it must also be rich enough in order to allow a discrimination of the performances of each object in a limited number of grades (Bordogna et al., 1997). The linguistic label s_α has the following characteristics:

(1) The set is ordered: $s_\alpha > s_\beta$, if $\alpha > \beta$;

(2) The negation operator is defined: $\text{neg}(s_\alpha) = s_\beta$ such that $\alpha + \beta = \tau$.

For example, a set of seven linguistic labels S_1 (Bordogna et al., 1997) could be (Figure 1.1)

$$\begin{aligned} S_1 =\{&s_0 = \text{none},\quad s_1 = \text{very low},\quad s_2 = \text{low},\quad s_3 = \text{medium},\\ &s_4 = \text{high},\quad s_5 = \text{very high},\quad s_6 = \text{perfect}\}\end{aligned}$$

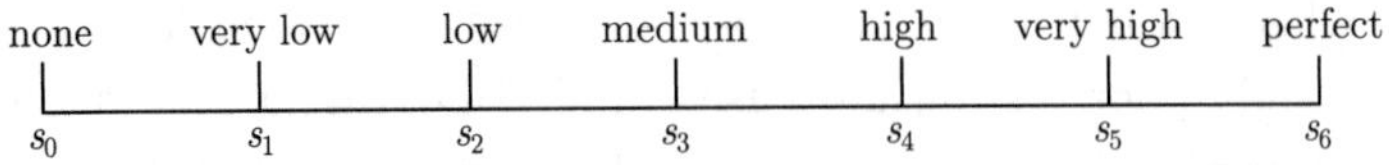

Figure 1.1 Additive linguistic evaluation scale S_1 ($\tau = 6$) (Xu, 2009c)

In the process of aggregating information, the aggregated result may not match any of the original linguistic labels in the additive linguistic evaluation scale S_1. In order to preserve all the given information, Dai et al. (2008) extended the discrete linguistic evaluation scale S_1 to a continuous linguistic evaluation scale:

$$\bar{S}_1 = \{s_\alpha | \alpha \in [0, q]\} \tag{1.3}$$

where q ($q > \tau$) is a sufficiently large positive integer. If $s_\alpha \in S_1$, then s_α is called an original linguistic label; otherwise, s_α is termed an extended (or virtual) linguistic label. The extended linguistic labels also meet the characteristics (1) and (2) described above.

In general, a decision maker uses the original linguistic labels to evaluate objects (or alternatives), and the virtual linguistic labels can only appear in calculations.

Based on the extended additive linguistic evaluation scale $\bar{S}_1$, Dai et al. (2008) introduced some operational laws:

Definition 1.1 (Dai et al., 2008) Let $s_\alpha, s_\beta \in \bar{S}$ and $\lambda \in [0, 1]$. Then

(1) $s_\alpha \oplus s_\beta = s_{\alpha+\beta}$;

(2) $\lambda\, s_\alpha = s_{\lambda\,\alpha}$.

However, in the process of operations, if we take $s_2 =$ low and $s_4 =$ high, then

$$s_2 \oplus s_4 = s_6 \tag{1.4}$$

i.e., the aggregated result of the linguistic labels "low" and "high" is "perfect". This is not in accordance with actual situations.

To overcome the issue above, Dai et al. (2008) improved the additive linguistic evaluation scale above, and gave a subscript-symmetric linguistic evaluation scale:

$$S_2 = \{s_\alpha | \alpha = -\tau, \cdots, -1, 0, 1, \cdots, \tau\} \tag{1.5}$$

where the mid linguistic label s_0 represents an assessment of "indifference", and with the rest of the linguistic labels being placed symmetrically around it. In particular, $s_{-\tau}$ and s_τ are the lower and upper limits of linguistic labels used by a decision maker in practical applications, τ is a positive integer, and S_2 satisfies the following conditions (Xu, 2005c):

(1) If $\alpha > \beta$, then $s_\alpha > s_\beta$;

(2) The negation operator is defined: $\text{neg}(s_\alpha) = s_{-\alpha}$, especially, $\text{neg}(s_0) = s_0$.

For example, when $\tau = 3$, S_2 can be taken as (Figure 1.2)

$$\begin{aligned} S_2 = \ &\{s_{-3} = \text{none},\ \ s_{-2} = \text{very low},\ \ s_{-1} = \text{low},\ \ s_0 = \text{medium}, \\ &s_1 = \text{high},\ \ s_2 = \text{very high},\ \ s_3 = \text{perfect}\} \end{aligned}$$

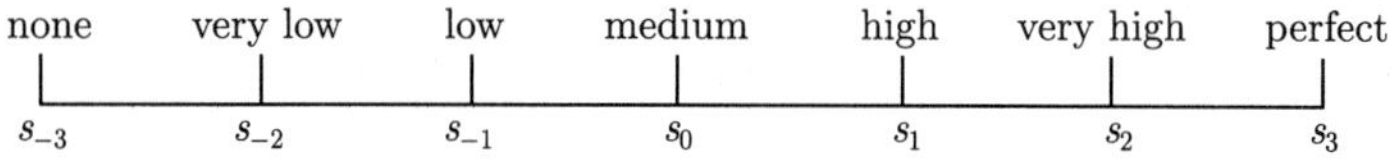

Figure 1.2 Additive linguistic evaluation scale S_2 ($\tau = 3$)

For the convenience of calculations and in order to preserve all the given information, Xu (2005c) extended the discrete linguistic evaluation scale S_2 to a continuous linguistic evaluation scale $\bar{S}_2 = \{s_\alpha | \alpha \in [-q, q]\}$, where q ($q > \tau$) is a sufficiently large positive integer. The operational laws of linguistic labels in $\bar{S}_2$ are similar to (1) and (2) in Definition 1.1. In this case, if we take $s_{-2} =$ low and $s_2 =$ high, then we have

$$s_{-2} \oplus s_2 = s_0 \tag{1.6}$$

i.e., the aggregated result of the linguistic labels "low" and "high" is "medium", which is clearly more in accord with actual situations than the case described in (1.4).

The linguistic labels in the above linguistic evaluation scales are uniformly and symmetrically distributed. However, in real-life, the unbalanced linguistic information may appear due to the nature of the linguistic variables used in the problems (Herrera and Herrera-Viedma, 2003). Therefore, to develop unbalanced linguistic label sets is an interesting and important research topic. In the following, we shall pay attention to this issue:

Xu (2000) gave a simulation-based evaluation of four common numerical evaluation scales (1-9 scale (Saaty, 1980), 9/9-9/1 scale (Wang and Ma, 1993), 10/10-18/2 scale (Wang and Ma, 1993) and exponential scale (Shu and Liang, 1990)) from different angles (Table 1.1). The results show that 10/10-18/2 scale is of the best performance.

Table 1.1 Four common numerical scale (Xu, 2000)

	1-9 scale	Exponential scale	9/9–9/1 scale	10/10–18/2 scale
Equal importance	1	$9^{(0)}$ (1)	9/9 (1)	10/10 (1)
Moderate importance of one over another	3	$9^{(1/9)}$ (1.277)	9/7 (1.286)	12/8 (5)
Essential or strong importance	5	$9^{(3/9)}$ (2.08)	9/5 (6)	14/6 (2.333)
Very strong importance	7	$9^{(6/9)}$ (4.327)	9/3 (3)	16/4 (4)
Extreme importance	9	$9^{(9/9)}$ (9)	9/1 (9)	18/2 (9)
	k	$9^{(k/9)}$ (4.327)	$9/(10-k)$	$(9+k)/(11-k)$

In the process of practical applications, such as selecting projects for different kinds of funding policies, and evaluating university faculty for tenure and promotion, we find that as the unbalanced linguistic information appears, the absolute value of the deviation between the indices of two adjoining linguistic labels should increase as the absolute value of the indices of the linguistic labels steadily increase. Motivated by this idea and based on the 10/10-18/2 scale, we introduce a linguistic evaluation scale from the viewpoints of simplicity, feasibility and practicability (Dai et al., 2008; Xu, 2009c), shown as below:

$$S_3 = \left\{ s_\alpha \middle| \alpha = -\frac{2(\tau-1)}{\tau+2-\tau}, -\frac{2(\tau-1-1)}{\tau+2-(\tau-1)}, \cdots, 0, \cdots, \right. \\ \left. \frac{2(\tau-1-1)}{\tau+2-(\tau-1)}, \frac{2(\tau-1)}{\tau+2-\tau} \right\} \tag{1.7}$$

The right part (after s_0) of S_3 can be written as

$$S_3^+ = \left\{ s_\alpha \middle| \alpha = \frac{\tau+i}{\tau+2-i} - 1 = \frac{2(i-1)}{\tau+2-i}, \quad i = 2, \cdots, \tau-1, \tau \right\} \tag{1.8}$$

and the left part (before s_0) of S_3 can be written as

$$S_3^- = \left\{ s_\alpha \middle| \alpha = -\frac{2(i-1)}{\tau+2-i}, \quad i = \tau, \tau-1, \cdots, 2 \right\} \tag{1.9}$$

Obviously, (1.7) can be simplified as

$$S_3 = \left\{ s_\alpha \middle| \alpha = -(\tau-1), \ -\frac{2}{3}(\tau-2), \cdots, \ 0, \cdots, \ \frac{2}{3}(\tau-2), \ \tau-1 \right\} \tag{1.10}$$

where the mid linguistic label s_0 represents an assessment of "indifference", and with the rest of the linguistic labels being placed symmetrically around it. Especially, $s_{-(\tau-1)}$ and $s_{(\tau-1)}$ are the lower and upper limits of linguistic labels used by a decision maker in practical applications. τ is a positive integer, and the cardinality value of S_3 is $2\tau-1$. The linguistic labels in S_3 have the following characteristics:

(1) If $\alpha > \beta$, then $s_\alpha > s_\beta$;

(2) The negation operator is defined: $\text{neg}(s_\alpha) = s_{-\alpha}$, especially, $\text{neg}(s_0) = s_0$.

For example, if $\tau = 1$, then we can take S_3 as (Figure 1.3)

$$S_3 = \{s_{-1} = \text{low}, \quad s_0 = \text{medium}, \quad s_1 = \text{high}\}$$

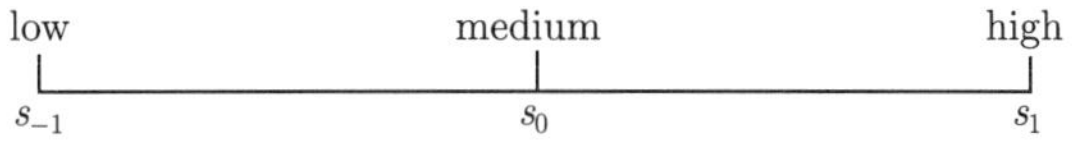

Figure 1.3 Additive linguistic evaluation scale S_3 ($\tau = 1$) (Xu, 2009c)

When $\tau = 4$, S_3 could be (Figure 1.4)

$$S_3 = \{s_{-3} = \text{none}, \ s_{-4/3} = \text{very low}, \ s_{-1/2} = \text{low}, \ s_0 = \text{medium}, \ s_{1/2} = \text{high}, \ s_{4/3} = \text{very high}, \ s_3 = \text{perfect}\}$$

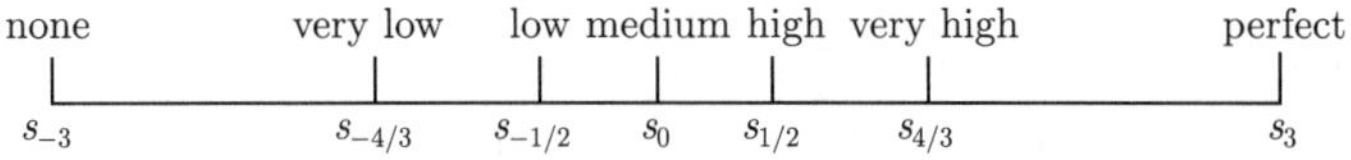

Figure 1.4 Additive linguistic evaluation scale S_3 ($\tau = 4$) (Xu, 2009c)

When $\tau = 5$, S_3 could be (Figure 1.5)

$$S_3 = \{s_{-4} = \text{none}, \ s_{-2} = \text{very low}, \ s_{-1} = \text{low}, \ s_{-0.4} = \text{slightly low}, \ s_0 = \text{medium},$$

$$s_{0.4} = \text{slightly high}, \quad s_1 = \text{high}, \quad s_2 = \text{very high}, \quad s_4 = \text{perfect}\}$$

Similarly, we extend the discrete linguistic evaluation scale S_3 into a continuous linguistic evaluation scale $\bar{S}_3 = \{s_\alpha | \alpha \in [-q, q]\}$, where q $(q > \tau - 1)$ is a sufficiently large positive integer. The operational laws of linguistic labels in $\bar{S}_3$ are similar to (1) and (2) in Definition 1.1.

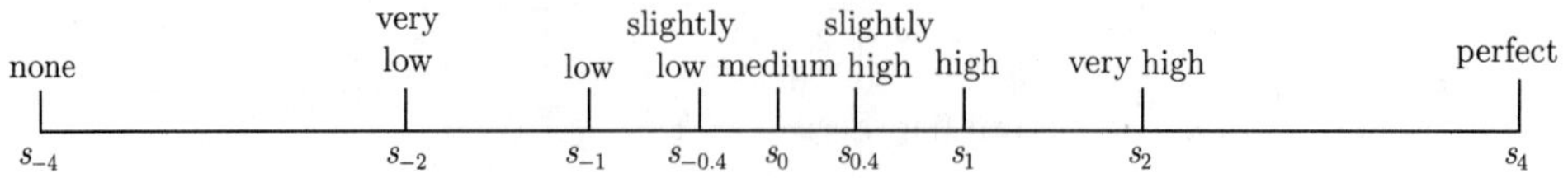

Figure 1.5 Additive linguistic evaluation scale S_3 $(\tau = 5)$ (Xu, 2009c)

Theorem 1.1 (Dai et al., 2008) For the linguistic evaluation scale S_3, the absolute value of the deviation between the indices of two adjoining linguistic labels should increase as the absolute values of the indices of the linguistic labels steadily increase.

Proof We first consider the linguistic label set S_3^+. Let s_{α_1} and s_{α_2} be two adjoining linguistic labels in S_3^+. Without loss of generality, let the indices of the linguistic labels s_{α_1} and s_{α_2} be

$$\alpha_1 = \frac{2(i-1)}{\tau + 2 - i}, \quad \alpha_2 = \frac{2(i+1-1)}{\tau + 2 - (i+1)}$$

respectively, then

$$\begin{aligned} |\alpha_1 - \alpha_2| &= \frac{2(i+1-1)}{\tau+2-(i+1)} - \frac{2(i-1)}{\tau+2-i} = \frac{2i}{\tau - i + 1} - \frac{2(i-1)}{\tau+2-i} \\ &= \frac{2(\tau+1)}{(\tau+2-i)(\tau+1-i)} \end{aligned} \tag{1.11}$$

From (1.11), we can see that with the increase of i, the numerator of (1.11) remains unchanged, but its denominator decreases, and thus, the absolute value of the deviation, $|\alpha_1 - \alpha_2|$, also increases correspondingly. We can prove the other cases in a similar way.

In the above, we have introduced three types of additive linguistic evaluation scales S_i $(i = 1, 2, 3)$, where S_1 and S_2 are the uniform additive linguistic evaluation scales, while S_3 is an unbalanced additive linguistic evaluation.

In the next section, we shall introduce some multiplicative linguistic evaluation scales.

1.2 Multiplicative Linguistic Evaluation Scales

Xu (2004e) defined a multiplicative linguistic evaluation scale as follows:

$$S_4 = \left\{ s_\alpha | \alpha = \frac{1}{\tau}, \cdots, \frac{1}{2}, 1, 2, \cdots, \tau \right\} \tag{1.12}$$

where s_α is a linguistic label. In particular, $s_{1/\tau}$ and s_τ are the lower and upper limits of the linguistic labels used by the decision makers in actual applications, τ is a positive integer, and S_4 has the following characteristics:

(1) If $\alpha > \beta$, then $s_\alpha > s_\beta$;

(2) The reciprocal operator is defined: $\text{rec}(s_\alpha) = s_\beta$, such that $\alpha\beta = 1$, especially, $\text{rec}(s_1) = s_1$.

For example, when $\tau = 4$, S_4 could be (Figure 1.6)

$$\begin{aligned} S_4 = \{ & s_{1/4} = \text{none}, \quad s_{1/3} = \text{very low}, \quad s_{1/2} = \text{low}, \quad s_1 = \text{medium}, \\ & s_2 = \text{high}, \quad s_3 = \text{very high}, \quad s_4 = \text{perfect} \} \end{aligned}$$

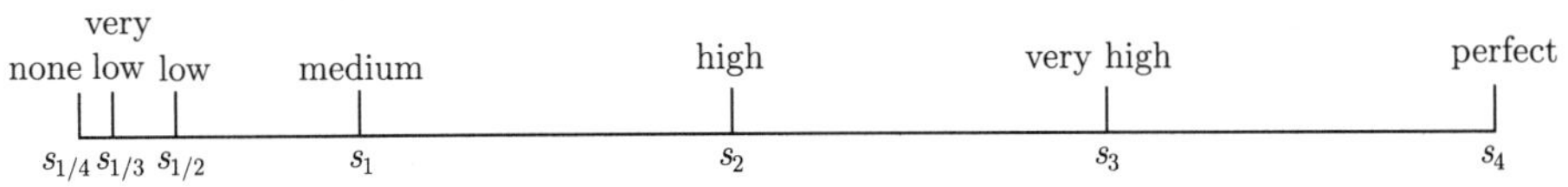

Figure 1.6 Multiplicative linguistic evaluation scale S_4 ($\tau = 4$)

The multiplicative linguistic evaluation scale S_4 has the following property:

Theorem 1.2 Let S_4 be defined as (1.12), and let the right and left parts of S_4 be defined as

$$S_4^+ = \{ s_\alpha | \alpha = 1, 2, \cdots, \tau - 1, \tau \} \tag{1.13}$$

and

$$S_4^- = \left\{ s_\alpha \middle| \alpha = \frac{1}{\tau}, \frac{1}{\tau - 1}, \cdots, \frac{1}{2}, 1 \right\} \tag{1.14}$$

respectively. Then

(1) The absolute value of the deviation between the indices of each two adjoining linguistic labels in S_4^+ is a constant;

(2) The absolute value of the deviation between the indices of two adjoining linguistic labels in S_4^+ should increase as the absolute values of the indices of the linguistic labels steadily increase.

Proof (1) Now we rewrite (1.13) as

$$S_4^+ = \{ s_{\alpha_i} | \alpha_i = i, \ i = 1, 2, \cdots, \tau \} \tag{1.15}$$

Then

$$|\alpha_{i+1} - \alpha_i| = (i + 1) - i = 1, \ i = 1, 2, \cdots, \tau - 1 \tag{1.16}$$

Hence, for any $i = 1, 2, \cdots, \tau - 1$, $|\alpha_{i+1} - \alpha_i|$ is a constant.

(2) We rewrite (1.14) as

$$S_4^- = \left\{ s_{\alpha_i} \middle| \alpha_i = \frac{1}{\tau - (i-1)},\ i = 1, 2, \cdots, \tau \right\} \tag{1.17}$$

Then

$$|\alpha_{i+1} - \alpha_i| = \frac{1}{\tau - (i+1-1)} - \frac{1}{\tau - (i-1)} = \frac{1}{(\tau - i + 1)(\tau - i)},$$
$$i = 1, 2, \cdots, \tau - 1 \tag{1.18}$$

We can see from (1.18) that $|\alpha_{i+1} - \alpha_i|$ increases with the increase of i. This completes the proof.

Xu (2009a) defined another multiplicative linguistic evaluation scale:

$$S_5 = \left\{ s_\alpha \middle| \alpha = \frac{1}{\tau}, \frac{2}{\tau}, \cdots, \frac{\tau - 1}{\tau}, 1, \frac{\tau}{\tau - 1}, \cdots, \frac{\tau}{2}, \tau \right\} \tag{1.19}$$

where $s_{1/\tau}$ and s_τ are the lower and upper limits of the linguistic labels used by the decision makers in actual applications, τ is a positive integer, and S_5 has the following characteristics:

(1) If $\alpha > \beta$, then $s_\alpha > s_\beta$;

(2) The reciprocal operator is defined: $\mathrm{rec}(s_\alpha) = s_\beta$, such that $\alpha\beta = 1$, especially, $\mathrm{rec}(s_1) = s_1$.

The multiplicative linguistic evaluation scale S_5 has the following property:

Theorem 1.3 (Xu, 2009a) Let S_5 be defined as (1.19), and let the right and left parts of S_5 be defined as

$$S_5^+ = \left\{ s_\alpha \middle| \alpha = 1, \frac{\tau}{\tau - 1}, \cdots, \frac{\tau}{2}, \tau \right\} \tag{1.20}$$

and

$$S_5^- = \left\{ s_\alpha \middle| \alpha = \frac{1}{\tau}, \frac{2}{\tau}, \cdots, \frac{\tau - 1}{\tau}, 1 \right\} \tag{1.21}$$

respectively. Then

(1) The greater the value of the index of a linguistic label in S_5^+, the greater the value of the deviation between the indices of this linguistic label and its adjoining linguistic label in S_5^+;

(2) The greater the value of the index of a linguistic label in S_5^-, the smaller the value of the deviation between the indices of this linguistic label and its adjoining linguistic label in S_5^-.

Proof (1) For convenience of description, we rewrite S_5^+ as

$$S_5^+ = \left\{ s_{\alpha_i} \middle| \alpha_i = \frac{\tau}{\tau - (i-1)},\ i = 1, 2, \cdots, \tau \right\} \tag{1.22}$$

Then

$$\frac{\alpha_i}{\alpha_{i-1}} = \frac{\dfrac{\tau}{\tau-(i-1)}}{\dfrac{\tau}{\tau-[(i-1)-1]}} = \frac{\tau-(i-2)}{\tau-(i-1)}, \quad i=1,2,\cdots,\tau \tag{1.23}$$

It follows from (1.23) that the value of α_i/α_{i-1} gets larger as i steadily increases.

(2) We rewrite S_5^- as

$$S_5^- = \left\{ s_{\alpha_i} \middle| \alpha_i = \frac{i}{\tau}, \ i=1,2,\cdots,\tau \right\} \tag{1.24}$$

Then

$$\frac{\alpha_{i+1}}{\alpha_i} = \frac{\dfrac{i+1}{\tau}}{\dfrac{i}{\tau}} = \frac{i+1}{i}, \quad i=1,2,\cdots,\tau-1 \tag{1.25}$$

Thus, the value of α_{i+1}/α_i gets smaller as i steadily increases. This completes the proof.

From (1) and (2), we know that the further a linguistic label in S_5 from s_1, the greater the value of the deviation between the indices of multiplicative linguistic label and its adjoining linguistic label, which is in accordance with the most practical situations such as medical diagnosis, personnel examination, and military system efficiency evaluation, etc.

If we replace (1.23) with

$$\alpha_i - \alpha_{i-1} = \frac{\tau}{\tau-(i-1)} - \frac{\tau}{\tau-[(i-1)-1]} = \frac{\tau}{(\tau-i+1)(\tau-i+2)}$$
$$i=1,2,\cdots,\tau \tag{1.26}$$

and replace (1.25) with

$$\alpha_{i+1} - \alpha_i = \frac{i+1}{\tau} - \frac{i}{\tau} = \frac{1}{\tau}, \quad i=1,2,\cdots,\tau-1 \tag{1.27}$$

then the results of (1.26) and (1.27) show that the multiplicative linguistic evaluation scale S_5 distributes linearly the labels on the scale when they are before linguistic label s_1 and not linearly when they are after s_1, which is inconsistent with Theorem 1.3. That is because both (1.26) and (1.27) are unsuitable for measuring the multiplicative linguistic labels, but generally suitable for measuring the additional linguistic labels, while both (1.23) and (1.25) are generally suitable for measuring the multiplicative linguistic labels, but unsuitable for measuring the additional linguistic labels.

For example, when $\tau = 4$, S_5 could be (Figure 1.7) (Xu, 2009a)

$$S_5 = \{s_{1/4} = \text{none}, \ s_{1/2} = \text{very low}, \ s_{3/4} = \text{low}, \ s_1 = \text{medium},$$
$$s_{4/3} = \text{high}, \ s_2 = \text{very high}, \ s_4 = \text{perfect}\}$$

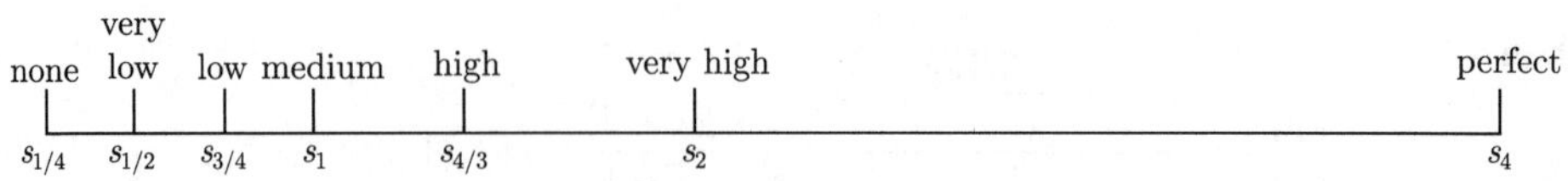

Figure 1.7 Multiplicative linguistic evaluation scale S_5 ($\tau = 4$) (Xu, 2009a)

Then in this case, by (1.23) and (1.25), all the values of the deviation between each pair of adjoining multiplicative linguistic labels in S_5 are as follows (Xu, 2009):

(1) The value of the deviation between "very low" and "none" is 2;

(2) The value of the deviation between "low" and "very low" is 3/2;

(3) The value of the deviation between "medium" and "low" is 4/3;

(4) The value of the deviation between "high" and "medium" is 4/3;

(5) The value of the deviation between "very high" and "high" is 3/2;

(6) The value of the deviation between "perfect" and "very high" is 2.

In the process of calculations, in order to preserve all the information, we extend the discrete multiplicative linguistic evaluation scale S_5 to a continuous multiplicative linguistic evaluation scale $\bar{S}_5 = \{s_\alpha | \alpha \in [1/q, q]\}$ (in a similar way, we can make an extension for the multiplicative linguistic evaluation scale S_4), where q ($q \geqslant \tau$) is a sufficiently large positive integer. If $s_\alpha \in S_5$, then s_α is termed an original multiplicative linguistic label; otherwise, s_α is termed a virtual multiplicative linguistic label.

Let s_α, $s_\beta \in \bar{S}_5$ and $\lambda \in [0, 1]$. Then their operational laws can be defined as follows (Xu, 2009a):

(1) $s_\alpha \otimes s_\beta = s_{\alpha\beta}$;

(2) $s_\alpha^\lambda = s_{\alpha^\lambda}$.

The linguistic evaluation scales and their operational laws introduced above make a good basis and prerequisite for decision making with linguistic information. In practical applications, we can select a proper one according to actual situations.

References

Bordogna G, Fedrizzi M, Passi G. 1997. A linguistic modeling of consensus in group decision making based on OWA operator. IEEE Transactions on Systems, Man, and Cybernetics, 27: 126-132.

Carlsson C, Fullér R. 2000. Multi-objective linguistic optimization. Fuzzy Sets and Systems, 115: 5-10.

Chen S J, Hwang C L. 1992. Fuzzy Multiple Attribute Decision Making: Methods and Applications. Berlin: Springer-Verlag.

Cordón O, Herrera F, Zwir I. 2002. Linguistic modeling by hierarchical systems of linguistic rules. IEEE Transactions on Fuzzy Systems, 10: 2-20.

Dai Y Q, Xu Z S, Da Q L. 2008. New evaluation scale of linguistic information and its application. Chinese Journal of Management Science, 16(2): 145-149.

Fan Z P, Ma J, Zhang Q. 2002. An approach to multiple attribute decision making based on fuzzy preference information on alternatives. Fuzzy Sets and Systems, 131: 101-106.

Fan Z P, Wang X R. 2004. Approach to solve assignment problems with linguistic assessment information. Journal of Systems Engineering, 19(1): 14-19.

Herrera F, Herrera-Viedma E, Chiclana F. 2001a. Multiperson decision making based on multiplicative preference relations. European Journal of Operational Research, 129: 372-385.

Herrera F, Herrera-Viedma E, Chiclana F. 2003. A study of the origin and uses of the ordered weighted geometric operator in multicriteria decision making. International Journal of Intelligent Systems, 18: 689-707.

Herrera F, Herrera-Viedma E, Martínez L. 2000. A fusion approach for managing multi-granularity linguistic term sets in decision making. Fuzzy Sets and Systems, 114: 43-58.

Herrera F, Herrera-Viedma E, Verdegay J L. 1995. A sequential selection process in group decision making with a linguistic assessment approach. Information Sciences, 85: 223-239.

Herrera F, Herrera-Viedma E, Verdegay J L. 1996a. A model of consensus in group decision making under linguistic assessments. Fuzzy Sets and Systems, 78: 73-87.

Herrera F, Herrera-Viedma E, Verdegay J L. 1996b. Direct approach processes in group decision making using linguistic OWA operators. Fuzzy Sets and Systems, 79: 175-190.

Herrera F, Herrera-Viedma E, Verdegay J L. 1997. A rational consensus model in group decision making using linguistic assessments. Fuzzy Sets and Systems, 88: 31-49.

Herrera F, Herrera-Viedma E. 1997. Aggregation operators for linguistic weighted information. IEEE Transactions on Systems, Man, and Cybernetics-Part A, 27: 646-656.

Herrera F, Herrera-Viedma E. 2000a. Choice functions and mechanisms for linguistic preference relations. European Journal of Operational Research, 120: 144-161.

Herrera F, Herrera-Viedma E. 2000b. Linguistic decision analysis: steps for solving decision problems under linguistic information. Fuzzy Sets and Systems, 115: 67-82.

Herrera F, Herrera-Viedma E. 2003. Linguistic preference modeling: foundation models and new trends. *In*: Proceedings of 4th International Workshop on Preferences and Decisions, Trento, Italy: 47-51.

Herrera F, López E, Mendaña, et al. 2001b. A linguistic decision model for personnel management solved with a linguistic biobjective genetic algorithm. Fuzzy Sets and Systems, 118: 47-64.

Herrera F, Martínez L, Sánchez P J. 2005. Managing non-homogeneous information in group decision making. European Journal of Operational Research, 166: 115-132.

Herrera F, Martínez L. 2000a. A 2-tuple fuzzy linguistic representation model for computing with words. IEEE Transactions on Fuzzy Systems, 8: 746-752.

Herrera F, Martínez L. 2000b. An approach for combining linguistic and numerical information based on the 2-tuple fuzzy linguistic representation model in decision-making. International Journal of Uncertainty, Fuzziness and Knowledge-Based Systems, 8: 539-562.

Herrera F, Martínez L. 2001a. The 2-tuple linguistic computational model: advantages of its linguistic description, accuracy and consistency. International Journal of Uncertainty, Fuzziness and Knowledge-Based Systems, 9 (Suppl.): 33-48.

Herrera F, Martínez L. 2001b. A model based on linguistic 2-tuples for dealing with multigranular hierarchical linguistic contexts in multi-expert decision-making. IEEE Transactions on Systems, Man, and Cybernetics, 31: 227-233.

Herrera F, Verdegay J L. 1993. Linguistic assessments in group decision. *In*: Proceedings of the First European Congress on Fuzzy and Intelligent Technologies, Aachen: 941-948.

Herrera-Viedma E, Cordon O, Luque M, et al. 2003. A model of fuzzy linguistic IRS based on multi-granular linguistic information. International Journal of Approximate Reasoning, 34: 221-239.

Herrera-Viedma E, Herrera F, Martínez L, et al. 2004. Incorporating filtering techniques in a fuzzy linguistic multi-agent model for gathering of information on the web. Fuzzy Sets and Systems, 148: 61-83.

Herrera-Viedma E, Martínez L, Mata F, et al. 2005. A consensus support systems model for group decision making problems with multigranular linguistic preference relations. IEEE Transactions on Fuzzy Systems, 13: 644-658.

Herrera-Viedma E, Peis E. 2003. Evaluating the informative quality of documents in SGML format from judgments by means of fuzzy linguistic techniques based on computing with words. Information Processing and Management, 39: 233-249.

Herrera-Viedma E. 2001. Modeling the retrieval process of an information retrieval system using an ordinal fuzzy linguistic approach. Journal of the American Society for Information Science and Technology, 52: 460-475.

Levrat E, Voisin A, Bombardier S, et al. 1997. Subjective evaluation of car seat comfort with fuzzy set techniques. International Journal of Intelligent Systems, 12: 891-913.

Martínez L, Liu J, Yang J B, et al. 2005. A multigranular hierarchical linguistic model for design evaluation based on safety and cost analysis. International Journal of Intelligent Systems, 20: 1161-1194.

Saaty T L. 1980. The Analytic Hierarchy Process. New York: McGraw-Hill.

Shu K, Liang Z H. 1990. The exponential scale method in AHP. Systems Engineering–Theory & Practice, 10(1): 6-8.

Torra V. 1996. Negation functions based semantics for ordered linguistic labels. International Journal of Intelligent Systems, 11: 975-988.

Torra V. 2001. Aggregation of linguistic labels when semantics is based on antonyms. International Journal of Intelligent Systems, 16: 513-524.

Wang H, Ma D. 1993. Scale evaluation and new scale methods. Systems Engineering–Theory & Practice, 13(5): 24-26.

Wang R C, Chuu S J. 2004. Group decision-making using a fuzzy linguistic approach for evaluating the flexibility in a manufacturing system. European Journal of Operational Research, 154: 563-572.

Wang X R, Fan Z P. 2003. Method for group decision making based on two-tuple linguistic information processing. Journal of Management Sciences in China, 6(5): 1-5.

Xu Z S, Da Q L. 2002. Multi-attribute decision making based on fuzzy linguistic assessments. Journal of Southeast University, 32: 656–658.

Xu Z S, Da Q L. 2003. Method based on fuzzy linguistic scale and fuzzy induced ordered weighted geometric averaging (FIOWGA) operator for decision making problems. Journal of Southeast University, 19: 88-91.

Xu Z S. 2000. A simulation-based evaluation of several scales in the analytic hierarchy process. Systems Engineering-Theory & Practice, 20(7): 58-62.

Xu Z S. 2004a. Uncertain linguistic aggregation operators based approach to multiple attribute group decision making under uncertain linguistic environment. Information Sciences, 168: 171-184.

Xu Z S. 2004b. On method of multi-attribute group decision making under pure linguistic information. Control and Decision, 19(7): 778-781.

Xu Z S. 2004c. Method based on fuzzy linguistic assessments and GIOWA operator in multi-attribute group decision making. Journal of Systems Science and Mathematical Sciences, 24: 218-224.

Xu Z S. 2004d. Uncertain Multiple Attribute Decision Making: Methods and Applications. Beijing: Tsinghua University Press.

Xu Z S. 2004e. EOWA and EOWG operators for aggregating linguistic labels based on linguistic preference relations. International Journal of Uncertainty, Fuzziness and Knowledge-Based Systems, 12: 791-810.

Xu Z S. 2004f. A method based on linguistic aggregation operators for group decision making with linguistic preference relations. Information Sciences, 166: 19-30.

Xu Z S. 2005a. An approach to pure linguistic multiple attribute decision making under uncertainty. International Journal of Information Technology and Decision Making, 4: 197-206.

Xu Z S. 2005b. An approach to group decision making based on incomplete linguistic preference relations. International Journal of Information Technology and Decision Making, 4: 153-160.

Xu Z S. 2005c. Deviation measures of linguistic preference relations in group decision making. Omega, 33: 249-254.

Xu Z S. 2005d. Maximizing deviations procedure for multiple attribute decision making under linguistic environment. Lecture Series on Computer and Computational Sciences, 2: 150-154.

Xu Z S. 2006a. A note on linguistic hybrid arithmetic averaging operator in group decision making with linguistic information. Group Decision and Negotiation, 15: 581-591.

Xu Z S. 2006b. On generalized induced linguistic aggregation operators. International Journal of General Systems, 35: 17-28.

Xu Z S. 2006c. Induced uncertain linguistic OWA operators applied to group decision making. Information Fusion, 7: 231-238.

Xu Z S. 2006d. A direct approach to group decision making with uncertain additive linguistic preference relations. Fuzzy Optimization and Decision Making, 5: 23-35.

Xu Z S. 2006e. An approach based on the uncertain LOWG and the induced uncertain LOWG operators to group decision making with uncertain multiplicative linguistic preference relations. Decision Support Systems, 41: 488-499.

Xu Z S. 2006f. Goal programming models for multiple attribute decision making under linguistic setting. Journal of Management Sciences in China, 9: 9-17.

Xu Z S. 2006g. Incomplete linguistic preference relations and their fusion. Information Fusion, 7: 331-337.

Xu Z S. 2006h. A practical procedure for group decision making under incomplete multiplicative linguistic preference relations. Group Decision and Negotiation, 15: 593-604.

Xu Z S. 2006i. Multiple attribute decision making based on different types of linguistic information. Journal of Southeast University, 22(1): 134-136.

Xu Z S. 2007a. A method for multiple attribute decision making with incomplete weight information in linguistic setting. Knowledge-Based Systems, 20: 719-725.

Xu Z S. 2007b. Linguistic aggregation operators: an overview. *In*: Bustince H, Herrera F, Montero J. eds. Fuzzy Sets and Their Extensions: Representation, Aggregation and Models. Heidelberg: Springer. 163-181.

Xu Z S. 2007c. A survey of preference relations. International Journal of General Systems, 36: 179-203.

Xu Z S. 2007d. Dynamic linguistic preference relations and their use in multi-period decision making. *In*: International Conference on Management Science and Engineering. Proceedings of 2007 International Conference on Management Science and Engineering. Harbin, China, 345-350.

Xu Z S. 2007e. An interactive procedure for linguistic multiple attribute decision making with incomplete weight information. Fuzzy Optimization and Decision Making, 6: 7-27.

Xu Z S. 2008. Group decision making based on multiple types of linguistic preference relations. Information Sciences, 178: 452-467.

Xu Z S. 2009a. Multi-period multi-attribute group decision making under linguistic assessments. International Journal of General Systems, 38: 823-850.

Xu Z S. 2009b. Correlated linguistic information aggregation. International Journal of Uncertainty, Fuzziness and Knowledge-Based System, 17: 633-647.

Xu Z S. 2009c. An interactive approach to multiple attribute group decision making with multigranular uncertain linguistic information. Group Decision and Negotiation, 18: 119-145.

Xu Z S. 2010. Interactive group decision making procedure based on uncertain multiplicative linguistic preference relations. Journal of Systems Engineering and Electronics, 21: 408-415.

Yager R R. 1992. Applications and extensions of OWA aggregations. International Journal of Man-Machine Studied, 37: 103-132.

Yager R R. 1995. An approach to ordinal decision making. International Journal of Approximate Reasoning, 12: 237-261.

Zadeh L A, Kacprzyk J, eds. 1999a. Computing with Words in Information/Intelligent Systems 1: Foundations. Heidelberg: Physica-Verlag.

Zadeh L A, Kacprzyk J, eds. 1999b. Computing with Words in Information/Intelligent Systems 2: Applications. Heidelberg: Physica-Verlag.

Chapter 2
Linguistic Aggregation Operators

When a problem is solved using linguistic information, it implies the need for computing with words (Zadeh and Kacprzyk, 1999a; 1999b). Thus, how to fuse the input linguistic information is an interesting and important research topic. Linguistic aggregation operators are a powerful tool to deal with linguistic information. Over the last decades, many scholars have focused their investigation on linguistic aggregation techniques and various linguistic aggregation operators have been proposed (Xu, 2007b), including the linguistic max and min operators (Xu and Da 2003b; Yager, 1998; 1996; 1995; 1992), linguistic median operator (Yager, 1998; 1996), Linguistic weighted median operator (Yager, 1998; 1996), linguistic max-min weighted averaging operator (Yager, 1995), extension principle-based linguistic aggregation operator (Bonissone and Decker, 1986; Bordogna and Passi, 1993; Chang and Chen, 1994; Chen, 1997; Degani and Bortolan, 1988; Delgado et al., 1998; Lee, 1996), symbol-based linguistic aggregation operator (Delgado et al., 1993a), 2-tuple arithmetic mean operator (Herrera and Martínez, 2000a), 2-tuple weighted averaging operator (Herrera and Martínez, 2000a), 2-tuple OWA (orded weighted averaging) operator (Herrera and Martínez, 2000a), linguistic weighted OWA operator (Torra, 1997), linguistic averaging operator (Xu, 2006b; 2005b; 2004d; 2004f), linguistic weighted disjunction operator (Herrera and Herrera-Viedma, 2000b; 1997), linguistic weighted conjunction operator (Herrera and Herrera-Viedma, 2000b; 1997), linguistic weighted averaging operator (Herrera and Herrera-Viedma, 1997; Xu, 2006b; 2005b; 2004d), ordinal OWA operator (Bordogna et al., 1997; Yager, 1995; 1992), ordinal linguistic aggregation operator (Delgado et al., 1993b), ordinal hybrid aggregation operator (Xu, 2004d), linguistic OWA operator (Delgado et al., 1998; Herrera and Herrera-Viedma, 2000a; Herrera et al., 2001b; 1997; 1996b; 1995; Xu, 2006a; 2006b; 2004b; 2004e; Xu and Da, 2003b), inverse-linguistic OWA operator (Herrera and Herrera-Viedma, 1997; Herrera et al., 2001b), linguistic hybrid aggregation operator (Xu, 2006a), induced linguistic OWA operator (Xu, 2006b), uncertain linguistic averaging operator (Xu, 2006b; 2004a), uncertain linguistic weighted averaging operator (Xu, 2006b), uncertain linguistic OWA operator (Xu, 2006b; 2004a), induced uncertain linguistic OWA operator (Xu, 2006b; 2006c), uncertain linguistic hybrid aggregation operator (Xu,

2004a), dynamic linguistic weighted averaging operator (Xu, 2007d), dynamic linguistic weighted geometric operator (Xu, 2009a), linguistic correlated averaging operator and linguistic correlated geometric operator (Xu, 2009b), etc. These linguistic aggregation operators have been studied and applied in a wide variety of areas, such as engineering (Levrat et al., 1997; Xu, 2004d), decision making (Bordogna et al., 1997; Chen and Hwang, 1992; Delgado et al., 2002; 1998; 1994; 1993a; 1993b; Herrera and Martínez, 2001a; Herrera et al., 2000; 1997; 1996a; 1996b; 1995; Herrera and Herrera-Viedma, 2003; 2000a; 2000b; 1997; Herrera and Martínez, 2001a; 2000b; Huynh and Nakamori, 2005; Xu, 2007a; 2006a; 2006c; 2006d; 2006e; 2006f; 2006g; 2006h; 2006i; 2005a; 2005b; 2005c; 2004a; 2004b; 2004c; 2004d; 2004e; 2004f; Tong and Bonissone, 1980; Yager, 1995; Yager and Kacprzyk, 1997), information retrieval (Bordogna and Passi, 1993; Delgado et al., 2002; Herrera-Viedma, 2001; Herrera-Viedma et al., 2003; Herrera-Viedma and Peis, 2003; Kostek, 2003), marketing (Herrera et al., 2002; Yager et al., 1994), scheduling (Adamopoulos and Pappis, 1996), biotechnology (Chang and Chen, 1994), materials selection (Chen, 1997), software system (Lee, 1996), personnel management (Herrera et al., 2001b), educational grading system (Law, 1996), medical diagnosis (Becker, 2001; Degani and Bortolan, 1988), artificial intelligence (Zadeh and Kacprzyk, 1999a; 1999b), supply chain management and maintenance service (Xu, 2004d), etc. Xu (2007b) provided a comprehensive survey of the existing main linguistic aggregation operators.

2.1 Linguistic Aggregation Operators Based on Linear Ordering

Let $(a_1, a_2, \cdots, a_n)$ be a collection of linguistic arguments, and $a_j \in S_1$ $(j = 1, 2, \cdots, n)$. Yager (1998; 1996; 1995; 1992) introduced a linguistic max operator:

$$\mathrm{LM}_1(a_1, a_2, \cdots, a_n) = \max_j\{a_j\} \tag{2.1}$$

and a linguistic min (LM_2) operator:

$$\mathrm{LM}_2(a_1, a_2, \cdots, a_n) = \min_j\{a_j\} \tag{2.2}$$

Yager (1998; 1996) developed a linguistic median (LM_3) operator:

$$\mathrm{LM}_3(a_1, a_2, \cdots, a_n) = \begin{cases} b_{\frac{n+1}{2}}, & n \text{ is odd} \\ b_{\frac{n}{2}}, & n \text{ is even} \end{cases} \tag{2.3}$$

where b_j is the jth biggest of a_i $(i = 1, 2, \cdots, n)$.

These three operators are the simplest linguistic aggregation operators, which are usually used to develop some other operators for aggregating linguistic information.

Example 2.1 Let $S_1 = \{s_0, s_1, s_2, s_3, s_4, s_5, s_6\}$, $(a_1, a_2, a_3, a_4, a_5) = (s_0, s_2, s_1, s_5, s_3)$ be a collection of linguistic arguments in S_1. Then

$$\text{LM}_1(s_0, s_2, s_1, s_5, s_3) = \max\{s_0, s_2, s_1, s_5, s_3\} = s_5$$

$$\text{LM}_2(s_0, s_2, s_1, s_5, s_3) = \min\{s_0, s_2, s_1, s_5, s_3\} = s_0$$

Reordering the linguistic arguments $(s_0, s_2, s_1, s_5, s_3)$ in descending order, we have

$$b_1 = s_5, \quad b_2 = s_3, \quad b_3 = s_2, \quad b_4 = s_1, \quad b_5 = s_0$$

and then

$$b_{\frac{5+1}{2}} = b_3 = s_2$$

Thus

$$\text{LM}_3(s_0, s_2, s_1, s_5, s_3) = b_3 = s_2$$

Yager and Rybalov (1997) further considered the problem of weighted median aggregation as below:

Suppose that $((w_1, a_1), (w_2, a_2), \cdots, (w_n, a_n))$ is a collection of pairs, where a_i is a linguistic label and w_i is its associated weight, with the condition:

$$w_i \geqslant 0, \ \ i = 1, 2, \cdots, n, \ \ \sum_{i=1}^{n} w_i = 1 \tag{2.4}$$

and assume that $a_i \, (i = 1, 2, \cdots, n)$ are reodered such that b_j is the jth largest of $a_i \, (i = 1, 2, \cdots, n)$. Then $((u_1, b_1), (u_2, b_2), \cdots, (u_n, b_n))$ is the reordered collection of $((w_1, a_1), (w_2, a_2), \cdots, (w_n, a_n))$, where u_j is the weight that is associated with a_i that becomes b_j. For example, if $b_j = a_5$, then $u_j = w_5$. Let

$$z_i = \sum_{j=1}^{i} u_j \tag{2.5}$$

if

$$\text{LWM}((u_1, b_1), (u_2, b_2), \cdots, (u_n, b_n)) = b_k \tag{2.6}$$

then LWM is called a linguistic weighted median (LWM) operator, where k satisfies

$$z_{k-1} < 0.5 \quad \text{and} \quad z_k \geqslant 0.5 \tag{2.7}$$

The linguistic weighted median b_k is the ordered value of the arguments $a_i \, (i = 1, 2, \cdots, n)$ for which the sum of the weights first crosses the value of 0.5, and k is called a cross over value.

Yager (1993) showed that the LWM operator is of idempotency, commutativity and monotonicity.

Example 2.2 Let $S_1 = \{s_0, s_1, s_2, s_3, s_4, s_5, s_6\}$, and let

$$((w_1, a_1), (w_2, a_2), (w_3, a_3), (w_4, a_4)) = ((0.1, s_6), (0.3, s_2), (0.2, s_3), (0.4, s_1))$$

be a collection of pairs. Then reorder the argunments a_i $(i = 1, 2, \cdots, n)$ in descending order. We have

$$((u_1, b_1), (u_2, b_2), (u_3, b_3), (u_4, b_4)) = ((0.1, s_6), (0.2, s_3), (0.3, s_2), (0.4, s_1))$$

Since

$$z_2 = u_1 + u_2 = 0.1 + 0.2 = 0.3 < 0.5$$

$$z_3 = u_1 + u_2 + u_3 = 0.1 + 0.2 + 0.3 = 0.6 > 0.5$$

then the cross over value $k = 3$, and $b_k = b_3 = s_2$. Accordingly,

$$\text{LWM}((u_1, b_1), (u_2, b_2), (u_3, b_3), (u_4, b_4)) = b_3 = s_2$$

Yager (1988) introduced an ordered weighted averaging (OWA) operator to provide a method for aggregating several numerical inputs that lie between the max and min operators. The fundamental aspect of the OWA operator is the re-ordering step. Using an ordinal form of the OWA operator, Yager (1992) defined the following:

Definition 2.1 (Yager, 1992) A mapping OOWA : $S_1^n \to S_1$ is called an ordinal OWA (OOWA) operator of dimension n, if it has an associated weighting vector $\boldsymbol{\omega} = (\omega_1, \omega_2, \cdots, \omega_n)^{\mathrm{T}}$, such that

(1) $\omega_j \in S_1$;

(2) If $j > i$, then $\omega_j \geqslant \omega_i$;

(3) $\max\limits_j\{\omega_j\} = s_\tau$.

Furthermore,

$$\text{OOWA}(a_1, a_2, \cdots, a_n) = \max_j\{\min\{\omega_j, b_j\}\} \tag{2.8}$$

where b_j is the jth largest of the linguistic arguments a_i $(i = 1, 2, \cdots, n)$.

Especially, if $\boldsymbol{\omega} = (s_\tau, s_\tau, \cdots, s_\tau)^{\mathrm{T}}$, where $\omega_i = s_\tau$ for all i, then the OOWA operator reduces to the LM_1 operator; if $\boldsymbol{\omega} = (s_0, \cdots, s_0, s_\tau)^{\mathrm{T}}$, where $\omega_i = s_0$ for all $i \neq n$ and $\omega_n = s_\tau$, then the OOWA operator reduces to the LM_2 operator; if $\boldsymbol{\omega} = (s_\alpha, \cdots, s_\alpha, s_\tau)^{\mathrm{T}}$, where $\omega_i = s_\alpha$ for all $i \neq n$ and $\omega_n = s_\tau$, then the OOWA operator reduces to the following:

$$\text{OOWA}(a_1, a_2, \cdots, a_n) = \max\{\min\{s_\alpha, \max_j(a_j)\}, \min_j(a_j)\} \tag{2.9}$$

which is called a linguistic max-min weighted averaging (LMMA) operator. This operator is analogous to the Hurwicz type criteria used in the numeric case, $\beta \max\limits_j(a_j) + (1 - \beta)\min\limits_j(a_j)$, where $\beta \in [0, 1]$.

The OOWA operator enjoys many useful properties (Yager, 1995; 1992):

(1) (Monotonicity) If $a_i \geqslant \dot{a}_i$ for all i, then

$$\mathrm{OOWA}(a_1, a_2, \cdots, a_n) \geqslant \mathrm{OOWA}(\dot{a}_1, \dot{a}_2, \cdots, \dot{a}_n) \tag{2.10}$$

(2) (Idempotency) If $a_i = a$ for all i, then

$$\mathrm{OOWA}(a_1, a_2, \cdots, a_n) = a \tag{2.11}$$

(3) (Commutativity) Let $(\dot{a}_1, \dot{a}_2, \cdots, \dot{a}_n)$ be any permutation of $(a_1, a_2, \cdots, a_n)$. Then

$$\mathrm{OOWA}(\dot{a}_1, \dot{a}_2, \cdots, \dot{a}_n) = \mathrm{OOWA}(a_1, a_2, \cdots, a_n) \tag{2.12}$$

(4) (Boundedness) For any the weighting vector $\boldsymbol{\omega} = (\omega_1, \omega_2, \cdots, \omega_n)^{\mathrm{T}}$, the following inequality holds:

$$\min_j(a_j) \leqslant \mathrm{OOWA}(a_1, a_2, \cdots, a_n) \leqslant \max_j(a_j) \tag{2.13}$$

Ovchinnikov (1999) investigated the invariance properties of the OOWA operator.

Example 2.3 Let $S_1 = \{s_0, s_1, s_2, s_3, s_4, s_5, s_6\}$. Assume that $(a_1, a_2, a_3, a_4, a_5) = (s_0, s_5, s_4, s_1, s_2)$ is a collection of linguistic arguments in S_1, and $\boldsymbol{\omega} = (s_0, s_2, s_3, s_4, s_6)^{\mathrm{T}}$, then after reordering the arguments $(a_1, a_2, a_3, a_4, a_5)$ in descending order, we have

$$(b_1, b_2, b_3, b_4, b_5) = (s_5, s_4, s_2, s_1, s_0)$$

Thus

$$\begin{aligned}\mathrm{OOWA}(s_0, s_5, s_4, s_1, s_2) =& \max\{\min\{s_0, s_5\}, \min\{s_2, s_4\}, \min\{s_3, s_2\},\\ & \min\{s_4, s_1\}, \min\{s_6, s_0\}\}\\ =& \max\{s_0, s_2, s_2, s_1, s_0\}\\ =& s_2\end{aligned}$$

Bordogna et al. (1997) proposed a model based on the OOWA operator for group decision making in a linguistic context, and presented a human-consistent definition that can be useful in a consensus reaching phase and a procedure for its computation. Yager (1995) developed a structure making use of the OOWA operator for the evaluation of decisions under ignorance in situations in which the payoff matrix contains linguistic information. In this structure, the decision maker's dispositional attitude is represented by the weighting vector. He also provided two measures for classifying the OOWA weighting vectors used in the aggregations. Yager et al. (1994) developed an OOWA operator-based approach to aggregating market research data based on fuzzy reasoning. This approach uses operations from fuzzy logic to construct the models and the principal of minimal entropy to choose between competing models.

They implemented this approach in a test case using economic attitudinal variables to predict the purchase of major consumer products.

Herrera and Herrera-Viedma (1997) provided the linguistic weighted disjunction and conjunction operators respectively:

Definition 2.2 (Herrera and Herrera-Viedma, 1997) Let $(a_1, a_2, \cdots, a_n)$ be a collection of linguistic arguments, and $\boldsymbol{w} = (w_1, w_2, \cdots, w_n)^{\mathrm{T}}$ be the weight vector of $(a_1, a_2, \cdots, a_n)$, $w_j \in S_1$ for any j. Then a linguistic weighted disjunction (LWD) operator is defined as

$$\mathrm{LWD}((w_1, a_1), (w_2, a_2), \cdots, (w_n, a_n)) = \max_j \min\{w_j, a_j\} \tag{2.14}$$

Definition 2.3 (Herrera and Herrera-Viedma, 1997) Let $(a_1, a_2, \cdots, a_n)$ be a collection of linguistic arguments, and $\boldsymbol{w} = (w_1, w_2, \cdots, w_n)^{\mathrm{T}}$ be the weight vector of $(a_1, a_2, \cdots, a_n)$, $w_j \in S_1$ for any j. Then a linguistic weighted conjunction (LWC) operator is defined as

$$\mathrm{LWC}((w_1, a_1), (w_2, a_2), \cdots, (w_n, a_n)) = \min_j \max\{\mathrm{neg}(w_j), a_j\} \tag{2.15}$$

Xu (2004d) developed an ordinal hybrid aggregation operator that reflects the importance degrees of both the linguistic arguments and its ordered position:

Definition 2.4 (Xu, 2004d) An ordinal hybrid aggregation (OHA) operator is a mapping OHA : $S_1^n \rightarrow S_1$, which has associated with it a linguistic weighting vector $\boldsymbol{\omega} = (\omega_1, \omega_2, \cdots, \omega_n)^{\mathrm{T}}$ with $\omega_j \in S_1$, such that

$$\mathrm{OHA}(a_1, a_2, \cdots, a_n) = \max_j \min\{\omega_j, b_j\} \tag{2.16}$$

where b_j is the jth largest of the linguistic weighted arguments $\bar{a}_i$ ($\bar{a}_i = \min\{w_i, a_i\}$, $i = 1, 2, \cdots, n$) and $\boldsymbol{w} = (w_1, w_2, \cdots, w_n)^{\mathrm{T}}$ is the linguistic weight vector of a_i ($i = 1, 2, \cdots, n$), with $w_i \in S_1$.

Especially, if $\boldsymbol{\omega} = (s_\tau, s_\tau, \cdots, s_\tau)^{\mathrm{T}}$, where $\omega_i = s_\tau$ for any i, then

$$\mathrm{OHA}(a_1, a_2, \cdots, a_n) = \max_j \min\{\omega_j, b_j\} = \max_i (\bar{a}_i) = \max_i \min\{w_i, a_i\} \tag{2.17}$$

Thus, in this case, the OHA operator reduces to the LWD operator.

If $\boldsymbol{w} = (s_\tau, s_\tau, \cdots, s_\tau)^{\mathrm{T}}$, where $w_i = s_\tau$ for any i, then

$$\bar{a}_i = \min\{w_i, a_i\} = a_i, \quad i = 1, 2, \cdots, n \tag{2.18}$$

In this case, the OHA operator reduces to the OOWA operator.

Example 2.4 Let $S_1 = \{s_0, s_1, s_2, s_3, s_4, s_5, s_6\}$ and assume that $(a_1, a_2, a_3, a_4, a_5, a_6) = (s_3, s_5, s_2, s_4, s_1, s_4)$ is a collection of linguistic labels in S_1, $\boldsymbol{w} = (s_2, s_3, s_5, s_3, s_4, s_1)^{\mathrm{T}}$ is the weight vector of the linguistic labels $(a_1, a_2, a_3, a_4, a_5, a_6)$. Then

$$\mathrm{LWD}((w_1, a_1), (w_2, a_2), \cdots, (w_6, a_6))$$

$$\begin{aligned}
&= \text{LWD}((s_2, s_3), (s_3, s_5), (s_5, s_2), (s_3, s_4), (s_4, s_1), (s_1, s_4)) \\
&= \max\{\min\{s_2, s_3\}, \min\{s_3, s_5\}, \min\{s_5, s_2\}, \min\{s_3, s_4\}, \\
&\quad \min\{s_4, s_1\}, \min\{s_1, s_4\}\} \\
&= \max\{s_2, s_3, s_2, s_3, s_1, s_1\} \\
&= s_3
\end{aligned}$$

$$\begin{aligned}
&\text{LWC}((w_1, a_1), (w_2, a_2), \cdots, (w_6, a_6)) \\
&= \min\{\max\{\text{neg}(s_2), s_3\}, \max\{\text{neg}(s_3), s_5\}, \max\{\text{neg}(s_5), s_2\}, \\
&\quad \max\{\text{neg}(s_3), s_4\}, \max\{\text{neg}(s_4), s_1\}, \max\{\text{neg}(s_1), s_4\}\} \\
&= \min\{\max\{s_4, s_3\}, \max\{s_3, s_5\}, \max\{s_1, s_2\}, \max\{s_3, s_4\}, \\
&\quad \max\{s_2, s_1\}, \max\{s_5, s_4\}\} \\
&= \min\{s_4, s_5, s_2, s_4, s_2, s_5\} \\
&= s_2
\end{aligned}$$

Since

$$\begin{aligned}
\bar{a}_1 &= \min\{w_1, a_1\} = \min\{s_2, s_3\} = s_2 \\
\bar{a}_2 &= \min\{w_2, a_2\} = \min\{s_3, s_5\} = s_3 \\
\bar{a}_3 &= \min\{w_3, a_3\} = \min\{s_5, s_2\} = s_2 \\
\bar{a}_4 &= \min\{w_4, a_4\} = \min\{s_3, s_4\} = s_3 \\
\bar{a}_5 &= \min\{w_5, a_5\} = \min\{s_4, s_1\} = s_1 \\
\bar{a}_6 &= \min\{w_6, a_6\} = \min\{s_1, s_4\} = s_1
\end{aligned}$$

then after reordering the linguistic weighted arguments $\bar{a}_i$ $(i = 1, 2, \cdots, 6)$ in descending order, we have

$$b_1 = b_2 = \bar{a}_2 = \bar{a}_4 = s_3, \quad b_3 = b_4 = \bar{a}_1 = \bar{a}_3 = s_2, \quad b_5 = b_6 = \bar{a}_5 = \bar{a}_6 = s_1$$

Let $\boldsymbol{\omega} = (s_1, s_2, s_3, s_3, s_2, s_1)^{\mathrm{T}}$ be the weighting vector associated with the OHA operator. Then

$$\begin{aligned}
\text{OHA}(a_1, a_2, \cdots, a_6) &= \max_j \min\{\omega_j, b_j\} \\
&= \max\{\min\{s_1, s_3\}, \min\{s_2, s_3\}, \min\{s_3, s_2\}, \min\{s_3, s_2\}, \\
&\quad \min\{s_2, s_1\}, \min\{s_1, s_1\}\} \\
&= \max\{s_1, s_2, s_2, s_2, s_1, s_1\} \\
&= s_2
\end{aligned}$$

2.2 Linguistic Aggregation Operators Based on Extension Principle and Symbols

Bonissone and Decker (1986), Bordogna and Passi (1993), Chang and Chen (1994), Chen (1997), Degani and Bortolan (1988), Delgado et al. (1993a), Lee (1996) studied the linguistic aggregation operators based on extension principle (Dubois and Prade, 1980). Herrera and Martínez (2000a) described an extension principle-based linguistic aggregation operator as follows:

$$S_1^n \xrightarrow{\tilde{F}} F(\mathbf{R}) \xrightarrow{\mathrm{app}_1(\cdot)} S_1 \tag{2.19}$$

where S_1^n symbolizes the n Cartesian product of S_1, $\tilde{F}$ is an aggregation operator based on extension principle, $F(\mathbf{R})$ is the set of fuzzy sets over the set of real numbers $\mathbf{R}$, $\mathrm{app}_1 : F(\mathbf{R}) \to S_1$ is a linguistic approximation function that returns a linguistic label from S_1 whose meaning is the closest to the obtained unlabeled fuzzy number, and S_1 is the initial linguistic label set.

Chang and Chen (1994) gave a decision algorithm based on extension principle based linguistic aggregation operator to solve the technology transfer strategy selection problem in the area of biotechnology management. The algorithm is based on the concepts of fuzzy set theory and the hierarchical structure analysis. The linguistic variables and fuzzy numbers were used to aggregate the decision makers' subjective assessments about criteria weightings and appropriateness of alternative transfer strategies versus selection criteria to obtain the final scores called fuzzy appropriateness indices. Chen (1997) utilized the extension principle-based linguistic aggregation operator to solve the tool steel materials selection problem in fuzzy environment where the importance weights of different criteria and the ratings of various alternatives under different criteria were assessed in linguistic terms represented by fuzzy numbers. Lee (1996) used the extension principle-based linguistic aggregation operator to build a structure model of risk in software development and proposed two algorithms to tackle the rate of aggregative risk in a fuzzy environment by fuzzy sets theory during any phase of the life cycle. Adamopoulos and Pappis (1996) proposed a fuzzy approach by using the extension principle-based linguistic aggregation operator to solve a single machine scheduling problem. In this approach, the system's variables are defined using linguistic terms. Each of these variables may take values described via triangular fuzzy numbers. Law (1996) built a structure model of a fuzzy educational grading system with the extension principle-based linguistic aggregation operator, and proposed an algorithm with it. They also discussed a method to build the membership functions of several linguistic values with different weights.

Delgado et al. (1993b) developed a symbol-based linguistic aggregation operator that makes compuations on the indices of the linguistic labels, which can be interpreted as (Herrera and Martínez, 2000a; 2000b)

$$S_1^n \xrightarrow{c} [0,\ \tau] \xrightarrow{\text{app}_2(\cdot)} \{0, \cdots, \tau\} \to S_1 \tag{2.20}$$

where c is a symbol-based linguistic aggregation operator, and $\text{app}_2(\cdot)$ is an approximation function used to obtain an index $\{0, \cdots, \tau\}$ associated to a linguistic label in $S_1 = \{s_0, s_1, \cdots, s_\tau\}$ from a value in $[0,\ \tau]$. For example, c is the convex combination of linguistic labels in S_1 and $\text{app}_2(\cdot)$ is the usual round operation.

Example 2.5 Let $S_1 = \{s_0, s_1, s_2, s_3, s_4\}$. Assume that

$$s_0 = \text{none},\quad s_1 = \text{low},\quad s_2 = \text{medium},\quad s_3 = \text{high},\quad s_4 = \text{perfect}$$

and their corresponding forms of interval numbers are denoted as

$$s_0 = [0, 0.2],\quad s_1 = [0.2, 0.4],\quad s_2 = [0.4, 0.6]$$

$$s_3 = [0.6, 0.8],\quad s_4 = [0.8, 1]$$

Moreover, let $(a_1, a_2, a_3, a_4) = (s_4, s_1, s_3, s_2)$ be a collection of linguistic arguments in S_1, and $\boldsymbol{w} = (0.2, 0.4, 0.3, 0.1)^{\mathrm{T}}$ be the weight vector of the linguistic arguments (a_1, a_2, a_3, a_4). We first denote the linguistic arguments (a_1, a_2, a_3, a_4) as interval numbers:

$$a_1 = [0.8, 1],\quad a_2 = [0.2, 0.4],\quad a_3 = [0.6, 0.8],\quad a_4 = [0.4, 0.6]$$

Then we utilize the operational laws of interval numbers (Xu and Da, 2002a; Xu and Zhai, 1992) and the uncertain weighted averaging operator (Xu and Da, 2002b) to aggregate the arguments a_i $(i = 1, 2, 3, 4)$:

$$\begin{aligned}\bar{a} =& 0.2 \times [0.8, 1] + 0.4 \times [0.2, 0.4] + 0.3 \times [0.6, 0.8] + 0.1 \times [0.4, 0.6] \\ =& [0.2 \times 0.8, 0.2 \times 1] + [0.4 \times 0.2, 0.4 \times 0.4] + [0.3 \times 0.6, 0.3 \times 0.8] \\ &+ [0.1 \times 0.4, 0.1 \times 0.6] \\ =& [0.16, 0.2] + [0.08, 0.16] + [0.18, 0.24] + [0.04, 0.06] \\ =& [0.16 + 0.08 + 0.18 + 0.04, 0.2 + 0.16 + 0.24 + 0.06] \\ =& [0.46, 0.66]\end{aligned}$$

Then we utilize the Euclidean distance measure to calculate the distance between $\bar{a}$ and the linguistic labels s_i $(i = 1, 2, 3, 4)$:

$$d(\bar{a}, s_0) = \sqrt{(0.46 - 0)^2 + (0.66 - 0.2)^2} = 0.46$$

$$d(\bar{a}, s_1) = \sqrt{(0.46 - 0.2)^2 + (0.66 - 0.4)^2} = 0.26$$

$$d(\bar{a}, s_2) = \sqrt{(0.46 - 0.4)^2 + (0.66 - 0.6)^2} = 0.06$$

$$d(\bar{a}, s_3) = \sqrt{(0.46 - 0.6)^2 + (0.66 - 0.8)^2} = 0.14$$

$$d(\bar{a}, s_4) = \sqrt{(0.46 - 0.8)^2 + (0.66 - 1)^2} = 0.34$$

Accordingly,

$$d(\bar{a}, s_2) < d(\bar{a}, s_3) < d(\bar{a}, s_2) < d(\bar{a}, s_4) < d(\bar{a}, s_0)$$

Since the distance between $\bar{a}$ and s_2 is the shortest one, we can take approximately $\bar{a}$ as the linguistic label "medium".

Herrera and Verdegay (1993) presented a linguistic OWA operator, which is based on the OWA operator (Yager, 1998) and the convex combination of linguistic labels (Delgado et al., 1993b).

Definition 2.5 (Herrera and Verdegay, 1993) Let $(a_1, a_2, \cdots, a_n)$ be a collection of linguistic labels to be aggregated. Then a linguistic OWA (LOWA_1) operator is defined as

$$\begin{aligned}\text{LOWA}_1(a_1, a_2, \cdots, a_n) =& \boldsymbol{\omega} \cdot \boldsymbol{b}^{\text{T}} = c^n\{\omega_k, b_k, k = 1, 2, \cdots, n\} \\ =& \omega_1 \otimes b_1 \oplus (1 - \omega_1) \otimes c^{n-1}\{\gamma_h, b_h, h = 2, \cdots, n\}\end{aligned} \tag{2.21}$$

where $\boldsymbol{\omega} = (\omega_1, \omega_2, \cdots, \omega_n)^{\text{T}}$ is a weighting vector, such that

$$\omega_i \in [0, 1],\ i = 1, 2, \cdots, n,\ \sum_{i=1}^{n} \omega_i = 1,\ \ \gamma_h = \frac{\omega_h}{\sum_{k=2}^{n} \omega_k},\ \ h = 2, \cdots, n \tag{2.22}$$

and $\boldsymbol{b} = (b_1, b_2, \cdots, b_m)$ is a vector related to $(a_1, a_2, \cdots, a_n)$, such that

$$\boldsymbol{b} = \sigma(a_1, a_2, \cdots, a_n) = (a_{\sigma(1)}, a_{\sigma(2)}, \cdots, a_{\sigma(n)}) \tag{2.23}$$

where $a_{\sigma(j)} \leqslant a_{\sigma(i)}$ for all $i \leqslant j$. $(a_{\sigma(1)}, a_{\sigma(2)}, \cdots, a_{\sigma(n)})$ is any permutation of $(a_1, a_2, \cdots, a_n)$, c^n is the convex combination operator of n linguistic labels and if $n = 2$. Then it is defined as

$$c^2\{\omega_i, b_i, i = 1, 2\} = \omega_1 \otimes s_j \oplus (1 - \omega_1) \otimes s_i = s_k, \quad s_j, s_i \in S_1,\ j \geqslant i \tag{2.24}$$

such that

$$k = \min\{\tau,\ i + \text{round}(\omega_1 \cdot (j - i))\} \tag{2.25}$$

where "round" is the usual round operation, $b_1 = s_j$ and $b_2 = s_i$.

If $\omega_j = 1$ and $\omega_i = 0$ for any $i \neq j$, then

$$c^n\{\omega_i, b_i, i = 1, 2, \cdots, n\} = b_j \tag{2.26}$$

Many approaches have been developed for determining the weighting vector $\boldsymbol{\omega} = (\omega_1, \omega_2, \cdots, \omega_n)^{\mathrm{T}}$, see Xu (2005d) for a detailed overview on this field.

Delgado et al. (1998), Herrera and Herrera-Viedma (2000a), Herrera et al. (1997; 1996a; 1996b; 1995) utilized the LOWA$_1$ operator to develop various approaches to group decision making under linguistic assessments, and presented some consensus models for the consensus reaching process in heterogeneous linguistic group decision making problems. Herrera et al. (2001b) established a linguistic decision model based on the LOWA$_1$ operator for promotion mix management solved with genetic algorithms, and demonstrated the usefulness of the model by solving a real problem from the business world. Herrera-Viedma and Peis (2003) presented a fuzzy linguistic evaluation method based on the LOWA$_1$ operator to characterize the information contained in SGML (standard generalized markup language)-based documents. The method generates linguistic recommendations from linguistic evaluation judgments provided by different recommenders on meaningful elements of document type definition (DTD).

Herrera and Herrera-Viedma (1997) proved that the LOWA$_1$ operator has the following properties:

(1) The LOWA$_1$ operator increases as the linguistic labels get larger;

(2) The LOWA$_1$ operator has the property of commutativity;

(3) The LOWA$_1$ operator is an "orand" operator, which aggregates the input arguments that lie between the max and min operators.

They also presented an extension of the LOWA$_1$ operator, called the inverse-LOWA$_1$ (I-LOWA$_1$) operator.

Definition 2.6 (Herrera and Herrera-Viedma, 1997) The I-LOWA$_1$ operator is a linguistic OWA operator, such that

$$\boldsymbol{b} = \sigma^{\mathrm{I}}(a_1, a_2, \cdots, a_n) = (a_{\sigma(1)}, a_{\sigma(2)}, \cdots, a_{\sigma(n)}) \tag{2.27}$$

where $a_{\sigma(i)} \leqslant a_{\sigma(j)}$ for all $i \leqslant j$. If $n = 2$, then

$$C^2\{\omega_i, b_i, i = 1, 2\} = \omega_1 \otimes s_j \oplus (1 - \omega_1) \otimes s_i = s_k, \quad s_j, s_i \in S_1, \ j \leqslant i \tag{2.28}$$

such that

$$k = \min\{\tau, \ i + \mathrm{round}(\omega_1 \cdot (j - i))\} \tag{2.29}$$

The I-LOWA$_1$ operator has some properties similar to those of the LOWA$_1$ operator.

Example 2.6 Let $S_1 = \{s_0, s_1, s_2, s_3, s_4\}$, and assume that $(a_1, a_2, a_3, a_4) = (s_2, s_4, s_1, s_3)$ is a collection of linguistic arguments in S_1. We first reorder the linguistic arguments (a_1, a_2, a_3, a_4) in descending order:

$$b_1 = a_2 = s_4, \quad b_2 = a_4 = s_3, \quad b_3 = a_1 = s_2, \quad b_4 = a_3 = s_1$$

Let $\boldsymbol{\omega} = (0.3, 0.4, 0.1, 0.2)^{\mathrm{T}}$ be the weighting vector associated with the LOWA$_1$ operator. Since

$$\begin{aligned}\mathrm{LOWA}_1(a_1, a_2, a_3, a_4) &= \boldsymbol{\omega} \cdot \boldsymbol{b}^{\mathrm{T}} = c^4\{\omega_k, b_k, k = 1, 2, 3, 4\} \\ &= 0.3 \otimes s_4 \oplus (1 - 0.3) \otimes c^{4-1}\{\gamma_h, b_h, h = 2, 3, 4\} \\ &= 0.3 \otimes s_4 \oplus 0.7 \otimes c^3\{\gamma_h, b_h, h = 2, 3, 4\}\end{aligned}$$

we utilize

$$\gamma_h = \frac{\omega_h}{\sum\limits_{k=2}^{4} \omega_k}, \quad h = 2, 3, 4$$

to calculate γ_h $(h = 2, 3, 4)$, and get

$$\gamma_2 = 0.571, \quad \gamma_3 = 0.143, \quad \gamma_4 = 0.286$$

Thus

$$c^3\{\gamma_h, b_h, h = 2, 3, 4\} = 0.571 \otimes s_3 \oplus 0.429 \otimes c^{3-1}\{\gamma_h', b_h, h = 3, 4\}$$

Then we utilize

$$\gamma_h' = \frac{\gamma_h}{\sum\limits_{k=3}^{4} \gamma_k}, \quad h = 3, 4$$

to calculate γ_h' $(h = 3, 4)$, and get

$$\gamma_3' = 0.333, \quad \gamma_4' = 0.667$$

Consequently,

$$c^2\{\gamma_h', b_h, h = 3, 4\} = 0.333 \otimes s_2 \oplus 0.667 \otimes s_1$$

According to (2.29), we have

$$k = \min\{4,\ 1 + \mathrm{round}(0.333 \times (2 - 1)\} = 1$$

i.e.,

$$c^2\{\gamma_h', b_h, h = 3, 4\} = s_1$$

Similarly, we have

$$c^3\{\gamma_h, b_h, h = 2, 3, 4\} = s_2$$

$$c^4\{\omega_k, b_k, k = 1, 2, 3, 4\} = s_3$$

i.e.,

$$\mathrm{LOWA}_1(a_1, a_2, a_3, a_4) = s_3$$

In what follows, we utilize the I-LOWA$_1$ operator to aggregate the linguistic arguments (a_1, a_2, a_3, a_4):

We first reorder the linguistic arguments (a_1, a_2, a_3, a_4) in descending order:

$$b_1 = a_3 = s_1, \quad b_2 = a_1 = s_2, \quad b_3 = a_4 = s_3, \quad b_4 = a_2 = s_4$$

Let $\boldsymbol{w} = (0.2, 0.4, 0.3, 0.1)^{\mathrm{T}}$ be the weighting vector associated with the I-LOWA$_1$ operator. Since

$$\begin{aligned}\text{I-LOWA}_1(a_1, a_2, a_3, a_4) &= \boldsymbol{\omega} \cdot \boldsymbol{b}^{\mathrm{T}} = c^4\{\omega_k, b_k, k = 1, 2, 3, 4\} \\ &= 0.2 \otimes s_1 \oplus (1 - 0.2) \otimes c^{4-1}\{\gamma_h, b_h, h = 2, 3, 4\} \\ &= 0.2 \otimes s_1 \oplus 0.8 \otimes c^3\{\gamma_h, b_h, h = 2, 3, 4\}\end{aligned}$$

we use

$$\gamma_h = \frac{\omega_h}{\sum_{k=2}^{4} \omega_k}, \quad h = 2, 3, 4$$

to calculate γ_h $(h = 2, 3, 4)$, and get

$$\gamma_2 = 0.500, \quad \gamma_3 = 0.375, \quad \gamma_4 = 0.125$$

Therefore,

$$c^3\{\gamma_h, b_h, h = 2, 3, 4\} = 0.500 \otimes s_2 \oplus 0.500 \otimes c^{3-1}\{\gamma'_h, b_h, h = 3, 4\}$$

Then we use

$$\gamma'_h = \frac{\gamma_h}{\sum_{k=3}^{4} \gamma_k}, \quad h = 3, 4$$

to calculate γ'_h $(h = 3, 4)$, and get

$$\gamma'_3 = 0.750, \quad \gamma'_4 = 0.250$$

Hence

$$c^2\{\gamma'_h, b_h, h = 3, 4\} = 0.750 \otimes s_3 \oplus 0.250 \otimes s_4$$

According to (2.25), we have

$$k = \min\{4,\ 4 + \text{round}(0.750 \times (3 - 4))\} = 3$$

i.e.,

$$c^2\{\gamma'_h, b_h, h = 3, 4\} = s_3$$

Similarly, we can obtain

$$c^3\{\gamma_h, b_h, h = 2, 3, 4\} = s_3$$

$$c^4\{\omega_k, b_k, k = 1, 2, 3, 4\} = s_3$$

i.e.,

$$\text{I-LOWA}_1(a_1, a_2, a_3, a_4) = s_3$$

Based on the LOWA_1 and I-LOWA_1 operators, Herrera and Herrera-Viedma (2000a) defined a linguistic weighted averaging operator using the concept of fuzzy majority (Zadeh, 1983) represented by fuzzy linguistic quantifiers and two families of linguistic connectives.

In the following, let us first review two families of linguistic connectives (Herrera and Herrera-Viedma, 2000a):

Let $a_i, w_i \in S_1$, where w_i is the weight of a_i. Then

(1) Linguistic conjunction operators $\text{LC}^{\rightarrow}$:

(i) The classical linguistic min operator:

$$\text{LC}_1^{\rightarrow}(w_i, a_i) = \min\{w_i, a_i\} \tag{2.30}$$

(ii) The nipotent linguistic min operator:

$$\text{LC}_2^{\rightarrow}(w_i, a_i) = \begin{cases} \min\{w_i, a_i\}, & w_i > \text{neg}(a_i) \\ s_0, & \text{otherwise} \end{cases} \tag{2.31}$$

(iii) The weakest linguistic conjunction:

$$\text{LC}_3^{\rightarrow}(w_i, a_i) = \begin{cases} \min\{w_i, a_i\}, & \max\{w_i, a_i\} = s_\tau \\ s_0, & \text{otherwise} \end{cases} \tag{2.32}$$

(2) Linguistic implication operators $\text{LI}^{\rightarrow}$:

(i) Kleene-Dienes' linguistic implication operator:

$$\text{LI}_1^{\rightarrow}(w_i, a_i) = \max\{\text{neg}(w_i), a_i\} \tag{2.33}$$

(ii) Gödel's linguistic implication operator:

$$\text{LI}_2^{\rightarrow}(w_i, a_i) = \begin{cases} s_\tau, & w_i \leqslant a_i \\ a_i & \text{otherwise} \end{cases} \tag{2.34}$$

(iii) Fodor's linguistic implication operator:

$$\text{LI}_3^{\rightarrow}(w_i, a_i) = \begin{cases} s_\tau, & w_i \leqslant a_i \\ \max\{\text{neg}(w_i),\ a_i\}, & \text{otherwise} \end{cases} \tag{2.35}$$

(iv) Lukasiewicz's linguistic implication operator:

$$\text{LI}_4^{\rightarrow}(w_i, a_i) = \begin{cases} s_\tau, & w_i \leqslant a_i \\ \text{neg}(w_i - a_i), & \text{otherwise} \end{cases} \tag{2.36}$$

Based on the linguistic connection functions, Herrera and Herrera-Viedma (2000a) defined the following concept:

Definition 2.7 (Herrera and Herrera-Viedma, 2000a) Let $(a_1, a_2, \cdots, a_n)$ be a collection of linguistic labels to be aggregated, $\boldsymbol{w} = (w_1, w_2, \cdots, w_n)^{\mathrm{T}}$ be the weight vector of $(a_1, a_2, \cdots, a_n)$, and $a_j, w_j \in S_1$ for any j. Then a linguistic weighted averaging (LWA_1) operator is defined as

$$\mathrm{LWA}_1((w_1, a_1), (w_2, a_2), \cdots, (w_n, a_n)) = f(g(w_1, a_1), g(w_2, a_2), \cdots, g(w_n, a_n)) \tag{2.37}$$

where f is a LOWA_1 (or I-LOWA_1) operator. If f is a LOWA_1 operator, then $g \in \mathrm{LC}^{\rightarrow}$; if LWA_1 is an I-LOWA_1 operator, then $g \in \mathrm{LI}^{\rightarrow}$, where

$$\mathrm{LC}^{\rightarrow} = \{\mathrm{LC}_1^{\rightarrow}, \mathrm{LC}_2^{\rightarrow}, \mathrm{LC}_3^{\rightarrow}\} \quad \text{and} \quad \mathrm{LI}^{\rightarrow} = \{\mathrm{LI}_1^{\rightarrow}, \mathrm{LI}_2^{\rightarrow}, \mathrm{LI}_3^{\rightarrow}\}$$

Example 2.7 Let $S_1 = \{s_0, s_1, s_2, s_3, s_4, s_5, s_6\}$, and assume that $(a_1, a_2, a_3, a_4, a_5) = (s_6, s_2, s_4, s_3, s_1)$ is a collection of linguistic arguments in S_1, $\boldsymbol{w} = (s_5, s_3, s_5, s_4, s_2)^{\mathrm{T}}$ is the weight vector of $(a_1, a_2, a_3, a_4, a_5)$. In order to illustrate (2.37), without loss of generality, we take f as a LOWA_1 operator and $g = \mathrm{LC}_1^{\rightarrow}$ (similarly, we can consider other cases).

Since

$$\mathrm{LC}_1^{\rightarrow}(w_1, a_1) = \min\{w_1, a_1\} = \min\{s_5, s_6\} = s_5$$

$$\mathrm{LC}_1^{\rightarrow}(w_2, a_2) = \min\{w_2, a_2\} = \min\{s_3, s_2\} = s_2$$

$$\mathrm{LC}_1^{\rightarrow}(w_3, a_3) = \min\{w_3, a_3\} = \min\{s_5, s_4\} = s_4$$

$$\mathrm{LC}_1^{\rightarrow}(w_4, a_4) = \min\{w_4, a_4\} = \min\{s_4, s_3\} = s_3$$

$$\mathrm{LC}_1^{\rightarrow}(w_5, a_5) = \min\{w_5, a_5\} = \min\{s_2, s_1\} = s_1$$

we reorder the linguistic arguments $\mathrm{LC}_1^{\rightarrow}(w_i, a_i)$ $(i = 1, 2, 3, 4, 5)$ in descending order:

$$b_1 = \mathrm{LC}_1^{\rightarrow}(w_1, a_1) = s_5, \quad b_2 = \mathrm{LC}_1^{\rightarrow}(w_3, a_3) = s_4$$

$$b_3 = \mathrm{LC}_1^{\rightarrow}(w_4, a_4) = s_3, \quad b_4 = \mathrm{LC}_1^{\rightarrow}(w_2, a_2) = s_2$$

$$b_5 = \mathrm{LC}_1^{\rightarrow}(w_5, a_5) = s_1$$

Let $\boldsymbol{\omega} = (0.1, 0.4, 0.2, 0.1, 0.2)^{\mathrm{T}}$ be the weighting vector associated with the LOWA_1 operator. Since

$$\begin{aligned}
&\mathrm{LWA}_1((w_1, a_1), (w_2, a_2), \cdots, (w_5, a_5)) \\
=&\mathrm{LOWA}_1(\mathrm{LC}_1^{\rightarrow}(w_1, a_1), \mathrm{LC}_1^{\rightarrow}(w_2, a_2), \cdots, \mathrm{LC}_1^{\rightarrow}(w_5, a_5)) \\
=&\boldsymbol{\omega} \cdot \boldsymbol{b}^{\mathrm{T}} = c^5\{\omega_k, b_k, k = 1, 2, 3, 4, 5\}
\end{aligned}$$

$$=0.1 \otimes s_5 \oplus (1-0.1) \otimes c^{5-1}\{\gamma_h, b_h, h=2,3,4,5\}$$
$$=0.1 \otimes s_5 \oplus 0.9 \otimes c^4\{\gamma_h, b_h, h=2,3,4,5\}$$

we use

$$\gamma_h = \frac{\omega_h}{\sum_{k=2}^{5} \omega_k}, \quad h=2,3,4,5$$

to calculate γ_h $(h=2,3,4,5)$, and get

$$\gamma_2 = 0.444, \quad \gamma_3 = 0.222, \quad \gamma_4 = 0.111, \quad \gamma_5 = 0.222$$

Thus

$$c^4\{\gamma_h, b_h, h=2,3,4,5\} = 0.444 \otimes s_4 \oplus (1-0.444) \otimes c^{4-1}\{\gamma_h', b_h, h=3,4,5\}$$

After that, we use

$$\gamma_h' = \frac{\gamma_h}{\sum_{k=3}^{5} \gamma_k}, \quad h=3,4,5$$

to calculate γ_h' $(h=3,4,5)$, and get

$$\gamma_3' = 0.400, \quad \gamma_4' = 0.200, \quad \gamma_5' = 0.400$$

Hence

$$c^3\{\gamma_h', b_h, h=3,4,5\} = 0.400 \otimes s_3 \oplus (1-0.400) \otimes c^{3-1}\{\gamma_h'', b_h, h=4,5\}$$

Furthermore, we utilize

$$\gamma_h'' = \frac{\gamma_h'}{\sum_{k=4}^{5} \gamma_k'}, \quad h=4,5$$

to calculate γ_h'' $(h=4,5)$, and get

$$\gamma_4'' = 0.333, \quad \gamma_5'' = 0.667$$

Therefore,

$$c^2\{\gamma_h'', b_h, h=4,5\} = 0.333 \otimes s_2 \oplus 0.667 \otimes s_1$$

According to (2.25), we have

$$k = \min\{6,\ 1 + \text{round}(0.333 \times (2-1))\} = 1$$

i.e.,

$$c^2\{\gamma_h'', b_h, h = 4, 5\} = s_1$$

Similarly, we get

$$c^3\{\gamma_h', b_h, h = 3, 4, 5\} = s_2$$

$$c^4\{\gamma_h, b_h, h = 2, 3, 4, 5\} = s_3$$

$$c^5\{\omega_k, b_k, k = 1, 2, 3, 4, 5\} = s_3$$

i.e.,

$$\text{LWA}_1((w_1, a_1), (w_2, a_2), \cdots, (w_5, a_5)) = s_3$$

Torra (1997) defined a weighted OWA operator that combines the advantages of the OWA operator and the ones of the weighted mean. In order to aggregate linguistic information, he extended the operator to linguistic environments and defined a linguistic weighted OWA operator:

Definition 2.8 (Torra, 1997) Let $(a_1, a_2, \cdots, a_n)$ be a collection of linguistic labels, $\boldsymbol{\omega} = (\omega_1, \omega_2, \cdots, \omega_n)^{\mathrm{T}}$ and $\boldsymbol{\xi} = (\xi_1, \xi_2, \cdots, \xi_n)^{\mathrm{T}}$ be the weighting vectors, which satisfy the following conditions:

$$\omega_i \geqslant 0, \quad i = 1, 2, \cdots, n, \quad \sum_{i=1}^{n} \omega_i = 1 \tag{2.38}$$

$$\xi_i \geqslant 0, \quad i = 1, 2, \cdots, n, \quad \sum_{i=1}^{n} \xi_i = 1 \tag{2.39}$$

and let LWOWA : $S_1^n \to S_1$. Then

$$\begin{aligned} &\text{LWOWA}(a_1, a_2, \cdots, a_n) \\ =&c^n\{\omega_k, b_k, k = 1, 2, \cdots, n\} \\ =&\omega_1 \otimes b_1 \oplus (1 - \omega_1) \otimes c^{n-1}\left\{\frac{\omega_h}{\sum\limits_{k=2}^{n} \omega_k}, b_h, h = 2, \cdots, n\right\} \end{aligned} \tag{2.40}$$

is called a linguistic weighted OWA (LWOWA) operator, where

$$(b_1, b_2, \cdots, b_n) = (a_{\sigma(1)}, a_{\sigma(2)}, \cdots, a_{\sigma(n)}) \tag{2.41}$$

$(a_{\sigma(1)}, a_{\sigma(2)}, \cdots, a_{\sigma(n)})$ is any permutation of $(a_1, a_2, \cdots, a_n)$, such that $a_{\sigma(j)} \leqslant a_{\sigma(i)}$ for any $i \leqslant j$. The convex combination c is defined according to Defintiion 2.5, and the weight ω_i is defined as

$$\omega_i = w^*\left(\sum_{j \leqslant i} \xi_{\sigma(j)}\right) - w^*\left(\sum_{j < i} \xi_{\sigma(j)}\right) \tag{2.42}$$

where w^* is a monotone increasing function that interpolates the points $\left(\frac{i}{n}, \sum_{j\leqslant i} w_j\right)$ together with the point (0, 0). w^* is required to straight line when the points can be interpolated in this way, $(\sigma(1), \sigma(2), \cdots, \sigma(n))$ is any permutation of $(1, 2, \cdots, n)$, such that $\xi_{\sigma(i-1)} \geqslant \xi_{\sigma(i)}$ $(i = 2, 3, \cdots, n)$.

Example 2.8 Let $S_1 = \{s_0, s_1, s_2, s_3, s_4, s_5, s_6\}$, and assume that $(a_1, a_2, a_3, a_4) = (s_5, s_3, s_6, s_2)$ is a collection of linguistic arguments in S_1, $\boldsymbol{\xi} = (0.3, 0.2, 0.1, 0.4)^{\mathrm{T}}$ and $\boldsymbol{\omega} = (0.25, 0.25, 0.25, 0.25)^{\mathrm{T}}$ are the weighting vectors. Since

$$\xi_{\sigma(1)} = 0.4, \quad \xi_{\sigma(2)} = 0.3, \quad \xi_{\sigma(3)} = 0.2, \quad \xi_{\sigma(4)} = 0.1$$

we get by (2.42) that $\omega_i = \xi_{\sigma(i)}$, i.e.,

$$\omega_1 = 0.4, \quad \omega_2 = 0.3, \quad \omega_3 = 0.2, \quad \omega_4 = 0.1$$

Also since

$$b_1 = a_{\sigma(1)} = a_3 = s_6, \quad b_2 = a_{\sigma(2)} = a_1 = s_5$$

$$b_3 = a_{\sigma(3)} = a_2 = s_3, \quad b_4 = a_{\sigma(4)} = a_4 = s_2$$

and by (2.40), we have

$$\begin{aligned}\text{LWOWA}(a_1, a_2, a_3, a_4) &= c^4\{\omega_k, b_k, k = 1, 2, 3, 4\} \\ &= 0.4 \otimes s_6 \oplus (1 - 0.4) \otimes C^{4-1}\left\{\frac{\omega_h}{\sum_{k=2}^{4} \omega_k}, b_h, h = 2, 3, 4\right\}\end{aligned}$$

Then we use

$$\gamma_h = \frac{\omega_h}{\sum_{k=2}^{4} \omega_k}, \quad h = 2, 3, 4$$

to calculate γ_h $(h = 2, 3, 4)$, and get

$$\gamma_2 = 0.500, \quad \gamma_3 = 0.333, \quad \gamma_4 = 0.167$$

Accordingly,

$$c^3\{\gamma_h, b_h, h = 2, 3, 4\} = 0.500 \otimes s_5 \oplus 0.500 \otimes c^{3-1}\{\gamma'_h, b_h, h = 3, 4\}$$

Furthermore, we use

$$\gamma'_h = \frac{\gamma_h}{\sum_{k=3}^{4} \gamma_k}, \quad h = 3, 4$$

to calculate γ'_h $(h = 3, 4)$, and get

$$\gamma'_3 = 0.666, \quad \gamma'_4 = 0.334$$

Consequently,

$$c^2\{\gamma'_h, b_h, h = 3, 4\} = 0.666 \otimes s_3 \oplus 0.667 \otimes s_2$$

According to (2.25), we get

$$k = \min\{6,\ 2 + \text{round}(0.666 \times (3 - 2))\} = 3$$

i.e.,

$$c^2\{\gamma'_h, b_h, h = 3, 4\} = s_3$$

Similarly, we have

$$c^3\{\gamma_h, b_h, h = 2, 3, 4\} = s_4$$

$$c^4\{\omega_k, b_k, k = 1, 2, 3, 4\} = s_5$$

i.e.,

$$\text{LWOWA}(a_1, a_2, a_3, a_4) = s_5$$

2.3 Linguistic Aggregation Operators Based on Linguistic 2-tuples

An important limitation for the extension principle-based linguistic aggregation operator and symbol-based linguistic aggregation operator appears because the computational techniques used present a common drawback, the "loss of information", that implies a lack of precision in the final results. To overcome this drawback, Herrera and Martínez (2000a) developed a fuzzy linguistic representation model, which represents the linguistic information with a pair of values called 2-tuple, $(s,\ \alpha)$, where $s \in S_1$ is a linguistic label and $\alpha \in [-0.5, 0.5)$ is a numerical value that represents the value of the symbolic translation.

Definition 2.9 (Herrera and Martínez, 2000a) Let β be the result of an aggregation of the indices of a collection of linguistic labels assessed in the linguistic evaluation scale S_1, i.e., the result of a symbolic aggregation operation. $\beta \in [0, \tau]$, being $\tau + 1$ the cardinality of S_1, and let $i = \text{round}(\beta)$ and $\alpha = \beta - i$, such that $i \in [0, \tau]$ and $\alpha \in [-0.5, 0.5)$. Then α is called a symbolic translation.

Herrera and Martínez (2000a) pointed out that the symbolic translation of a linguistic label, s_i, is a numerical value assessed in $[-0.5, 0.5)$ that supports the "difference of information" between a counting of information $\beta \in [0, \tau]$ obtained after a symbolic aggregation operation and the closest value $i = \text{round}(\beta)$ in $\{0, \cdots, \tau\}$ that indicates the index of the closest linguistic label in S_1.

Definition 2.10 (Herrera and Martínez, 2000a) The equivalent 2-tuple representation of linguistic label $s_i \in S_1$ is obtained by means of the function ϑ as

$$\vartheta : S_1 \to S_1 \times [-0.5, 0.5)$$

$$\vartheta(s_i) = (s_i, 0)/s_i \in S_1 \tag{2.43}$$

Example 2.9 Let $S_1 = \{s_0, s_1, s_2, s_3, s_4, s_5, s_6\}$. Then the 2-tuple representation of S_1 is as follows:

$$S_1 \times [-0.5, 0.5) = \{(s_0, 0), (s_1, 0), (s_2, 0), (s_3, 0), (s_4, 0), (s_5, 0), (s_6, 0)\}$$

Definition 2.11 (Herrera and Martínez, 2000a) Let S_1 be a linguistic evaluation scale and $\beta \in [0, \tau]$ be a value of symbolic aggregation operation. Then the 2-tuple $(s_i,\ \alpha)$ that expresses the equivalent information to β is obtained with the following function Δ:

$$\Delta : [0, \tau] \to S_1 \times [-0.5, 0.5)$$

$$\Delta(\beta) = \begin{cases} s_i, & i = \text{round}(\beta) \\ \alpha = \beta - i, & \alpha \in [-0.5, 0.5) \end{cases} \tag{2.44}$$

with "round" being the usual round operation, s_i has the closest index label to β and α is the value of the symbolic translation.

Example 2.10 Let $S_1 = \{s_0, s_1, s_2, s_3, s_4, s_5, s_6\}$ and $\beta = 5.5$ be the value of symbolic aggregation operation. Then

$$\Delta(\beta) = \Delta(5.5) = (s_6, -0.5)$$

For the linguistic evaluation scale S_1 and the 2-tuple $(s_i,\ \alpha)$, there is always a function Δ^{-1}, such that, from the 2-tuple $(s_i,\ \alpha)$ it returns its equivalent numerical value $\beta \in [0, \tau]$:

$$\Delta^{-1} : S_1 \times [-0.5,\ 0.5) \to [0, \tau]$$

$$\Delta^{-1}(s_i, \alpha) = i + \alpha = \beta \tag{2.45}$$

Example 2.11 Let $S_1 = \{s_0, s_1, s_2, s_3, s_4, s_5, s_6\}$ and a 2-tuple $(s_i, \alpha) = (s_4,\ 0.3)$. Then

$$\Delta^{-1}(s_i, \alpha) = \Delta^{-1}(s_4, 0.3) = 4 + 0.3 = 4.3$$

Herrera and Martínez (2000a) further introduced some 2-tuple aggregation operators as follows:

(1) 2-tuple comparison operator. The linguistic labels represented by 2-tuples can be compared with the following approach:

Let (s_k, α_1) and (s_l, α_2) be two 2-tuples. Then

(i) If $k < l$, then (s_k, α_1) is smaller than (s_l, α_2), i.e., $(s_k, \alpha_1) < (s_l, \alpha_2)$;

(ii) If $k = l$, then

(a) If $\alpha_1 = \alpha_2$, then (s_k, α_1) and (s_l, α_2) represent the same information, i.e., $(s_k, \alpha_1) = (s_l, \alpha_2)$;

(b) If $\alpha_1 < \alpha_2$, then (s_k, α_1) is smaller than (s_l, α_2), i.e., $(s_k, \alpha_1) < (s_l, \alpha_2)$;

(c) If $\alpha_1 > \alpha_2$, then (s_k, α_1) is larger than (s_l, α_2), i.e., $(s_k, \alpha_1) > (s_l, \alpha_2)$.

Example 2.12 Let $((s_4, -0.4), (s_3, 0.2), (s_4, 0.1), (s_3, 0), (s_5, -0.5))$ be a collection of 2-tuples. Then according to the 2-tuple comparison operator, we get

$$(s_5, -0.5) > (s_4, 0.1) > (s_4, -0.4) > (s_3, 0.2) > (s_3, 0)$$

(2) 2-tuple negation operator. The negation operator over a 2-tuple can be defined as

$$\text{neg}(s_i, \alpha) = \Delta(\tau - (\Delta^{-1}(s_i, \alpha))) \tag{2.46}$$

Example 2.13 Let $S_1 = \{s_0, s_1, s_2, s_3, s_4, s_5, s_6\}$ and $(s_i, \alpha) = (s_4, 0.2)$ be a 2-tuple. Then according to the 2-tuple negation operator, we have

$$\text{neg}(s_4, 0.2) = \Delta(6 - (\Delta^{-1}(s_4, 0.2))) = \Delta(6 - 4.2) = \Delta(1.8) = (s_2, -0.2)$$

(3) 2-tuple averaging operator. Let $((r_1, \alpha_1), (r_2, \alpha_2), \cdots, (r_n, \alpha_n))$ be a collection of 2-tuples. Then

$$\text{TAM}((r_1, \alpha_1), (r_2, \alpha_2), \cdots, (r_n, \alpha_n)) = \Delta\left(\sum_{i=1}^{n} \frac{1}{n}\Delta^{-1}(r_i, \alpha_i)\right) = \Delta\left(\frac{1}{n}\sum_{i=1}^{n} \beta_i\right) \tag{2.47}$$

is called a 2-tuple averaging operator.

Example 2.14 Let $((s_4, -0.4), (s_3, 0.2), (s_4, 0.1), (s_3, 0), (s_5, -0.5), (s_2, 0.3))$ be a collection of 2-tuples. Since

$$\Delta^{-1}(s_4, -0.4) = 3.6, \quad \Delta^{-1}(s_3, 0.2) = 3.2, \quad \Delta^{-1}(s_4, 0.1) = 4.1$$

$$\Delta^{-1}(s_3, 0) = 3, \quad \Delta^{-1}(s_5, -0.5) = 4.5, \quad \Delta^{-1}(s_2, 0.3) = 2.3$$

then according to the 2-tuple averaging operator (2.47), we get

$$\begin{aligned} &\text{TAM}((s_4, -0.4), (s_3, 0.2), (s_4, 0.1), (s_3, 0), (s_5, -0.5), (s_2, 0.3)) \\ =&\Delta\left(\frac{1}{6} \times (3.6 + 3.2 + 4.1 + 3 + 4.5 + 2.3)\right) \\ =&\Delta(3.45) = (s_3, 0.45) \end{aligned}$$

(4) 2-tuple weighted averaging operator. Let $((r_1, \alpha_1), (r_2, \alpha_2), \cdots, (r_n, \alpha_n))$ be a collection of 2-tuples, and $\boldsymbol{w} = (w_1, w_2, \cdots, w_n)^{\mathrm{T}}$ be their weight vector, which satisfies the condition:

$$w_i \geqslant 0, \quad i = 1, 2, \cdots, n, \quad \sum_{i=1}^{n} w_i = 1$$

Then

$$\text{TWA}((r_1, \alpha_1),\ (r_2, \alpha_2),\ \cdots, (r_n, \alpha_n)) = \Delta\left(\sum_{i=1}^{n} w_i \cdot \Delta^{-1}(r_i, \alpha_i)\right) = \Delta\left(\sum_{i=1}^{n} w_i \beta_i\right) \tag{2.48}$$

is called a 2-tuple weighted averaging operator. Especially, if $\boldsymbol{w} = (1/n, 1/n, \cdots, 1/n)^{\mathrm{T}}$, then the TWA operator reduces to the TAM operator.

Example 2.15 Let $((s_6, -0.4), (s_5, 0.2), (s_3, 0.1), (s_1, 0), (s_4, -0.5), (s_0, 0.3))$ be a collection of 2-tuples, and $\boldsymbol{w} = (0.20, 0.15, 0.10, 0.30, 0.15, 0.10)^{\mathrm{T}}$ be their weight vector. Since

$$\Delta^{-1}(s_6, -0.4) = 5.6, \quad \Delta^{-1}(s_5, 0.2) = 5.2, \quad \Delta^{-1}(s_3, 0.1) = 3.1$$

$$\Delta^{-1}(s_1, 0) = 1, \quad \Delta^{-1}(s_4, -0.5) = 3.5, \quad \Delta^{-1}(s_0, 0.3) = 0.3$$

then according to the 2-tuple weighted averaging operator (2.48), we get

$$\begin{aligned}
&\text{TWA}((s_6, -0.4),\ (s_5, 0.2),\ (s_3, 0.1), (s_1, 0), (s_4, -0.5), (s_0, 0.3))\\
=&\Delta\left(0.20 \times 5.6 + 0.15 \times 5.2 + 0.10 \times 3.1 + 0.30 \times 1 + 0.15 \times 3.5 + 0.10 \times 0.3\right)\\
=&\Delta(3.065) = (s_3, 0.065)
\end{aligned}$$

(5) 2-tuple OWA operator. Let $\{(r_1, \alpha_1),\ (r_2, \alpha_2),\ \cdots, (r_n, \alpha_n)\}$ be a set of n 2-tuples. Then

$$\text{TOWA}((r_1, \alpha_1),\ (r_2, \alpha_2),\ \cdots, (r_n, \alpha_n)) = \Delta\left(\sum_{j=1}^{n} \omega_j \beta_j^*\right) \tag{2.49}$$

is called a 2-tuple OWA (TOWA) operator, where $\boldsymbol{\omega} = (\omega_1, \omega_2, \cdots, \omega_n)^{\mathrm{T}}$ is the weighting vector associated with the TOWA operator, which satisfies the condition:

$$\omega_i \geqslant 0, \quad i = 1, 2, \cdots, n, \quad \sum_{i=1}^{n} \omega_i = 1$$

and β_j^* is the jth largest of the linguistic arguments β_i ($\beta_i = \Delta^{-1}(r_i, \alpha_i)$, $i = 1, 2, \cdots, n$).

Example 2.16 Let $((s_5, -0.2), (s_3, 0.3), (s_1, 0.2), (s_4, 0), (s_5, -0.4), (s_6, -0.3))$ be a collection of 2-tuples. Then

$$\Delta^{-1}(s_5, -0.2) = 4.8, \quad \Delta^{-1}(s_3, 0.3) = 3.3, \quad \Delta^{-1}(s_1, 0.2) = 1.2$$

$$\Delta^{-1}(s_4, 0) = 4, \quad \Delta^{-1}(s_5, -0.4) = 4.6, \quad \Delta^{-1}(s_6, -0.3) = 5.7$$

Thus

$$\beta_1^* = \Delta^{-1}(s_6, -0.3), \quad \beta_2^* = \Delta^{-1}(s_5, -0.2), \quad \beta_3^* = \Delta^{-1}(s_5, -0.4)$$

$$\beta_4^* = \Delta^{-1}(s_4, 0), \quad \beta_5^* = \Delta^{-1}(s_3, 0.3), \quad \beta_6^* = \Delta^{-1}(s_1, 0.2)$$

Assume that $\boldsymbol{\omega} = (0.10, 0.15, 0.25, 0.25, 0.15, 0.10)^{\mathrm{T}}$ is the weighting vector associated with the TOWA operator, then according to the 2-tuple TOWA operator (2.49), we get

$$\begin{aligned}&\mathrm{TOWA}((r_1, \alpha_1),\ (r_2, \alpha_2),\ \cdots, (r_6, \alpha_6))\\ =&\Delta(0.10 \times 5.7 + 0.15 \times 4.8 + 0.25 \times 4.6 + 0.25 \times 4 + 0.15 \times 3.3 + 0.10 \times 1.2)\\ =&\Delta(4.055) = (s_4, 0.055)\end{aligned}$$

2.4 Linguistic Aggregation Operators Computing with Words Directly

The operators introduced in Section 2.3 can deal with the 2-tuples without loss of information. But the transformation between a counting of information and a linguistic 2-tuple in the aggregation process is somewhat cumbersome, and thus, is a bit inconvenient in calculation. In this section, we shall introduce some operators which compute with words directly. These operators can not only avoid losing any linguistic information, but also are straightforward and very convenient in calculation, and thus, are practical in actual applications.

2.4.1 Linguistic Averaging Operator

Let $S_2 = \{s_\alpha | \alpha = -\tau, \cdots, -1, 0, 1, \cdots, \tau\}$ be an additive linguistic evaluation scale, and $\bar{S}_2 = \{s_\alpha | \alpha \in [-q, q]\}$ be the extended additive linguistic evaluation scale of S_2 (see Section 1.1), where q ($q > \tau$) is a sufficiently large positive integer. $s_{\alpha_j}(j = 1, 2, \cdots, n)$ is a collection of linguistic labels, and $s_{\alpha_j} \in S_2$ $(j = 1, 2, \cdots, n)$.

Based on the additve linguistic evaluation scale $\bar{S}_2$ and Definition 1.1, we have the following definition:

Definition 2.12 (Xu, 2004d) Let $s_\alpha, s_{\alpha_1}, s_{\alpha_2} \in \bar{S}_2$. Then

(1) $s_{\alpha_1} \oplus s_{\alpha_2} = s_{\alpha_2} \oplus s_{\alpha_1}$;

(2) $\lambda(s_{\alpha_1} \oplus s_{\alpha_2}) = \lambda s_{\alpha_1} \oplus \lambda s_{\alpha_2}$;

(3) $\lambda_1 s_\alpha \oplus \lambda_2 s_\alpha = (\lambda_1 + \lambda_2) s_\alpha$.

Proof (1) Since

$$s_{\alpha_1} \oplus s_{\alpha_2} = s_{\alpha_1+\alpha_2} = s_{\alpha_2+\alpha_1} = s_{\alpha_2} \oplus s_{\alpha_1}$$

then (1) holds.

(2) Since

$$\lambda(s_{\alpha_1} \oplus s_{\alpha_2}) = \lambda s_{\alpha_1+\alpha_2} = s_{\lambda(\alpha_1+\alpha_2)} = s_{\lambda\alpha_1+\lambda\alpha_2}$$
$$= s_{\lambda\alpha_1} \oplus s_{\lambda\alpha_2} = \lambda s_{\alpha_1} \oplus \lambda s_{\alpha_2}$$

then (2) holds.

(3) Since

$$\lambda_1 s_\alpha \oplus \lambda_2 s_\alpha = s_{\lambda_1\alpha} \oplus s_{\lambda_2\alpha} = s_{\lambda_1\alpha+\lambda_2\alpha}$$
$$= s_{(\lambda_1+\lambda_2)\alpha} = (\lambda_1+\lambda_2)s_\alpha$$

then (3) holds.

This completes the proof.

Definition 2.13 (Xu, 2006b) Let LA : $(\bar{S}_2)^n \to \bar{S}_2$. Then

$$\mathrm{LA}(s_{\alpha_1}, s_{\alpha_2}, \cdots, s_{\alpha_n}) = \frac{1}{n}(s_{\alpha_1} \oplus s_{\alpha_2} \oplus \cdots \oplus s_{\alpha_n}) = s_{\bar{\alpha}} \tag{2.50}$$

is called a linguistic averaging (LA) operator, where

$$\bar{\alpha} = \frac{1}{n}\sum_{j=1}^{n} \alpha_j \tag{2.51}$$

Example 2.17 Let (s_{-1}, s_0, s_4, s_1) be a collection of linguistic labels. Then

$$\mathrm{LA}(s_{-1}, s_0, s_4, s_1) = \frac{1}{4}(s_{-1} \oplus s_0 \oplus s_4 \oplus s_1) = s_1$$

Definition 2.14 (Xu, 2006b) Let $\mathrm{LWA}_2 : \bar{S}_2^n \to \bar{S}_2$. If

$$\mathrm{LWA}_2(s_{\alpha_1}, s_{\alpha_2}, \cdots, s_{\alpha_n}) = w_1 s_{\alpha_1} \oplus w_2 s_{\alpha_2} \oplus \cdots \oplus w_n s_{\alpha_n} = s_{\dot{\alpha}} \tag{2.52}$$

then LWA_2 is called a linguistic weighted averaging (LWA_2) operator, where

$$\dot{\alpha} = \sum_{j=1}^{n} w_j \alpha_j \tag{2.53}$$

and $\boldsymbol{w} = (w_1, w_2, \cdots, w_n)^{\mathrm{T}}$ is the weight vector of $s_{\alpha_i} (i = 1, 2, \cdots, n)$, $w_j \geqslant 0$ $(j = 1, 2, \cdots, n)$ and $\sum_{j=1}^{n} w_j = 1$. The LWA_2 operator is the extension of the WA operator (Harsanyi, 1955; Xu and Da, 2003b).

Especially, if $\boldsymbol{w} = (1/n, 1/n, \cdots, 1/n)^{\mathrm{T}}$, then the LWA_2 operator reduces to the LA operator.

The characteristic of the LWA_2 operator is that it can consider the importance degrees of the arguments in the process of aggregation.

Example 2.18 Let (s_{-2}, s_3, s_4, s_2) be a collection of linguistic labels, and $\boldsymbol{w} = (0.2, 0.3, 0.4, 0.1)^{\mathrm{T}}$ be their weight vector. Then

$$\begin{aligned}\text{LWA}_2(s_{-2}, s_3, s_4, s_2) &= 0.2 \times s_{-2} \oplus 0.3 \times s_3 \oplus 0.4 \times s_4 \oplus 0.1 \times s_2 \\ &= s_{-0.4} \oplus s_{0.9} \oplus s_{1.6} \oplus s_{0.2} \\ &= s_{2.3}\end{aligned}$$

Definition 2.15 (Xu, 2006b) Let $\text{LOWA}_2 : \bar{S}_2^n \to \bar{S}_2$, whose associated weighting vector is $\boldsymbol{\omega} = (\omega_1, \omega_2, \cdots, \omega_n)^{\text{T}}$, $\omega_i \geqslant 0$ $(i = 1, 2, \cdots, n)$ and $\sum_{i=1}^{n} \omega_i = 1$, such that

$$\text{LOWA}_2(s_{\alpha_1}, s_{\alpha_2}, \cdots, s_{\alpha_n}) = \omega_1 s_{\beta_1} \oplus \omega_2 s_{\beta_2} \oplus \cdots \oplus \omega_n s_{\beta_n} = s_{\bar{\beta}} \tag{2.54}$$

where

$$\bar{\beta} = \sum_{j=1}^{n} \omega_j \beta_j$$

and s_{β_j} $(j = 1, 2, \cdots, n)$ is the jth largest of the linguistic arguments s_{α_i} $(i = 1, 2, \cdots, n)$. Then LOWA_2 is called a linguistic OWA (LOWA_2) operator.

In particular, if $\boldsymbol{\omega} = (1, 0, \cdots, 0)^{\text{T}}$, where $\omega_1 = 1$, $\omega_i = 0$ $(i = 2, \cdots, n)$, then the LOWA_2 operator reduces to the LM_1 operator; if $\boldsymbol{\omega} = (0, \cdots, 0, 1)^{\text{T}}$, where $\omega_i = 0$ $(i = 1, 2, \cdots, n-1), \omega_n = 1$, then the LOWA_2 operator reduces to the LM_2 operator; if $\boldsymbol{\omega} = (1/n, 1/n, \cdots, 1/n)^{\text{T}}$, then the LOWA_2 operator reduces to the LA operator.

The fundamental aspect of the LOWA_2 operator is its reordering step, i.e., it needs to reorder the arguments in descending order and weights their ordered positions, and then aggregates the ordered arguments together with the corresponding position weights.

The LOWA_2 operator is the extension of the OWA operator (Yager, 1988) in linguistic environments. It has many desirable properties similar to the OWA operator, such as monotonicity, idempotency, commutativity and boundedness, and is a linguistic aggregation tool that lies between the max and min operators.

Example 2.19 Let $(s_2, s_4, s_1, s_3, s_{-1})$ be a collection of linguistic labels. After reordering these linguistic arguments in descending order, we have

$$s_{\beta_1} = s_4, \quad s_{\beta_2} = s_3, \quad s_{\beta_3} = s_2, \quad s_{\beta_4} = s_1, \quad s_{\beta_5} = s_{-1}$$

Let $\boldsymbol{\omega} = (0.1, 0.2, 0.4, 0.2, 0.1)^{\text{T}}$ be a weighting vector associated with the LOWA_2 operator. Then

$$\begin{aligned}\text{LOWA}_2\ (s_2, s_4, s_1, s_3, s_{-1}) &= 0.1 \times s_4 \oplus 0.2 \times s_3 \oplus 0.4 \times s_2 \oplus 0.2 \times s_1 \oplus 0.1 \times s_{-1} \\ &= s_{0.4} \oplus s_{0.6} \oplus s_{0.8} \oplus s_{0.2} \oplus s_{-0.1} \\ &= s_{1.9}\end{aligned}$$

From Definitions 2.14 and 2.15, we can see that the LWA_2 operator only weights the argument itself, but ignores the importance of the ordered position of the argument, while the $LOWA_2$ operator only weights the ordered position of the argument, but ignores the importance of the argument. Both the operators have their own drawbacks. To solve this issue, Xu (2006a) introduced a linguistic hybrid aggregation operator.

Definition 2.16 (Xu, 2006a) A linguistic hybrid aggregation (LHA) operator is a mapping LHA : $\bar{S}_2^n \to \bar{S}_2$, which has an associated vector $\boldsymbol{\omega} = (\omega_1, \omega_2, \cdots, \omega_n)^{\mathrm{T}}$, $\omega_i \geqslant 0$ $(i = 1, 2, \cdots, n)$ and $\sum_{i=1}^{n} \omega_i = 1$, such that

$$\mathrm{LHA}(s_{\alpha_1}, s_{\alpha_2}, \cdots, s_{\alpha_n}) = \omega_1 s_{\beta_1} \oplus \omega_2 s_{\beta_2} \oplus \cdots \oplus \omega_n s_{\beta_n} \tag{2.55}$$

where s_{β_j} is the jth largest of the linguistic weighted arguments $\bar{s}_{\alpha_i} (i = 1, 2, \cdots, n)$ $(\bar{s}_{\alpha_i} = n w_i s_{\alpha_i},\ i = 1, 2, \cdots, n)$, $\boldsymbol{w} = (w_1, w_2, \cdots, w_n)^{\mathrm{T}}$ is the weight vector of the linguistic arguments $s_{\alpha_i} (i = 1, 2, \cdots, n)$, $w_i \geqslant 0$ $(i = 1, 2, \cdots, n)$, $\sum_{i=1}^{n} w_i = 1$, and n is the balancing coefficient, which plays a role of balance (in such a case, if the vector $(w_1, w_2, \cdots, w_n)^{\mathrm{T}}$ approaches $(1/n, 1/n, \cdots, 1/n)^{\mathrm{T}}$, then the vector $(nw_1 s_{\alpha_1}, nw_2 s_{\alpha_2}, \cdots, nw_n s_{\alpha_n})^{\mathrm{T}}$ approaches $(s_{\alpha_1}, s_{\alpha_2}, \cdots, s_{\alpha_n})$).

In particular, if $w = (1/n, 1/n, \cdots, 1/n)^{\mathrm{T}}$, then the LHA operator reduces to the LWA_2 operator; if $\boldsymbol{w} = (1/n, 1/n, \cdots, 1/n)^{\mathrm{T}}$, then the LHA operator reduces to the $LOWA_2$ operator.

It is clear that the LHA operator generalizes both the LWA_2 and $LOWA_2$ operators, and reflects the importance degrees of both the given linguistic argument and its ordered position.

Example 2.20 Let $(s_3, s_{-2}, s_0, s_{-3}, s_{-1}, s_4)$ be a collection of linguistic arguments, and $\boldsymbol{w} = (0.15, 0.10, 0.20, 0.25, 0.10, 20)^{\mathrm{T}}$ be the weight vector. We first calculate the linguistic weighted arguments:

$$\bar{s}_{\alpha_1} = 6 \times 0.15 \times s_3 = s_{2.7}, \quad \bar{s}_{\alpha_2} = 6 \times 0.10 \times s_{-2} = s_{-1.2}$$

$$\bar{s}_{\alpha_3} = 6 \times 0.20 \times s_0 = s_0, \quad \bar{s}_{\alpha_4} = 6 \times 0.25 \times s_{-3} = s_{-4.5}$$

$$\bar{s}_{\alpha_5} = 6 \times 0.10 \times s_{-1} = s_{-0.6}, \quad \bar{s}_{\alpha_6} = 6 \times 0.20 \times s_4 = s_{4.8}$$

After reordering the arguments $\bar{s}_{\alpha_i} (i = 1, 2, \cdots, 6)$ in descending order, we get

$$s_{\beta_1} = \bar{s}_{\alpha_6} = s_{4.8}, \quad s_{\beta_2} = \bar{s}_{\alpha_1} = s_{2.7}, \quad s_{\beta_3} = \bar{s}_{\alpha_3} = s_0$$

$$s_{\beta_4} = \bar{s}_{\alpha_5} = s_{-0.6}, \quad s_{\beta_5} = \bar{s}_{\alpha_2} = s_{-1.2}, \quad s_{\beta_6} = \bar{s}_{\alpha_4} = s_{-4.5}$$

Assume that $\boldsymbol{\omega} = (0.10, 0.15, 0.25, 0.25, 0.15, 0.10)^{\mathrm{T}}$ is the weighting vector associated with the LHA operator. Then

$$\begin{aligned}\text{LHA}(s_3, s_{-2}, s_0, s_{-3}, s_{-1}, s_4) =& 0.1 \times s_{4.8} \oplus 0.15 \times s_{2.7} \oplus 0.25 \times s_0 \oplus 0.25 \times s_{-0.6} \\ & \oplus 0.15 \times s_{-1.2} \oplus 0.10 \times s_{-4.5} \\ =& s_{0.48} \oplus s_{0.405} \oplus s_0 \oplus s_{-0.15} \oplus s_{-0.18} \oplus s_{-0.45} \\ =& s_{0.105}\end{aligned}$$

Yager and Filev (1999) introduced the concept of induced OWA (IOWA) operator, which takes as its argument pairs, called OWA pairs, in which one component is used to induce an ordering over the second components which are then aggregated. The IOWA operators, however, can only be used in situations in which the aggregated arguments are the exact numerical values. Xu (2006b) introduced an induced linguistic OWA (ILOWA) operator, which can be used to aggregate linguistic arguments and is defined as follows:

Definition 2.17 (Xu, 2006b) An induced linguistic OWA (ILOWA) operator is defined as

$$\text{ILOWA}(\langle u_1, s_{\alpha_1}\rangle, \langle u_2, s_{\alpha_2}\rangle \cdots, \langle u_n, s_{\alpha_n}\rangle) = \omega_1 s_{\gamma_1} \oplus \omega_2 s_{\gamma_2} \oplus \cdots \oplus \omega_n s_{\gamma_n} = s_{\bar{\gamma}} \quad (2.56)$$

where

$$\bar{\gamma} = \sum_{i=1}^{n} \omega_i \gamma_i \quad (2.57)$$

and $\boldsymbol{\omega} = (\omega_1, \omega_2, \cdots, \omega_n)^{\mathrm{T}}$ is the weighting vector, such that $\omega_i \geqslant 0$ $(i = 1, 2, \cdots, n)$ and $\sum_{i=1}^{n} \omega_i = 1$. s_{γ_j} is the s_{α_i} value of the OWA pair $\langle u_i,\ s_{\alpha_i}\rangle$ having the jth largest of u_i $(i = 1, 2, \cdots, n)$, and u_i in $\langle u_i,\ s_{\alpha_i}\rangle$ is referred to as the order inducing variable and s_{α_i} as the linguistic label.

Especially, if $\boldsymbol{\omega} = (1/n,\ 1/n, \cdots,\ 1/n)^{\mathrm{T}}$, then the ILOWA operator reduces to the LA operator; if $u_i = s_{\alpha_i}$ for all i, then the LOWA operator reduces to the LOWA$_2$ operator; if $u_i =$ No. i for all i, where No. i is the ordered position of the linguistic label s_i, then the ILOWA operator reduces to the LWA$_2$ operator.

However, if there is a tie between the OWA pairs $\langle u_i,\ s_{\alpha_i}\rangle$ and $\langle u_j,\ s_{\alpha_j}\rangle$ with respect to the order inducing variables, then in this case, we can follow the policy presented by Yager and Filev (1999). That is, we replace the arguments of the tied objects by the average of the arguments of the tied objects, i.e., we replace the argument component of each of $\langle u_i,\ s_{\alpha_i}\rangle$ and $\langle u_j,\ s_{\alpha_j}\rangle$ by their average $\dfrac{s_{\alpha_i} \oplus s_{\alpha_j}}{2}$. Thus, we get $\left\langle u_i,\ \dfrac{s_{\alpha_i} \oplus s_{\alpha_j}}{2}\right\rangle$ and $\left\langle u_j,\ \dfrac{s_{\alpha_i} \oplus s_{\alpha_j}}{2}\right\rangle$. Similarly, if k items are tied, then we replace these by k replica's of their average.

Example 2.21 Given a collection of the OWA pairs $\langle u_i,\ s_{\alpha_i}\rangle (i = 1, 2, 3, 4, 5)$:

$$(\langle 0.15,\ s_2\rangle, \langle 0.1,\ s_3\rangle, \langle 0.2,\ s_4\rangle, \langle 0.3,\ s_{-1}\rangle, \langle 0.25,\ s_{-3}\rangle)$$

Suppose that the weighting vector associated with the ILOWA operator is $\boldsymbol{\omega} = (0.15, 0.20, 0.30, 0.20, 0.15)^{\mathrm{T}}$. We first utilize the values of u_i $(i = 1, 2, 3, 4, 5)$ to rank the OWA pairs $\langle u_i,\ s_{\alpha_i}\rangle$ $(i = 1, 2, 3, 4, 5)$. Since

$$u_4 > u_5 > u_3 > u_1 > u_2$$

we get the ordered OWA pairs:

$$\langle 0.3,\ s_{-1}\rangle, \quad \langle 0.25,\ s_{-3}\rangle, \quad \langle 0.2,\ s_4\rangle, \quad \langle 0.15,\ s_2\rangle, \quad \langle 0.1,\ s_3\rangle$$

i.e.,

$$s_{\gamma_1} = s_{-1}, \quad s_{\gamma_2} = s_{-3}, \quad s_{\gamma_3} = s_4, \quad s_{\gamma_4} = s_2, \quad s_{\gamma_5} = s_3$$

Therefore,

$$\begin{aligned}
&\text{ILOWA}\left(\langle u_1, s_{\alpha_1}\rangle, \langle u_2, s_{\alpha_2}\rangle, \langle u_3, s_{\alpha_3}\rangle, \langle u_4, s_{\alpha_4}\rangle, \langle u_5, s_{\alpha_5}\rangle\right)\\
=&0.15 \times s_{-1} \oplus 0.20 \times s_{-3} \oplus 0.30 \times s_4 \oplus 0.20 \times s_2 \oplus 0.15 \times s_3\\
=&s_{-0.15} \oplus s_{-0.6} \oplus s_{1.2} \oplus s_{0.4} \oplus s_{0.45}\\
=&s_{1.30}
\end{aligned}$$

If we replace $\langle \mu_1,\ s_{\alpha_1}\rangle = \langle 0.15,\ s_2\rangle$ and $\langle \mu_4,\ s_{\alpha_4}\rangle = \langle 0.25,\ s_{-3}\rangle$ in the above as $\langle \mu_1,\ s_{\alpha_1}\rangle = \langle 0.2,\ s_2\rangle$ and $\langle \mu_4,\ s_{\alpha_4}\rangle = \langle 0.20,\ s_{-3}\rangle$ respectively, then $u_1 = \mu_3 = \mu_5$. In the process of aggregation, we average $s_{\alpha_1} = s_2$, $s_{\alpha_3} = s_4$ and $s_{\alpha_5} = s_{-3}$, i.e., $(s_2 \oplus s_4 \oplus s_{-3})/3 = s_1$. Thus, we get

$$\begin{aligned}
\left\langle u_1,\ \frac{s_{\alpha_1} \oplus s_{\alpha_3} \oplus s_{\alpha_5}}{3}\right\rangle = \left\langle u_3,\ \frac{s_{\alpha_1} \oplus s_{\alpha_3} \oplus s_{\alpha_5}}{3}\right\rangle = \left\langle u_5,\ \frac{s_{\alpha_1} \oplus s_{\alpha_3} \oplus s_{\alpha_5}}{3}\right\rangle\\
= \langle 0.2,\ s_1\rangle
\end{aligned}$$

i.e.,

$$s_{\gamma_1} = s_{-1}, \quad s_{\gamma_2} = s_{-3}, \quad s_{\gamma_2} = s_{\gamma_3} = s_{\gamma_4} = s_1, \quad s_{\gamma_5} = s_3$$

Consequently,

$$\begin{aligned}
&\text{ILOWA}\left(\langle u_1, s_{\alpha_1}\rangle, \langle u_2, s_{\alpha_2}\rangle, \langle u_3, s_{\alpha_3}\rangle, \langle u_4, s_{\alpha_4}\rangle, \langle u_5, s_{\alpha_5}\rangle\right)\\
=&0.15 \times s_{-1} \oplus 0.20 \times s_1 \oplus 0.30 \times s_1 \oplus 0.20 \times s_1 \oplus 0.15 \times s_3\\
=&s_{-0.15} \oplus s_{0.2} \oplus s_{0.3} \oplus s_{0.2} \oplus s_{0.45}\\
=&s_1
\end{aligned}$$

The IOWA operator, which essentially aggregates objects that are pairs, provides a very general family of aggregation operators. Particularly noteworthy is its ability to provide for aggregations in environments that mix linguistic and numeric variables.

However, in some situations, when we need to provide more information about the objects, i.e., each object may consist of three components, a direct locater, an indirect locater and a prescribed value, it is unsuitable to use the IOWA operator as an aggregation tool. Motivated by this issue, in the following, we introduce a more general aggregation technique called the generalized IOWA (GIOWA) operator (Xu, 2004c):

Definition 2.18 (Xu, 2004c) A generalized induced LOWA (GILOWA) operator is given by

$$\begin{aligned}&\text{GILOWA}\left(\langle v_1, u_1, s_{\alpha_1}\rangle, \langle v_2, u_2, s_{\alpha_2}\rangle, \cdots, \langle v_n, u_n, s_{\alpha_n}\rangle\right)\\&=\omega_1 s_{\beta_1} \oplus \omega_2 s_{\beta_2} \oplus \cdots \oplus \omega_n s_{\beta_n}\end{aligned} \tag{2.58}$$

where $\boldsymbol{\omega} = (\omega_1, \omega_2, \cdots, \omega_n)^{\mathrm{T}}$ is the associated weighting vector with $\omega_i \geqslant 0$ $(i = 1, 2, \cdots, n)$ and $\sum_{i=1}^{n} \omega_i = 1$. The object $\langle v_i, u_i, s_{\alpha_i}\rangle$ consists of three components, where the first component v_i represents the importance or character of the second component u_i, and the second component u_i is used to induce an ordering through the first component v_i over the third components s_{α_i}, which are then aggregated. Here, s_{β_j} is the s_{α_i} value of the object having the jth largest of v_i $(i = 1, 2, \cdots, n)$. In discussing these objects $\langle v_i, u_i, s_{\alpha_i}\rangle$ $(i = 1, 2, \cdots, n)$, because of their roles, we shall refer to v_i as the direct order inducing variable, u_i as the indirect order inducing variable, and s_{α_i} as the argument variable.

In particular, if $v_i = u_i$ for all i, then the GILOWA operator reduces to the ILOWA operator; if $v_i = s_{\alpha_i}$ for all i, then the GILOWA operator reduces to the LOWA_2 operator; if $v_i =$ No. i for all i, where No. i is the ordered position of s_{α_i}, then the GILOWA operator reduces to the LWA_2 operator; if $\boldsymbol{\omega} = (1/n,\ 1/n, \cdots,\ 1/n)^{\mathrm{T}}$, then the GILOWA operator reduces to the LA operator.

Especially, if there exist two objects $\langle v_i, u_i,\ s_{\alpha_i}\rangle$ and $\left\langle v_j, u_j,\ s_{\alpha_j}\right\rangle$ such that $v_i = v_j$, then we can replace the arguments s_{α_i} and s_{α_j} of the tied objects $\langle v_i, u_i,\ s_{\alpha_i}\rangle$ and $\left\langle v_j, u_j,\ s_{\alpha_j}\right\rangle$ by the average $\dfrac{s_{\alpha_i} \oplus s_{\alpha_j}}{2}$ of the arguments of the tied objects, and get the corresponding objects $\left\langle v_i, u_i, \dfrac{s_{\alpha_i} \oplus s_{\alpha_j}}{2}\right\rangle$ and $\left\langle v_j, u_j, \dfrac{s_{\alpha_i} \oplus s_{\alpha_j}}{2}\right\rangle$. If k items are tied, we replace these by k replica's of their average.

Example 2.22 Given a collection of the objects $\langle v_i, u_i, s_{\alpha_i}\rangle$ $(i = 1, 2, 3, 4)$:

$$(\langle \text{No.3, Johnson}, s_0\rangle, \quad \langle \text{No.2, Brown}, s_2\rangle, \quad \langle \text{No.1, Smith}, s_4\rangle,$$

$$\langle \text{No.4, Anderson}, s_{-2}\rangle)$$

Suppose that the weighting vector associated with the ILOWA operator is $\boldsymbol{w} = (0.15, 0.30, 0.40, 0.15)^{\mathrm{T}}$. We first rank the objects $\langle v_i, u_i,\ s_{\alpha_i}\rangle$ $(i = 1, 2, 3, 4)$ according to the ranking of v_i $(i = 1, 2, 3, 4)$. Since

$$v_3 > v_2 > v_1 > v_4$$

we get the ordered OWA pairs:

$$\langle \text{No.1, Smith, } s_4 \rangle$$

$$\langle \text{No.2, Brown, } s_2 \rangle$$

$$\langle \text{No.3, Johnson, } s_0 \rangle$$

$$\langle \text{No.4, Anderson, } s_{-2} \rangle$$

i.e.,

$$s_{\gamma_1} = s_4, \quad s_{\gamma_2} = s_2, \quad s_{\gamma_3} = s_0, \quad s_{\gamma_4} = s_{-2}$$

Accordingly,

$$\begin{aligned} & \text{GILOWA}\left(\langle v_1, u_1, s_{\alpha_1}\rangle, \langle v_2, u_2, s_{\alpha_2}\rangle, \langle v_3, u_3, s_{\alpha_3}\rangle, \langle v_4, u_4, s_{\alpha_4}\rangle\right) \\ =& 0.15 \times s_4 \oplus 0.30 \times s_2 \oplus 0.40 \times s_0 \oplus 0.15 \times s_{-2} \\ =& s_{0.6} \oplus s_{0.6} \oplus s_0 \oplus s_{-0.3} \\ =& s_{0.9} \end{aligned}$$

The linguistic aggregation operators introduced above only consider situations where all the aggregated linguistic arguments are independent. However, in some actual situations, the considered linguistic arguments may be correlative. For example, Grabisch (1995) and Torra (2003) gave the following classical example: "We are to evaluate a set of students in relation to three subjects: {mathematics, physics, literature}, we want to give more importance to science-related subjects than to literature, but on the other hand, we want to give some advantage to students that are good both in literature and in any of the science related subjects." Here, the evaluation information is very suitable to be expressed in linguistic arguments. Thus, the investigation on the approach to aggregating the input agruments in such cases is necessary. Aiming to this problem, Xu (2009b) developed a linguistic information aggregation operator called the linguistic correlated averaging operator, which is based on the idea of the well-known Choquet integral (Choquet, 1953).

Let the linguistic labels $s_{\alpha_i} \in \bar{S}_2$ $(i = 1, 2, \cdots, n)$, and $\mu(\{s_{\alpha_i}\})$ $(i = 1, 2, \cdots, n)$ be the weights of the linguistic labels $s_{\alpha_i} (i = 1, 2, \cdots, n)$, where μ is a fuzzy measure. Similar to Wang and klir (1992), Xu (2009b) defined μ as follows:

Definition 2.19 (Xu, 2009b) A fuzzy measure μ on the extended additive linguistic evaluation scale $\bar{S}_2$ is a set function $\mu : \bar{S}_2 \rightarrow [0, 1]$ satisfying the following axioms:

(1) $\mu(\varnothing) = 0$, $\mu(\bar{S}_2) = 1$;

(2) $A \subseteq B$ implies $\mu(A) \leqslant \mu(B)$ for all $A, B \subseteq \bar{S}_2$;

(3) $\mu(A\cup B)=\mu(A)+\mu(B)+\lambda\mu(A)\mu(B)$ for all $A,B\subseteq \bar{S}_2$ and $A\cap B=\varnothing$, where $\lambda\in(-1,\infty)$.

Especially, if $\lambda=0$, then (3) reduces to the axiom of additive measure:

$$\mu(A\cup B)=\mu(A)+\mu(B)\quad \text{for all}\quad A,B\subseteq \bar{S}_2\quad \text{and}\quad A\cap B=\varnothing \tag{2.59}$$

In this case, all the additional linguistic labels in $\bar{S}_2$ are independent, and we have

$$\mu(A)=\sum_{s_{\alpha_i}\in A}\mu\left(\{s_{\alpha_i}\}\right)\quad \text{for all } A\subseteq \bar{S}_2 \tag{2.60}$$

Based on Definition 2.19 and the Choquet integral (Choquet, 1953), Xu (2009b) gave an operator for aggregating linguistic labels together with their correlative weights:

Definition 2.20 (Xu, 2009b) Let μ be a fuzzy measure on $\bar{S}_2$, and s_{α_i} $(i=1,2,\cdots,n)$ be n additional linguistic labels in $\bar{S}_2$. Then a linguistic correlated averaging (LCA) operator is defined as

$$(C_1)\int s_\alpha \mathrm{d}\mu=\mathrm{LCA}\left(s_{\alpha_1},s_{\alpha_2},\cdots,s_{\alpha_n}\right)=\bigoplus_{i=1}^{n}\left\{[\mu(A_{\sigma(i)})-\mu(A_{\sigma(i-1)})]s_{\alpha_{\sigma(i)}}\right\} \tag{2.61}$$

where $(C_1)\int s_\alpha d\mu$ is a notation of Choquet integral, $(\sigma(1),\sigma(2),\cdots,\sigma(n))$ is a permutation of $(1,2,\cdots,n)$ such that $s_{\alpha_{\sigma(1)}}\geqslant s_{\alpha_{\sigma(2)}}\geqslant\cdots\geqslant s_{\alpha_{\sigma(n)}}$, $A_{\sigma(k)}=\{s_{\alpha_{\sigma(j)}}|j\leqslant k\}$, when $k\geqslant 1$ and $A_{\sigma(0)}=\varnothing$.

In what follows, we discuss some special cases of the LCA operator:

(1) If (2.60) holds, then

$$w_i=\mu\left(\{s_{\alpha_{\sigma(i)}}\}\right)=\mu(A_{\sigma(i)})-\mu(A_{\sigma(i-1)}),\quad i=1,2,\cdots,n \tag{2.62}$$

where $\boldsymbol{w}=(w_1,w_2,\cdots,w_n)^{\mathrm{T}}$, $w_i\geqslant 0$ $(i=1,2,\cdots,n)$ and $\sum_{i=1}^{n}w_i=1$. Thus, the LCA operator reduces to the linguistic weighted averaging (LWA_2) operator (2.52). Especially, if $w_i=\dfrac{1}{n}$ for all $i=1,2,\cdots,n$, then the LWA operator reduces to the linguistic averaging (LA) operator (2.50).

(2) If

$$\mu(A)=\sum_{i=1}^{|A|}\omega_i\quad \text{for all } A\subseteq \bar{S}_2 \tag{2.63}$$

where $|A|$ is the number of the additional linguistic labels in the set A, then

$$\omega_i=\mu(A_{\sigma(i)})-\mu(A_{\sigma(i-1)}),\quad i=1,2,\cdots,n \tag{2.64}$$

where $\boldsymbol{\omega}=(\omega_1,\omega_2,\cdots,\omega_n)^{\mathrm{T}}$, $\omega_i\geqslant 0$ $(i=1,2,\cdots,n)$ and $\sum_{i=1}^{n}\omega_i=1$. In this case, the

LCA operator reduces to the linguistic ordered weighted averaging ($LOWA_2$) operator (2.54). Especially, if $\mu(A) = \frac{|A|}{n}$ for all $A \subseteq \bar{S}_2$, then both the LCA and $LOWA_2$ operators reduce to the LA operator (2.50).

(3) If

$$\mu(A) = f\left(\sum_{s_{\alpha_i} \in A} \mu\left(\{s_{\alpha_i}\}\right)\right) \quad \text{for all } A \subseteq \bar{S}_2 \tag{2.65}$$

where f is a basic unit-interval monotonic (BUM) function $f: [0,1] \to [0,1]$ and is monotonic with the following properties:

(i) $f(0) = 0$;

(ii) $f(1) = 1$;

(iii) $f(x) \geqslant f(y)$ for $x > y$,

then we let

$$w_i = \mu(A_{\sigma(i)}) - \mu(A_{\sigma(i-1)}) = f\left(\sum_{j \leqslant i} \mu\left(\{s_{\alpha_{\sigma(i)}}\}\right)\right) - f\left(\sum_{j<i} \mu\left(\{s_{\alpha_{\sigma(i)}}\}\right)\right)$$
$$i = 1, 2, \cdots, n \tag{2.66}$$

where $\boldsymbol{w} = (w_1, w_2, \cdots, w_n)^{\mathrm{T}}$, $w_i \geqslant 0$ $(i = 1, 2, \cdots, n)$ and $\sum_{i=1}^{n} w_i = 1$. In this case, the LCA operator reduces to a linguistic weighted ordered weighted averaging (LWOWA) operator similar to (2.40). Especially, if $\mu\left(\{s_{\alpha_i}\}\right) = \frac{1}{n}$ for all $i = 1, 2, \cdots, n$, then the LWOWA operator reduces to the $LOWA_2$ operator (2.54).

Example 2.23 (Xu, 2009b) We are to evaluate a set of four students $X = \{x_1, x_2, x_3, x_4\}$ in relation to a set of three attributes: $G = \{G_1, G_2, G_3\} = \{$mathematics, physics, literature$\}$, we want to give more importance to science-related subjects than to literature, but on the other hand, we want to give some advantage to students that are good both in literature and in any of the science related subjects. Let

$$\mu(\varnothing) = 0, \quad \mu(G) = \mu\left(\{G_1, G_2, G_3\}\right) = 1, \quad \mu\left(\{G_1\}\right) = \mu\left(\{G_2\}\right) = 0.45$$

$$\mu\left(\{G_3\}\right) = 0.3, \quad \mu(\{G_1, G_2\}) = 0.5, \quad \mu(\{G_1, G_3\}) = \mu(\{G_2, G_3\}) = 0.9$$

The evaluation information about the four students x_i $(i = 1, 2, 3, 4)$ under the three attributes: ① mathematics; ② physics; ③ literature, are expressed by the additional linguistic labels in S_2, shown as in Table 2.1.

For convenience of depiction, we denote the additional linguistic labels in Table 2.1 as a_{ij} $(i = 1, 2, \cdots, 5;\ j = 1, 2, 3)$, i.e., $a_{11} = s_1$, $a_{12} = s_3$, $a_{13} = s_4$, $a_{21} = s_2$, $a_{22} = s_2$, $a_{23} = s_3$, $a_{31} = s_4$, $a_{32} = s_3$, $a_{33} = s_0$, $a_{41} = s_2$, $a_{42} = s_3$, $a_{43} = s_3$.

Table 2.1 The evaluation information about students (Xu, 2009b)

	mathematics	physics	literature
x_1	s_{-1}	s_3	s_4
x_2	s_0	s_2	s_3
x_3	s_4	s_3	s_{-2}
x_4	s_{-2}	s_4	s_2

Then we rearrange the additional linguistic labels corresponding to each student in descending order:

$$a_{\sigma(11)} = a_{13} = s_4, \quad a_{\sigma(12)} = a_{12} = s_3, \quad a_{\sigma(13)} = a_{11} = s_{-1}$$

$$a_{\sigma(21)} = a_{23} = s_3, \quad a_{\sigma(22)} = a_{22} = s_2, \quad a_{\sigma(23)} = a_{21} = s_0$$

$$a_{\sigma(31)} = a_{31} = s_4, \quad a_{\sigma(32)} = a_{32} = s_3, \quad a_{\sigma(33)} = a_{33} = s_{-2}$$

$$a_{\sigma(41)} = a_{42} = s_4, \quad a_{\sigma(42)} = a_{43} = s_2, \quad a_{\sigma(43)} = a_{41} = s_{-2}$$

By the LCA operator, we calculate the overall evaluation information corresponding to each student:

$$\begin{aligned} a_1 =& \mathrm{LCA}\,(a_{11}, a_{12}, a_{13}) \\ =& [\mu(\{G_3\}) - \mu(\{\phi\})]a_{13} \oplus [\mu(\{G_2, G_3\}) - \mu(\{G_3\})]a_{12} \\ & \oplus [\mu(\{G_1, G_2, G_3\}) - \mu(\{G_2, G_3\})]a_{11} \\ =& (0.3 - 0)\, s_4 \oplus (0.9 - 0.3)\, s_3 \oplus (1 - 0.9)\, s_{-1} = s_{2.9} \\ a_2 =& \mathrm{LCA}\,(a_{21}, a_{22}, a_{23}) \\ =& [\mu(\{G_3\}) - \mu(\{\phi\})]a_{23} \oplus [\mu(\{G_2, G_3\}) - \mu(\{G_3\})]a_{22} \\ & \oplus [\mu(\{G_1, G_2, G_3\}) - \mu(\{G_2, G_3\})]a_{21} \\ =& (0.3 - 0)\, s_3 \oplus (0.9 - 0.3)\, s_2 \oplus (1 - 0.9)\, s_0 = s_{2.1} \\ a_3 =& \mathrm{LCA}\,(a_{31}, a_{32}, a_{33}) \\ =& [\mu(\{G_1\}) - \mu(\{\phi\})]a_{31} \oplus [\mu(\{G_1, G_2\}) - \mu(\{G_1\})]a_{32} \\ & \oplus [\mu(\{G_1, G_2, G_3\}) - \mu(\{G_1, G_2\})]a_{33} \\ =& (0.45 - 0)\, s_4 \oplus (0.5 - 0.45)\, s_3 \oplus (1 - 0.5)\, s_{-2} = s_{0.95} \\ a_4 =& \mathrm{LCA}\,(a_{41}, a_{42}, a_{43}) \\ =& [\mu(\{G_2\}) - \mu(\{\phi\})]a_{42} \oplus [\mu(\{G_2, G_3\}) - \mu(\{G_2\})]a_{43} \\ & \oplus [\mu(\{G_1, G_2, G_3\}) - \mu(\{G_2, G_3\})]a_{41} \\ =& (0.45 - 0)\, s_4 \oplus (0.9 - 0.45)\, s_2 \oplus (1 - 0.9)\, s_{-2} = s_{2.5} \end{aligned}$$

Since $a_1 > a_4 > a_2 > a_3$, then the ranking of the four students x_i $(i = 1, 2, 3, 4)$ is

$$x_1 \succ x_4 \succ x_2 \succ x_3$$

and then x_1 is the best one.

Sometimes, however, the input linguistic arguments may not match any of the original linguistic labels, and they may be located between two of them. For example, when evaluating the "design" of a car, an expert may provide his/her opinion with "between 'fair' and 'good' ". To solve this issue, Xu (2004a; 2004d) introduced the concept of uncertain linguistic variable and defined some operational laws of uncertain linguistic variables.

Definition 2.21 (Xu, 2004a; 2004d) Let $\tilde{s} = [s_\alpha, s_\beta]$, where $s_\alpha, s_\beta \in \bar{S}_2$, s_α and s_β be the lower and upper limits of $\tilde{s}$. Then $\tilde{s}$ is called an uncertain linguistic variable.

For convenience, let $\tilde{S}_2$ be the set of all the uncertain linguistic variables, which is on the basis of the extended additive linguistic evaluation scale $\bar{S}_2$.

Definition 2.22 (Xu, 2004a) Let $\tilde{s}=[s_\alpha, s_\beta]$, $\tilde{s}_1=[s_{\alpha_1}, s_{\beta_1}]$ and $\tilde{s}_2=[s_{\alpha_2}, s_{\beta_2}] \in \tilde{S}_2$. Then

(1) $\tilde{s}_1 \oplus \tilde{s}_2 = [s_{\alpha_1}, s_{\beta_1}] \oplus [s_{\alpha_2}, s_{\beta_2}] = [s_{\alpha_1} \oplus s_{\alpha_2},\ s_{\beta_1} \oplus s_{\beta_2}]$;

(2) $\lambda \tilde{s} = [\lambda s_\alpha, \lambda s_\beta]$ $(\lambda \in [0,1])$.

In order to compare any two uncertain linguistic variables $\tilde{s}_1 = [s_{\alpha_1}, s_{\beta_1}]$ and $\tilde{s}_2 = [s_{\alpha_2}, s_{\beta_2}]$, Xu (2004a) gave a simple formula:

$$p(\tilde{s}_1 \geqslant \tilde{s}_2) = \min\left\{\max\left\{\frac{\beta_1 - \alpha_2}{\text{len}(\tilde{s}_1) + \text{len}(\tilde{s}_2)}, 0\right\}, 1\right\} \tag{2.67}$$

where $\text{len}(\tilde{s}_1) = \beta_1 - \alpha_1$, $\text{len}(\tilde{s}_2) = \beta_2 - \alpha_2$, $p(\tilde{s}_1 \geqslant \tilde{s}_2)$ is called the degree of possibility of $\tilde{s}_1 \geqslant \tilde{s}_2$. Xu (2006e; 2004d) gave another two equivalent forms of (2.67):

$$p(\tilde{s}_1 \geqslant \tilde{s}_2) = \max\left\{1 - \max\left\{\frac{\beta_2 - \alpha_1}{\text{len}(\tilde{s}_1) + \text{len}(\tilde{s}_2)}, 0\right\}, 0\right\} \tag{2.68}$$

and

$$p(\tilde{s}_1 \geqslant \tilde{s}_2) = \frac{\max\{0,\ \text{len}(\tilde{s}_1) + \text{len}(\tilde{s}_2) -\ \max\{\beta_2 - \alpha_1,\ 0\}\}}{\text{len}(\tilde{s}_1) + \text{len}(\tilde{s}_2)} \tag{2.69}$$

Especially,

(1) If one of $\tilde{s}_1$ and $\tilde{s}_2$ reduces to a linguistic variable, without loss of generality, suppose that $\tilde{s}_1$ reduces to a linguistic variable, i.e., $\text{len}(\tilde{s}_1) = 0$. In such a case, $\alpha_1 = \beta_1$, then the degree of possibility of $\tilde{s}_1 \geqslant \tilde{s}_2$ is defined as

$$p(\tilde{s}_1 \geqslant \tilde{s}_2) = \frac{\max\{0,\ \text{len}(\tilde{s}_2) -\ \max\{\beta_2 - \alpha_1,\ 0\}\}}{\text{len}(\tilde{s}_2)} \tag{2.70}$$

(2) If both $\tilde{s}_1$ and $\tilde{s}_2$ reduce to the linguistic variables, i.e., $\text{len}(\tilde{s}_1) + \text{len}(\tilde{s}_2) = 0$, then the degree of possibility of $\tilde{s}_1 >= \tilde{s}_2$ is defined as

$$p(\tilde{s}_1 \geqslant \tilde{s}_2) = \begin{cases} 1, & \tilde{s}_1 > \tilde{s}_2 \\ \dfrac{1}{2}, & \tilde{s}_1 = \tilde{s}_2 \\ 0, & \tilde{s}_1 < \tilde{s}_2 \end{cases} \tag{2.71}$$

The formulas (2.67)~(2.69) provide an approach to ranking uncertain linguistic arguments, and reflect the possibility degree of the event $\tilde{s}_1 \geqslant \tilde{s}_2$ occurs.

The possibility degree $p(\tilde{s}_1 >= \tilde{s}_2)$ has the following properties:

Theorem 2.1 (Xu, 2004d) Let $\tilde{s}_1 = [s_{\alpha_1}, s_{\beta_1}]$, $\tilde{s}_2 = [s_{\alpha_2}, s_{\beta_2}]$ and $\tilde{s}_3 = [s_{\alpha_3}, s_{\beta_3}]$ be three uncertain linguistic variables. Then

(1) $0 \leqslant p(\tilde{s}_1 \geqslant \tilde{s}_2) \leqslant 1$;

(2) $p(\tilde{s}_1 \geqslant \tilde{s}_2) = 1$ if and only if $\beta_2 \leqslant \alpha_1$;

(3) $p(\tilde{s}_1 \geqslant \tilde{s}_2) = 0$ if and only if $\beta_1 \leqslant \alpha_2$;

(4) $p(\tilde{s}_1 \geqslant \tilde{s}_2) + p(\tilde{s}_2 \geqslant \tilde{s}_1) = 1$, especially, $p(\tilde{s}_1 \geqslant \tilde{s}_1) = 1/2$;

(5) $p(\tilde{s}_1 \geqslant \tilde{s}_2) \geqslant 1/2$ if and only if $\alpha_1 + \beta_1 \geqslant \alpha_2 + \beta_2$, especially, $p(\tilde{s}_1 \geqslant \tilde{s}_2) = 1/2$ if and only if $\alpha_1 + \beta_1 = \alpha_2 + \beta_2$;

(6) $p(\tilde{s}_1 \geqslant \tilde{s}_2) \geqslant 1/2$ and $p(\tilde{s}_2 \geqslant \tilde{s}_3) \geqslant 1/2$, then $p(\tilde{s}_1 \geqslant \tilde{s}_3) \geqslant 1/2$.

Proof (1) Since

$$\max\{0,\ \mathrm{len}(\tilde{s}_1) + \mathrm{len}(\tilde{s}_2) - \max\{\beta_2 - \alpha_1,\ 0\}\} \geqslant 0$$

we get by (2.69) that $p(\tilde{s}_1 \geqslant \tilde{s}_2) \geqslant 0$. Also since $\max\{\beta_2 - \alpha_1,\ 0\} \geqslant 0$, then

$$\mathrm{len}(\tilde{s}_1) + \mathrm{len}(\tilde{s}_2) - \max\{\beta_2 - \alpha_1,\ 0\} \leqslant \mathrm{len}(\tilde{s}_1) + \mathrm{len}(\tilde{s}_2)$$

We can get from (2.69) that $p(\tilde{s}_1 \geqslant \tilde{s}_2) \leqslant 1$. Thus, $0 \leqslant p(\tilde{s}_1 \geqslant \tilde{s}_2) \leqslant 1$.

(2) If $\beta_2 \leqslant \alpha_1$, then $\max\{\beta_2 - \alpha_1,\ 0\} = 0$. Thus, by (2.69), we get $p(\tilde{s}_1 \geqslant \tilde{s}_2) = 1$. And vice versa.

(3) If $\beta_1 \leqslant \alpha_2$, then

$$\beta_2 - \alpha_1 \geqslant \beta_2 - \alpha_1 - (\alpha_2 - \beta_1) = (\beta_1 - \alpha_1) + (\beta_2 - \alpha_2) = \mathrm{len}(\tilde{s}_1) + \mathrm{len}(\tilde{s}_2)$$

i.e.,

$$\mathrm{len}(\tilde{s}_1) + \mathrm{len}(\tilde{s}_2) - \max(\beta_2 - \alpha_1,\ 0) \leqslant 0$$

Thus, by (2.69), we know that $p(\tilde{s}_1 \geqslant \tilde{s}_2) = 0$. And vice versa.

(4) Since

$$p(\tilde{s}_1 \geqslant \tilde{s}_2) + p(\tilde{s}_2 \geqslant \tilde{s}_1) = \frac{\max\{0,\ \mathrm{len}(\tilde{s}_1) + \mathrm{len}(\tilde{s}_2) - \max\{\beta_2 - \alpha_1,\ 0\}\}}{\mathrm{len}(\tilde{s}_1) + \mathrm{len}(\tilde{s}_2)} + \frac{\max\{0,\ \mathrm{len}(\tilde{s}_1) + \mathrm{len}(\tilde{s}_2) - \max\{\beta_1 - \alpha_2,\ 0\}\}}{\mathrm{len}(\tilde{s}_1) + \mathrm{len}(\tilde{s}_2)} \tag{2.72}$$

we have

(i) If $\beta_1 \leqslant \alpha_2$, then $p(\tilde{s}_1 \geqslant \tilde{s}_2) = 0$ and $p(\tilde{s}_2 \geqslant \tilde{s}_1) = 1$. Thus, $p(\tilde{s}_1 \geqslant \tilde{s}_2) + p(\tilde{s}_2 \geqslant \tilde{s}_1) = 1$.

(ii) If $\beta_2 \leqslant \alpha_1$, then $p(\tilde{s}_2 \geqslant \tilde{s}_1) = 0$ and $p(\tilde{s}_1 \geqslant \tilde{s}_2) = 1$. Thus, $p(\tilde{s}_1 \geqslant \tilde{s}_2) + p(\tilde{s}_2 \geqslant \tilde{s}_1) = 1$.

(iii) If $\alpha_1 \leqslant \alpha_2 < \beta_1 \leqslant \beta_2$, then $\beta_1 - \alpha_2 > 0$ and $\beta_2 - \alpha_1 > 0$. Therefore,

$$\text{len}\,(\tilde{s}_1) + \text{len}\,(\tilde{s}_2) - \max\{\beta_2 - \alpha_1,\ 0\} = \beta_1 - \alpha_1 + \beta_2 - \alpha_2 - \beta_2 + \alpha_1 = \beta_1 - \alpha_2 \quad (2.73)$$

and

$$\text{len}\,(\tilde{s}_1) + \text{len}\,(\tilde{s}_2) - \max\{\beta_1 - \alpha_2,\ 0\} = \beta_1 - \alpha_1 + \beta_2 - \alpha_2 - \beta_1 + \alpha_2 = \beta_2 - \alpha_1 \quad (2.74)$$

Accordingly, from (2.72)~(2.74), we have

$$\begin{aligned} p(\tilde{s}_1 \geqslant \tilde{s}_2) + p(\tilde{s}_2 \geqslant \tilde{s}_1) &= \frac{\beta_1 - \alpha_2 + \beta_2 - \alpha_1}{\text{len}(\tilde{s}_1) + \text{len}(\tilde{s}_2)} = \frac{(\beta_1 - \alpha_1) + (\beta_2 - \alpha_2)}{\text{len}(\tilde{s}_1) + \text{len}(\tilde{s}_2)} \\ &= \frac{\text{len}(\tilde{s}_1) + \text{len}(\tilde{s}_2)}{\text{len}(\tilde{s}_1) + \text{len}(\tilde{s}_2)} = 1 \end{aligned} \quad (2.75)$$

(iv) If $\alpha_1 < \alpha_2 \leqslant \beta_2 < \beta_1$, then (2.73)~(2.75) hold.

(v) If $\alpha_2 < \alpha_1 \leqslant \beta_1 < \beta_2$, then (2.73)~(2.75) hold.

(vi) If $\alpha_2 \leqslant \alpha_1 < \beta_2 \leqslant \beta_1$, then (2.73)~(2.75) hold.

We can see from (i)~(vi) that $p(\tilde{s}_1 \geqslant \tilde{s}_2) + p(\tilde{s}_2 \geqslant \tilde{s}_1) = 1$. Especially, if $\tilde{s}_1 = \tilde{s}_2$, i.e., $s_{a_1} = s_{a_2}$ and $s_{b_1} = s_{b_2}$, then $p\,(\tilde{s}_1 \geqslant \tilde{s}_1) + p\,(\tilde{s}_1 \geqslant \tilde{s}_1) = 1$. Thus, $p(\tilde{s}_1 \geqslant \tilde{s}_1) = 1/2$.

(1) If $\beta_2 > \alpha_1$, then by $p(\tilde{s}_1 \geqslant \tilde{s}_2) \geqslant 1/2$, we have

$$\frac{\text{len}(\tilde{s}_1) + \text{len}(\tilde{s}_2) - \max\{\beta_2 - \alpha_1,\ 0\}}{\text{len}(\tilde{s}_1) + \text{len}(\tilde{s}_2)} \geqslant \frac{1}{2} \quad (2.76)$$

It follows from (2.73) that

$$\frac{\beta_1 - \alpha_2}{(\beta_1 - \alpha_1) + (\beta_2 - \alpha_2)} \geqslant \frac{1}{2} \quad (2.77)$$

which can be simplified as $\alpha_1 + \beta_1 \geqslant \alpha_2 + \beta_2$. If $\beta_2 \leqslant \alpha_1$, then $p(\tilde{s}_1 \geqslant \tilde{s}_2) = 1$. Consequently, the condition $p(\tilde{s}_1 \geqslant \tilde{s}_2) \geqslant 1/2$ holds. In this case, $\beta_1 \geqslant \alpha_1 \geqslant \beta_2 \geqslant \alpha_2$, and then $\alpha_1 + \beta_1 \geqslant \alpha_2 + \beta_2$. And vice versa. If $\alpha_1 + \beta_1 \geqslant \alpha_2 + \beta_2$, then both (2.76) and (2.77) hold. Therefore, $p(\tilde{s}_1 \geqslant \tilde{s}_2) \geqslant 1/2$. Similarily, we can prove that $p(\tilde{s}_1 \geqslant \tilde{s}_2) = 1/2$ if and only if $\alpha_1 + \beta_1 = \alpha_2 + \beta_2$.

(2) If $p(\tilde{s}_1 \geqslant \tilde{s}_2) \geqslant 1/2$ and $p(\tilde{s}_2 \geqslant \tilde{s}_3) \geqslant 1/2$, then $\alpha_1 + \beta_1 \geqslant \alpha_2 + \beta_2$ and $\alpha_2 + \beta_2 \geqslant \alpha_3 + \beta_3$. Thus, $\alpha_1 + \beta_1 \geqslant \alpha_3 + \beta_3$, and then $p(\tilde{s}_1 \geqslant \tilde{s}_3) \geqslant 1/2$. This completes the proof.

Based on Definition 2.22, Xu (2006b; 2004a; 2004d) developed some uncertain linguistic aggregation operators:

Definition 2.23 (Xu, 2004a) Let ULA : $\tilde{S}_2^n \to \tilde{S}_2$. Then

$$\mathrm{ULA}(\tilde{s}_1, \tilde{s}_2, \cdots, \tilde{s}_n) = \frac{1}{n}(\tilde{s}_1 \oplus \tilde{s}_2 \oplus \cdots \oplus \tilde{s}_n) \tag{2.78}$$

is called an uncertain linguistic averaging (ULA) operator.

Example 2.24 Let $\tilde{s}_1 = [s_{-1}, s_0]$, $\tilde{s}_2 = [s_{-2}, s_{-1}]$, $\tilde{s}_3 = [s_2, s_3]$ and $\tilde{s}_4 = [s_1, s_3]$ be a collection of uncertain linguistic variables. Then

$$\begin{aligned}\mathrm{ULA}(\tilde{s}_1, \tilde{s}_2, \tilde{s}_3, \tilde{s}_4) &= \frac{1}{4}([s_{-1}, s_0] \oplus [s_{-2}, s_{-1}] \oplus [s_2, s_3] \oplus [s_1, s_3]) \\ &= [s_0, s_{1.25}]\end{aligned}$$

Definition 2.24 (Xu, 2004a) Let ULWA : $\tilde{S}_2^n \to \tilde{S}_2$. If

$$\mathrm{ULWA}(\tilde{s}_1, \tilde{s}_2, \cdots, \tilde{s}_n) = w_1\tilde{s}_1 \oplus w_2\tilde{s}_2 \oplus \cdots \oplus w_n\tilde{s}_n \tag{2.79}$$

where $\boldsymbol{w} = (w_1, w_2, \cdots, w_n)^{\mathrm{T}}$ is the weighting vector of $\tilde{s}_i$ $(i = 1, 2, \cdots, n)$, with $w_i \geqslant 0$ $(i = 1, 2, \cdots, n)$ and $\sum_{i=1}^{n} w_i = 1$, then the function ULWA is called an uncertain linguistic weighted averaging (ULWA) operator.

In particular, If $\boldsymbol{w} = (1/n, 1/n, \cdots, 1/n)^{\mathrm{T}}$, then the ULWA operator reduces to the ULA operator.

Example 2.25 Let $\tilde{s}_1 = [s_{-1}, s_0]$, $\tilde{s}_2 = [s_{-2}, s_{-1}]$, $\tilde{s}_3 = [s_2, s_3]$ and $\tilde{s}_4 = [s_1, s_3]$ be a collection of uncertain linguistic variables, and $\boldsymbol{w} = (0.3, 0.2, 0.3, 0.2)^{\mathrm{T}}$ be their weight vector. Then

$$\begin{aligned}\mathrm{ULWA}(\tilde{s}_1, \tilde{s}_2, \tilde{s}_3, \tilde{s}_4) &= 0.3 \times [s_{-1}, s_0] \oplus 0.2 \times [s_{-2}, s_{-1}] \oplus 0.3 \times [s_2, s_3] \oplus 0.2 \times [s_1, s_3] \\ &= [s_{-0.3}, s_0] \oplus [s_{-0.4}, s_{-0.2}] \oplus [s_{0.6}, s_{0.9}] \oplus [s_{0.2}, s_{0.6}] \\ &= [s_{0.1}, s_{1.3}]\end{aligned}$$

Definition 2.25 (Xu, 2004a) Let ULOWA : $\tilde{S}_2^n \to \tilde{S}_2$. If

$$\mathrm{ULOWA}(\tilde{s}_1, \tilde{s}_2, \cdots, \tilde{s}_n) = \omega_1\tilde{s}_{\sigma_1} \oplus \omega_2\tilde{s}_{\sigma_2} \oplus \cdots \oplus \omega_n\tilde{s}_{\sigma_n} \tag{2.80}$$

then the function ULOWA is called an uncertain linguistic ordered weighted averaging (ULOWA) operator, where $\tilde{s}_{\sigma_j}$ is the jth largest of the uncertain linguistic arguments $\tilde{s}_i$ $(i = 1, 2, \cdots, n)$, $\boldsymbol{\omega} = (\omega_1, \omega_2, \cdots, \omega_n)^{\mathrm{T}}$ is the weighting vector associated with the ULOWA operator, $\omega_i \geqslant 0$ $(i = 1, 2, \cdots, n)$ and $\sum_{i=1}^{n} \omega_i = 1$.

In particular, if $\boldsymbol{\omega} = (1/n, 1/n, \cdots, 1/n)^{\mathrm{T}}$, then the ULOWA operatror reduces to the ULA operator.

In order to rank the uncertain linguistic variables $\tilde{s}_i$ $(i = 1, 2, \cdots, n)$, we first use (2.67) to compare $\tilde{s}_i$ and $\tilde{s}_j$, and let $p_{ij} = p\ (\tilde{s}_i \geqslant \tilde{s}_j)$. Then we construct the possibility degree matrix $\boldsymbol{P} = (p_{ij})_{n\times n}$, where

$$p_{ij} \geqslant 0, \quad p_{ij} + p_{ji} = 1, \quad p_{ii} = \frac{1}{2}, \quad i, j = 1, 2, \cdots, n \tag{2.81}$$

Clearly, the possibility degree matrix $\boldsymbol{P} = (p_{ij})_{n\times n}$ is a complementary judgment matrix (Chiclana et al., 2001; 1998; Fan et al., 2002; Ma et al., 2006; Xu, 2007c; 2004g; 2004h; 2001a; 2001b; 2000; Xu and Chen, 2008; Xu and Da, 2005; 2002b):

After that, we utilize a simple ranking formula given by Xu (2001b):

$$\zeta_i = \frac{1}{n(n-1)}\left(\sum_{j=1}^{n} p_{ij} + \frac{n}{2} - 1\right), \quad i = 1, 2, \cdots, n \tag{2.82}$$

to derive the priority vector $\boldsymbol{\zeta} = (\zeta_1, \zeta_2, \cdots, \zeta_n)^{\mathrm{T}}$ from the possibility degree matrix $\boldsymbol{P}$, and employ ζ_i $(i = 1, 2, \cdots, n)$ to rank the uncertain linguistic variables $\tilde{s}_i$ $(i = 1, 2, \cdots, n)$.

Example 2.26 Let $\tilde{s}_1 = [s_{-2}, s_1]$, $\tilde{s}_2 = [s_{-2}, s_0]$, $\tilde{s}_3 = [s_{-3}, s_0]$ and $\tilde{s}_4 = [s_{-2}, s_2]$ be a collection of uncertain linguistic variables. In order to rank the uncertain linguistic variables $\tilde{s}_i$ $(i = 1, 2, 3, 4)$, we first use (2.67) to compare $\tilde{s}_i$ and $\tilde{s}_j$, and construct the possiblity degree matrix:

$$\boldsymbol{P} = \begin{pmatrix} 0.5 & 0.6 & 0.667 & 0.429 \\ 0.4 & 0.5 & 0.600 & 0.333 \\ 0.333 & 0.400 & 0.5 & 0.286 \\ 0.571 & 0.667 & 0.714 & 0.5 \end{pmatrix}$$

Then we utilize (2.82) to derive the priority vector from the possibility degree matrix $\boldsymbol{P}$:

$$\boldsymbol{\zeta} = (0.266, 0.236, 0.210, 0.288)^{\mathrm{T}}$$

and employ the priority vector $\boldsymbol{\zeta}$ to rank the uncertain linguistic variables $\tilde{s}_i$ $(i = 1, 2, 3, 4)$:

$$\tilde{s}_{\sigma_1} = \tilde{s}_4 = [s_{-2}, s_2], \quad \tilde{s}_{\sigma_2} = \tilde{s}_1 = [s_{-2}, s_1],$$

$$\tilde{s}_{\sigma_3} = \tilde{s}_2 = [s_{-2}, s_0], \quad \tilde{s}_{\sigma_4} = \tilde{s}_3 = [s_{-3}, s_0]$$

Assume that $\boldsymbol{\omega} = (0.2, 0.3, 0.3, 0.2)^{\mathrm{T}}$ is the weighting vector associated with the ULOWA operator. Then

$$\mathrm{ULOWA}(\tilde{s}_1, \tilde{s}_2, \tilde{s}_3, \tilde{s}_4) = 0.2 \times [s_{-2}, s_2] \oplus 0.3 \times [s_{-2}, s_{-1}] \oplus 0.3$$

$$\times [s_{-2}, s_0] \oplus 0.2 \times [s_{-3}, s_0]$$
$$=[s_{-0.4}, s_{0.4}] \oplus [s_{-0.6}, s_{-0.3}] \oplus [s_{-0.6}, s_0] \oplus [s_{-0.6}, s_0]$$
$$=[s_{-1.6}, s_{0.1}]$$

Definition 2.26 (Xu, 2004a) Let a mapping ULHA : $\tilde{S}_2^n \rightarrow \tilde{S}_2$. If

$$\text{ULHA}(\tilde{s}_1, \tilde{s}_2, \cdots, \tilde{s}_n) = \omega_1 \tilde{s}_{\beta_1} \oplus \omega_2 \tilde{s}_{\beta_2} \oplus \cdots \oplus \omega_n \tilde{s}_{\beta_n} \tag{2.83}$$

then ULHA is called an uncertain linguistic hybrid averaging (ULHA) operator, where $\boldsymbol{\omega} = (\omega_1, \omega_2, \cdots, \omega_n)^{\mathrm{T}}$ is the weighting vector, $\omega_i \geqslant 0$ $(i = 1, 2, \cdots, n)$ and $\sum_{i=1}^{n} \omega_i = 1$. $\tilde{s}_{\beta_j}$ is the jth largest of the weighted uncertain linguistic variables $\tilde{s}_i'$ ($\tilde{s}_i' = n w_i \tilde{s}_i$ $(i = 1, 2, \cdots, n)$, $\boldsymbol{w} = (w_1, w_2, \cdots, w_n)^{\mathrm{T}}$ is the weight vector of $\tilde{s}_i$ $(i = 1, 2, \cdots, n)$, $w_i \geqslant 0$ $(i = 1, 2, \cdots, n)$, $\sum_{i=1}^{n} w_i = 1$, and n is the balancing coefficient.

Especially, if $\boldsymbol{\omega} = (1/n, 1/n, \cdots, 1/n)^{\mathrm{T}}$, then the ULHA operator reduces to the ULWA operator; if $\boldsymbol{w} = (1/n, 1/n, \cdots, 1/n)^{\mathrm{T}}$, then the ULHA operator reduces to the ULOWA operator.

Example 2.27 Let $\tilde{s}_1 = [s_{-3}, s_{-2}]$, $\tilde{s}_2 = [s_{-2}, s_{-1}]$, $\tilde{s}_3 = [s_2, s_3]$, $\tilde{s}_4 = [s_2, s_4]$ and $\tilde{s}_5 = [s_1, s_3]$ be a collection of uncertain linguistic variables, and let $\boldsymbol{w} = (0.20, 0.15, 0.25, 0.30, 0.10)^{\mathrm{T}}$ be their weight vector. We first calculate the weighted uncertain linguistic variables:

$$\tilde{s}_1' = 5w_1\tilde{s}_1 = 5 \times 0.20 \times [s_{-3}, s_{-2}] = [s_{-3}, s_{-2}]$$

$$\tilde{s}_2' = 5w_2\tilde{s}_2 = 5 \times 0.15 \times [s_{-2}, s_{-1}] = [s_{-1.5}, s_{-0.75}]$$

$$\tilde{s}_3' = 5w_3\tilde{s}_3 = 5 \times 0.25 \times [s_2, s_3] = [s_{2.50}, s_{3.75}]$$

$$\tilde{s}_4' = 5w_4\tilde{s}_4 = 5 \times 0.30 \times [s_2, s_4] = [s_3, s_6]$$

$$\tilde{s}_5' = 5w_5\tilde{s}_5 = 5 \times 0.10 \times [s_1, s_3] = [s_{0.5}, s_{1.5}]$$

In order to rank the uncertain linguistic varibles $\tilde{s}_i'$ $(i = 1, 2, 3, 4, 5)$, we first use (2.67) to compare $\tilde{s}_i'$ and $\tilde{s}_j'$, and construct the possibility degree matrix:

$$\boldsymbol{P} = \begin{pmatrix} 0.5 & 0 & 0 & 0 & 0 \\ 1 & 0.5 & 0 & 0 & 0 \\ 1 & 1 & 0.5 & 0.176 & 1 \\ 1 & 1 & 0.824 & 0.5 & 1 \\ 1 & 1 & 0 & 0 & 0.5 \end{pmatrix}$$

Then we employ (2.82) to derive the priority vector from the possibility degree matrix $\boldsymbol{P}$:

$$\boldsymbol{\zeta} = (0.100, 0.150, 0.259, 0.291, 0.200)^{\mathrm{T}}$$

and utilize the priority vector $\boldsymbol{\zeta}$ to rank the uncertain linguistic variables $\tilde{s}'_i\,(i = 1, 2, 3, 4, 5)$:

$$\tilde{s}_{\sigma_1} = \tilde{s}'_4 = [s_3, s_6], \quad \tilde{s}_{\sigma_2} = \tilde{s}'_3 = [s_{2.50}, s_{3.75}], \quad \tilde{s}_{\sigma_3} = \tilde{s}'_5 = [s_{0.5}, s_{1.5}]$$

$$\tilde{s}_{\sigma_4} = \tilde{s}'_2 = [s_{-1.5}, s_{-0.75}], \quad \tilde{s}_{\sigma_5} = \tilde{s}'_1 = [s_{-3}, s_{-2}]$$

Suppose that $\boldsymbol{\omega} = (0.15, 0.2, 0.3, 0.2, 0.15)^{\mathrm{T}}$ is the weighting vector associated with the ULHA operator. Then

$$\begin{aligned}\text{ULHA}(\tilde{s}_1, \tilde{s}_2, \tilde{s}_3, \tilde{s}_4, \tilde{s}_5) =& 0.15 \times [s_3, s_6] \oplus 0.20 \times [s_{2.50}, s_{3.75}] \oplus 0.3 \times [s_{0.5}, s_{1.5}] \\ & \oplus 0.20 \times [s_{-1.5}, s_{-0.75}] \oplus 0.15 \times [s_{-3}, s_{-2}] \\ =& [s_{-0.35}, s_{1.65}]\end{aligned}$$

Definition 2.27 (Xu, 2006b) An induced uncertain LOWA (IULOWA) operator is given as

$$\text{IULOWA}\left(\langle u_1, \tilde{s}_1\rangle, \langle u_2, \tilde{s}_2\rangle, \cdots, \langle u_n, \tilde{s}_n\rangle\right) = \omega_1 \tilde{s}_{\sigma_1} \oplus \omega_2 \tilde{s}_{\sigma_2} \oplus \cdots \oplus \omega_n \tilde{s}_{\sigma_n} \tag{2.84}$$

where $\boldsymbol{\omega} = (\omega_1, \omega_2, \cdots, \omega_n)^{\mathrm{T}}$ is the weighting vector, such that $\omega_i \geqslant 0\,(i = 1, 2, \cdots, n)$ and $\sum\limits_{j=1}^{n} \omega_j = 1$. $\tilde{s}_{\sigma_j}$ is the $\tilde{s}_i$ value of the pairs $\langle u_i, \tilde{s}_i\rangle\,(i = 1, 2, \cdots, n)$ having the jth largest u_i value. The term u_i is referred as the order inducing variable, and $\tilde{s}_i$ is referred as the argument variable.

In particular, if $u_i = \tilde{s}_i$ for all i, then the IULOWA operator reduces to the ULOWA operator; if $u_i =$ No. i for all i, where No. i is the ordered position of $\tilde{s}_i$, then the IULOWA operator reduces to the ULWA operator; if $\boldsymbol{\omega} = (1/n, 1/n, \cdots, 1/n)^{\mathrm{T}}$, then the IULOWA operator reduces to the ULA operator.

If there is a tie between $\langle u_i, \tilde{s}_i\rangle$ and $\langle u_j, \tilde{s}_j\rangle$ with respect to order inducing variables, then in the process of aggregation, we can average $\tilde{s}_i$ and $\tilde{s}_j$, and get the corresponding pairs $\left\langle u_i, \dfrac{\tilde{s}_i \oplus \tilde{s}_j}{2}\right\rangle$ and $\left\langle u_j, \dfrac{\tilde{s}_i \oplus \tilde{s}_j}{2}\right\rangle$. If k items are tied, then we replace these by k replica's of their average.

Example 2.28 Given a collection of uncertain linguistic pairs $\langle u_i, \tilde{s}_i\rangle\,(i = 1, 2, 3, 4)$:

$$\left(\langle -1,\ [s_2, s_3]\rangle, \langle -3,\ [s_2, s_4]\rangle, \langle -2,\ [s_1, s_3]\rangle, \langle -4,\ [s_{-1}, s_0]\rangle\right)$$

We first use the values u_i $(i = 1, 2, 3, 4)$ to rank the uncertain linguistic pairs $\langle u_i, \tilde{s}_i\rangle$ $(i = 1, 2, \cdots, n)$. Since

$$u_1 > u_3 > u_2 > u_4$$

we get the ordered uncertain linguistic pairs:

$$(\langle -1, [s_2, s_3]\rangle, \langle -2, [s_1, s_3]\rangle, \langle -3, [s_2, s_4]\rangle, \langle -4, [s_{-1}, s_0]\rangle)$$

i.e.,

$$\tilde{s}_{\sigma_1} = [s_2, s_3], \quad \tilde{s}_{\sigma_2} = [s_1, s_3], \quad \tilde{s}_{\sigma_3} = [s_2, s_4], \quad \tilde{s}_{\sigma_4} = [s_{-1}, s_0]$$

Assume that the weighting vector associated with the IULOWA operator is $\boldsymbol{w} = (0.10, 0.40, 0.40, 0.10)^{\mathrm{T}}$. Then

$$\begin{aligned}&\text{IULOWA}(\langle u_1, \tilde{s}_1\rangle, \langle u_2, \tilde{s}_2\rangle, \langle u_3, \tilde{s}_3\rangle, \langle u_4, \tilde{s}_4\rangle)\\ =&0.10 \times [s_2, s_3] \oplus 0.40 \times [s_1, s_3] \oplus 0.40 \times [s_2, s_4] \oplus 0.10 \times [s_{-1}, s_0]\\ =&[s_{0.2}, s_{0.3}] \oplus [s_{0.4}, s_{1.2}] \oplus [s_{0.8}, s_{1.6}] \oplus [s_{-0.1}, s_0]\\ =&[s_{1.3}, s_{3.1}]\end{aligned}$$

If we replace (2.58) with

$$\begin{aligned}&\text{GIULOWA}(\langle v_1, u_1, \tilde{s}_1\rangle, \langle v_2, u_2, \tilde{s}_2\rangle, \cdots, \langle v_n, u_n, \tilde{s}_n\rangle)\\ =&\omega_1\tilde{s}_{\sigma_1} \oplus \omega_2\tilde{s}_{\sigma_2} \oplus \cdots \oplus \omega_n\tilde{s}_{\sigma_n}\end{aligned} \tag{2.85}$$

then by Definition 2.18, we get a generalized induced uncertain LOWA (GIULOWA) operator.

In particular, if $v_i = u_i$ for any i, then the GIULOWA operator reduces to the IULOWA operator; if $v_i = \tilde{s}_i$ for all i, then the GIULOWA operator reduces to the ULOWA operator; if $v_i =$ No. i for all i, where No. i is the ordered position of $\tilde{s}_i$, then the GIULOWA operator reduces to the ULWA operator; if $\boldsymbol{\omega} = (1/n, 1/n \cdots, 1/n)^{\mathrm{T}}$, then the GIULOWA operator reduces to the ULA operator.

Example 2.29 Given a collection of four objects $\langle v_i, u_i, \tilde{s}_i\rangle$ $(i = 1, 2, 3, 4)$:

$$(\langle 0.2, \text{Johnson}, [s_0, s_1]\rangle, \langle 0.4, \text{Brown}, [s_2, s_3]\rangle, \langle 0.3, \text{Smith}, [s_{-3}, s_{-1}]\rangle,$$
$$\langle 0.1, \text{Anderson}, [s_{-2}, s_0]\rangle)$$

We first use the values v_i $(i = 1, 2, 3, 4)$ to rank the objects $\langle v_i, u_i, s_{\alpha_i}\rangle$ $(i = 1, 2, 3, 4)$:

$$\langle 0.4, \text{Brown}, [s_2, s_3]\rangle$$
$$\langle 0.3, \text{Smith}, [s_{-3}, s_{-1}]\rangle$$
$$\langle 0.2, \text{Johnson}, [s_0, s_1]\rangle$$

$$\langle 0.1, \text{ Anderson}, \ [s_{-2}, s_0]\rangle)$$

i.e.,

$$\tilde{s}_{\sigma_1} = [s_2, s_3], \quad \tilde{s}_{\sigma_2} = [s_{-3}, s_{-1}], \quad \tilde{s}_{\sigma_3} = [s_0, s_1], \quad \tilde{s}_{\sigma_4} = [s_{-2}, s_0]$$

Assume that the weighting vector associated with the GIULOWA operator is $\boldsymbol{\omega} = (0.20, 0.40, 0.30, 0.10)^{\mathrm{T}}$. Then we have

$$\begin{aligned} &\text{GIULOWA}\left(\langle v_1, u_1, \tilde{s}_1\rangle, \langle v_2, u_2, \tilde{s}_2\rangle, \langle v_3, u_3, \tilde{s}_3\rangle, \langle v_4, u_4, \tilde{s}_4\rangle\right) \\ =&0.20 \times [s_2, s_3] \oplus 0.40 \times [s_{-3}, s_{-1}] \oplus 0.30 \times [s_0, s_1] \oplus 0.10 \times [s_{-2}, s_0] \\ =&[s_{0.4}, s_{0.6}] \oplus [s_{-1.2}, s_{-0.4}] \oplus [s_0, s_{0.3}] \oplus [s_{-0.2}, s_0] \\ =&[s_{-1}, s_{0.5}] \end{aligned}$$

Yager (2004) gave a continuous ordered weighted averaging (C-OWA) operator. Zhang and Xu (2005) extended the C-OWA operator to uncertain linguistic environment:

Definition 2.28 (Yager, 2004) Let $[a, b]$ be an interval number, and

$$f_{\varphi}([a,b]) = \int_0^1 \frac{\mathrm{d}\varphi(x)}{\mathrm{d}x}[b - x(b-a)]\mathrm{d}x = a\left[1 - \int_0^1 \varphi(x)\mathrm{d}x\right] + b\int_0^1 \varphi(x)\mathrm{d}x \tag{2.86}$$

where $\varphi : [0,1] \to [0,1]$ is a BUM function, and is monotonic with the following properties:

(1) $\varphi(0) = 0$;

(2) $\varphi(1) = 1$;

(3) $\varphi(x) \geqslant \varphi(y)$ if $x > y$,

then f is called a continuous OWA (C-OWA) operator.

Definition 2.29 (Zhang and Xu, 2005) Let $s_\alpha \in \bar{S}_2$ and $I(s_\alpha) = \alpha$. Then the function I is an operator of the subscripts of linguistic labels.

Definition 2.30 (Zhang and Xu, 2005) Let $[s_\alpha, s_\beta] \in \tilde{S}_2$ be an uncertain linguistic variable, and $lf_\varphi([s_\alpha, s_\beta]) = s_\gamma$, where

$$\gamma = f_\varphi([I(s_\alpha), I(s_\beta)]) = f_\varphi([\alpha, \beta]) \tag{2.87}$$

φ is a BUM function. Then lf is called an uncertain linguistic C-OWA (ULC-OWA) operator.

Theorem 2.2 (Zhang and Xu, 2005) (Boundedness) $s_\alpha \leqslant lf_\varphi([s_\alpha, s_\beta]) \leqslant s_\beta$ for any BUM function φ.

Proof Since $lf_\varphi([s_\alpha, s_\beta]) = s_{f_\varphi([\alpha,\beta])}$, where

$$f_\varphi([\alpha,\beta]) = \int_0^1 \frac{\mathrm{d}\varphi(x)}{\mathrm{d}x}[\beta - x(\beta-\alpha)]\mathrm{d}x = \int_0^1 [\beta - x(\beta - \alpha)]\mathrm{d}\varphi(x)$$

$$= [[\beta - x(\beta - \alpha)]\varphi(x)]_0^1 - \int_0^1 \varphi(x)\mathrm{d}[\beta - x(\beta - \alpha)]$$
$$= \alpha + (\beta - \alpha)\int_0^1 \varphi(x)\mathrm{d}x$$

and since $0 \leqslant \int_0^1 \varphi(x)\mathrm{d}x \leqslant 1$, we have

$$\alpha \leqslant f_\varphi([\alpha, \beta]) = \alpha + (\beta - \alpha)\int_0^1 \varphi(x)\mathrm{d}x \leqslant \beta$$

Thus, $s_\alpha \leqslant lf_\varphi([s_\alpha, s_\beta]) \leqslant s_\beta$. This completes the proof.

Example 2.30 (Zhang and Xu, 2005) Let $\tilde{s} = [s_{-1}, s_3]$ be an uncertain linguistic variable, and BUM be the function $\varphi(x) = x^2$. Then

$$f_\varphi([-1, 3]) = \int_0^1 \frac{\mathrm{d}x^2}{\mathrm{d}x}\{3 - x[3 - (-1)]\}\mathrm{d}x = \frac{1}{3}$$

and thus, $lf_\varphi([s_{-1}, s_3]) = s_{1/3}$.

Xu (2009b) extended the LCA operator (2.61) to uncertain linguistic environment:

Definition 2.31 (Xu, 2009b) Let μ be a fuzzy measure on $\tilde{S}_2$, and $\tilde{s}_i$ $(i = 1, 2, \cdots, n)$ be n uncertain additional linguistic variables in $\tilde{S}_2$. Then an uncertain linguistic correlated averaging (ULCA) operator is defined as

$$(C_2)\int \tilde{s}\,\mathrm{d}\mu = \mathrm{ULCA}\,(\tilde{s}_1, \tilde{s}_2, \cdots, \tilde{s}_n) = \bigoplus_{i=1}^{n}\left\{[\mu(\tilde{A}_{\sigma(i)}) - \mu(\tilde{A}_{\sigma(i-1)})]\tilde{s}_{\sigma(i)}\right\} \tag{2.88}$$

where $(C_2)\int \tilde{s}\,\mathrm{d}\mu$ is a notation of Choquet integral, $(\sigma(1), \sigma(2), \cdots, \sigma(n))$ is a permutation of $(1, 2, \cdots, n)$ such that $\tilde{s}_{\sigma(1)} \geqslant \tilde{s}_{\sigma(2)} \geqslant \cdots \geqslant \tilde{s}_{\sigma(n)}$, $\tilde{A}_{\sigma(k)} = \{\tilde{s}_{\sigma(j)} | j \leqslant k\}$, when $k \geqslant 1$ and $\tilde{A}_{\sigma(0)} = \varnothing$.

Now we discuss some special cases of the ULCA operator:

(1) If

$$\mu(\tilde{A}) = \sum_{\tilde{s}_i \in \tilde{A}} \mu(\{\tilde{s}_i\}) \quad \text{for all} \quad \tilde{A} \subseteq \tilde{S}_2 \tag{2.89}$$

then

$$w_i = \mu(\{\tilde{s}_{\sigma(i)}\}) = \mu(\tilde{A}_{\sigma(i)}) - \mu(\tilde{A}_{\sigma(i-1)}), \quad i = 1, 2, \cdots, n \tag{2.90}$$

where $\boldsymbol{w} = (w_1, w_2, \cdots, w_n)^{\mathrm{T}}$, $w_i \geqslant 0$ $(i = 1, 2, \cdots, n)$ and $\sum_{i=1}^{n} w_i = 1$. Thus, the ULCA operator reduces to the ULWA operator. Especially, if $w_i = \frac{1}{n}$ for all $i = 1, 2, \cdots, n$, then the ULWA operator reduces to the ULA operator.

(2) If

$$\mu(\tilde{A})=\sum_{i=1}^{|\tilde{A}|}\omega_i \quad \text{for all } \tilde{A}\subseteq\tilde{S}_2 \tag{2.91}$$

where $|\tilde{A}|$ is the number of the uncertain additional linguistic variables in the set $\tilde{A}$, then

$$\omega_i=\mu(\tilde{A}_{\sigma(i)})-\mu(\tilde{A}_{\sigma(i-1)}),\quad i=1,2,\cdots,n \tag{2.92}$$

where $\boldsymbol{\omega}=(\omega_1,\omega_2,\cdots,\omega_n)^{\mathrm{T}}$, $\omega_i\geqslant 0$ $(i=1,2,\cdots,n)$ and $\sum_{i=1}^{n}\omega_i=1$. In this case, the ULCA operator reduces to the ULOWA operator. Especially, if $\mu(\tilde{A})=\frac{|\tilde{A}|}{n}$ for all $\tilde{A}\subseteq\tilde{S}_2$, then both the ULCA and ULOWA operators reduce to the ULA operator.

(3) If

$$\mu(\tilde{A})=f\left(\sum_{\tilde{s}_i\in\tilde{A}}\mu\left(\{\tilde{s}_i\}\right)\right) \quad \text{for all } \tilde{A}\subseteq\tilde{S}_2 \tag{2.93}$$

where f is the BUM function $f:\ [0,1]\to[0,1]$, then we let

$$w_i=\mu(\tilde{A}_{\sigma(i)})-\mu(\tilde{A}_{\sigma(i-1)})=f\left(\sum_{j\leqslant i}\mu\left(\{\tilde{s}_{\sigma(i)}\}\right)\right)-f\left(\sum_{j<i}\mu\left(\{\tilde{s}_{\sigma(i)}\}\right)\right),$$
$$i=1,2,\cdots,n \tag{2.94}$$

where $\boldsymbol{w}=(w_1,w_2,\cdots,w_n)^{\mathrm{T}}$, $w_i\geqslant 0$ $(i=1,2,\cdots,n)$ and $\sum_{i=1}^{n}w_i=1$. In this case, the ULCA operator reduces to the following form:

$$\text{ULWOWA}\left(\tilde{s}_1,\tilde{s}_2,\cdots,\tilde{s}_n\right)=\bigoplus_{i=1}^{n}\left(w_i\tilde{s}_{\sigma(i)}\right) \tag{2.95}$$

which we call an uncertain linguistic weighted ordered weighted averaging (ULWOWA) operator. Especially, if $\mu\left(\{s_{\alpha_i}\}\right)=\frac{1}{n}$ for all $i=1,2,\cdots,n$, then the ULWOWA operator reduces to the ULOWA operator.

Example 2.31 (Xu, 2009b) If the evaluation information in Example 2.23 is represented by the uncertain additive linguistic variables, shown as in Table 2.2:

Table 2.2 The evaluation information about students (Xu, 2009b)

	mathematics	physics	literature
x_1	$[s_{-1},\ s_0]$	$[s_2,\ s_4]$	$[s_3,\ s_4]$
x_2	$[s_{-1},\ s_0]$	$[s_1,\ s_3]$	$[s_2,\ s_4]$
x_3	$[s_3,\ s_4]$	$[s_2,\ s_4]$	$[s_{-2},\ s_0]$
x_4	$[s_{-2},\ s_0]$	$[s_3,\ s_4]$	$[s_1,\ s_2]$

Then by (2.68), we rearrange the uncertain additive linguistic variables corresponding to each student in descending order:

$$\tilde{a}_{\sigma(11)} = \tilde{a}_{13} = [s_3, s_4], \quad \tilde{a}_{\sigma(12)} = \tilde{a}_{12} = [s_2, s_3], \quad \tilde{a}_{\sigma(13)} = \tilde{a}_{11} = [s_{-1}, s_0]$$

$$\tilde{a}_{\sigma(21)} = \tilde{a}_{23} = [s_2, s_4], \quad \tilde{a}_{\sigma(22)} = \tilde{a}_{22} = [s_1, s_3], \quad \tilde{a}_{\sigma(23)} = \tilde{a}_{21} = [s_{-1}, s_0]$$

$$\tilde{a}_{\sigma(31)} = \tilde{a}_{31} = [s_3, s_4], \quad \tilde{a}_{\sigma(32)} = \tilde{a}_{32} = [s_2, s_4], \quad \tilde{a}_{\sigma(33)} = \tilde{a}_{33} = [s_{-2}, s_0]$$

$$\tilde{a}_{\sigma(41)} = \tilde{a}_{42} = [s_3, s_4], \quad \tilde{a}_{\sigma(42)} = \tilde{a}_{43} = [s_1, s_2], \quad \tilde{a}_{\sigma(43)} = \tilde{a}_{41} = [s_{-2}, s_0]$$

and utilize the ULCA operator to calculate the overall evaluation information corresponding to each student:

$$\begin{aligned}
\tilde{a}_1 &= \text{ULCA}(\tilde{a}_{11}, \tilde{a}_{12}, \tilde{a}_{13}) \\
&= (0.3 - 0)[s_3, s_4] \oplus (0.9 - 0.3)[s_2, s_3] \oplus (1 - 0.9)[s_{-1}, s_0] = [s_2, s_3] \\
\tilde{a}_2 &= \text{ULCA}(\tilde{a}_{21}, \tilde{a}_{22}, \tilde{a}_{23}) \\
&= (0.3 - 0)[s_2, s_4] \oplus (0.9 - 0.3)[s_1, s_3] \oplus (1 - 0.9)[s_{-1}, s_0] = [s_{1.1}, s_3] \\
\tilde{a}_3 &= \text{ULCA}(\tilde{a}_{31}, \tilde{a}_{32}, \tilde{a}_{33}) \\
&= (0.45 - 0)[s_3, s_4] \oplus (0.5 - 0.45)[s_2, s_4] \oplus (1 - 0.5)[s_{-2}, s_0] = [s_{0.4}, s_2] \\
\tilde{a}_4 &= \text{ULCA}(\tilde{a}_{41}, \tilde{a}_{42}, \tilde{a}_{43}) \\
&= (0.45 - 0)[s_3, s_4] \oplus (0.9 - 0.45)[s_1, s_2] \oplus (1 - 0.9)[s_{-2}, s_0] = [s_{1.6}, s_{2.7}]
\end{aligned}$$

Using (2.68), we compare each pair of $\tilde{a}_i$ $(i = 1, 2, 3, 4)$, and construct a complementary matrix $\boldsymbol{P} = (p_{ij})_{4\times4}$:

$$\boldsymbol{P} = \begin{pmatrix} 0.5 & 0.655 & 1 & 0.667 \\ 0.345 & 0.5 & 0.743 & 0.467 \\ 0 & 0.257 & 0.5 & 0.148 \\ 0.333 & 0.533 & 0.852 & 0.5 \end{pmatrix}$$

Summing all elements in each line of the matrix $\boldsymbol{P}$, we have

$$p_1 = 2.822, \quad p_2 = 2.055, \quad p_3 = 0.905, \quad p_4 = 2.218$$

Then we rank $\tilde{a}_i$ $(i = 1, 2, 3, 4)$ in descending order in accordance with the values of p_i $(i = 1, 2, 3, 4)$: $\tilde{a}_1 > \tilde{a}_4 > \tilde{a}_2 > \tilde{a}_3$. Thus, the ranking of the four students x_i $(i = 1, 2, 3, 4)$ is

$$x_1 \succ x_4 \succ x_2 \succ x_3$$

All the operators above can only be used to deal with time independent linguistic arguments. However, if time is taken into account, for example, the argument information may be collected at different periods, then the aggregation operators and their associated weights should not be kept constant. As a result, Xu (2009a) first defined the notion of additive linguistic label at the time t.

Definition 2.32 (Xu, 2009a) Let t be a time variable. Then $s_\alpha(t)$ is called an additive linguistic label at the time t, where $s_\alpha(t) \in \bar{S}_2$ for any t. If $t = t_1, t_2, \cdots, t_p$, then $s_\alpha(t_k)\,(k = 1, 2, \cdots, p)$ denote p additive linguistic labels collected from p different periods.

Based on Definition 2.32, Xu (2009a) defined the concept of dynamic linguistic weighted averaging operator:

Definition 2.33 (Xu, 2009a) Let $s_\alpha(t_k)\ (k = 1, 2, \cdots, p)$ be p additive linguistic labels collected from p different periods $t_k\,(k = 1, 2, \cdots, p)$, and $\boldsymbol{w}(t) = (w(t_1), w(t_2), \cdots, w(t_p))^{\mathrm{T}}$ be the weight vector of the periods $t_k\,(k = 1, 2, \cdots, p)$. Then

$$\mathrm{DLWA}(s_\alpha(t_1), s_\alpha(t_2), \cdots, s_\alpha(t_p)) = w(t_1)s_\alpha(t_1) \oplus w(t_2)s_\alpha(t_2) \oplus \cdots \oplus w(t_p)s_\alpha(t_p) \tag{2.96}$$

is called a dynamic linguistic weighted averaging (DLWA) operator, where $w(t_k) \geqslant 0$ $(k = 1, 2, \cdots, p)$, and $\sum_{k=1}^{m} w(t_k) = 1$. Especially, if $\boldsymbol{w}(t) = (1/p, 1/p, \cdots, 1/p)^{\mathrm{T}}$, then (2.96) reduces to the dynamic linguistic averaging (DLA) operator:

$$\mathrm{DLA}\,(s_\alpha(t_1), s_\alpha(t_2), \cdots, s_\alpha(t_p)) = \frac{1}{p}\left[s_\alpha(t_1) \oplus s_\alpha(t_2) \oplus \cdots \oplus s_\alpha(t_p)\right] \tag{2.97}$$

Clearly, the key of the DLWA operator is to determine the weighting vector $\boldsymbol{w}(t) = (w(t_1), w(t_2), \cdots, w(t_p))^{\mathrm{T}}$ of the periods $t_k\,(k = 1, 2, \cdots, p)$.

In what follows, we utilize the BUM function based method (Yager, 2004; 1996) to determine the weighting vector $\boldsymbol{w}(t)$:

Let $f : [0, 1] \to [0, 1]$ be a BUM function. Then Yager (1988) defined the weighting vector $\boldsymbol{w}(t)$ as

$$w(t_k) = f\left(\frac{k}{p}\right) - f\left(\frac{k-1}{p}\right), \quad k = 1, 2, \cdots, p \tag{2.98}$$

Now we discuss the following two cases:

Case 1 If f is a monotone increasing function, then we can take

$$f(x) = \frac{\mathrm{e}^{\alpha x} - 1}{\mathrm{e}^{\alpha} - 1}, \quad \alpha > 0 \tag{2.99}$$

In such a case, by (2.98), we have

$$w(t_k) = \frac{\mathrm{e}^{\frac{\alpha k}{p}} - 1}{\mathrm{e}^{\alpha} - 1} - \frac{\mathrm{e}^{\frac{\alpha(k-1)}{p}} - 1}{\mathrm{e}^{\alpha} - 1} = \frac{\mathrm{e}^{\frac{\alpha k}{p}} - \mathrm{e}^{\frac{\alpha(k-1)}{p}}}{\mathrm{e}^{\alpha} - 1}$$

$$= \frac{e^{\frac{\alpha k}{p}}(1 - e^{-\frac{\alpha}{p}})}{e^{\alpha} - 1}, \quad k = 1, 2, \cdots, p \tag{2.100}$$

Let

$$F(x) = f(x) - f\left(x - \frac{1}{p}\right) = \frac{e^{\alpha x}\left(1 - e^{-\frac{\alpha}{p}}\right)}{e^{\alpha} - 1}, \quad x \geqslant \frac{1}{p} \tag{2.101}$$

Then

$$F'(x) = \frac{\alpha\left(1 - e^{-\frac{\alpha}{p}}\right) e^{\alpha x}}{e^{\alpha} - 1}, \quad x \geqslant \frac{1}{p} \tag{2.102}$$

Since $F'(x) > 0$ for all $\alpha > 0$, then $F(x)$ is a strictly monotone increasing function. Thus, from (2.100), it follows that

$$w(t_{k+1}) > w(t_k), \quad k = 1, 2, \cdots, p - 1 \tag{2.103}$$

i.e., the sequence $\{w(t_k)\}$ is a monotonic increasing sequence.

Yager (2007a) defined another function called centering function:

Definition 2.34 (Yager, 2007a) A function $g : [0, 1] \to \mathbf{R}$ is called centering function, if it satisfies the following conditions:

(1) $g(x) > 0$;

(2) g is symmetric about 0.5: $g(0.5 + z) = g(0.5 - z)$ for $z \in [0, 0.5]$;

(3) g is unimodal, i.e.,

(i) $g(x) < g(y)$ for $x < y \leqslant 0.5$;

(ii) $g(x) < g(y)$ for $x > y \geqslant 0.5$.

Yager (2007a) also gave the following interesting result:

Theorem 2.3 (Yager, 2007a) Let g be a centering function. If f is a function defined on the unit interval such that $f(x) = K\int_0^x g(y)\mathrm{d}y$, where $K = 1\Big/\int_0^1 g(y)\mathrm{d}y$, then

(1) f is a BUM function;

(2) The weights generated from f and (2.98) satisfy the conditions of a centered weighting vector for all p:

(i) (Inclusive) $w_j > 0$;

(ii) (Symmetric) $w_j = w_{p+1-j}$;

(iii) (Strongly decaying) if $i < j \leqslant \dfrac{p+1}{2}$, then $w_i < w_j$.

For example, if we take $g(x) = \sin(\pi x)$, then

$$f(x) = \frac{\int_0^x g(y)\mathrm{d}y}{\int_0^1 g(y)\mathrm{d}y} = \frac{\int_0^x \sin(\pi x)\mathrm{d}y}{\int_0^1 \sin(\pi x)\mathrm{d}y} = \frac{1}{2}(1 - \cos x) \tag{2.104}$$

Thus, by (2.98), we have

$$\begin{aligned} w(t_k) &= f\left(\frac{k}{p}\right) - f\left(\frac{k-1}{p}\right) \\ &= \frac{1}{2}\left[\cos\left(\frac{j-1}{p}\pi\right) - \cos\frac{j}{p}\pi\right], \quad k = 1, 2, \cdots, p \end{aligned} \tag{2.105}$$

Case 2 If f is a monotone decreasing function, then we can take

$$f(x) = \frac{e^{\alpha x} - 1}{e^{\alpha} - 1}, \quad \alpha < 0 \tag{2.106}$$

Using (2.106), we know that $F(x) = f(x) - f\left(x - \frac{1}{p}\right)$ $(\alpha < 0)$ is a strictly monotone decreasing function. Thus, it follows from (2.98) that

$$w(t_{k+1}) < w(t_k), \quad k = 1, 2, \cdots, p-1 \tag{2.107}$$

i.e., the sequence $\{w(t_k)\}$ is a monotonic decreasing sequence.

Xu (2009a) employed the minimum variability model (developed by Fullér and Majlender (2003)) to derive the time series weights $w(t_k)$ $(k = 1, 2, \cdots, p)$:

They first introduced the variance of the time series weight vector as

$$\begin{aligned} D^2(\boldsymbol{w}(t)) &= \sum_{k=1}^{p} \frac{1}{p}\left[w(t_k) - E(w(t_k))\right]^2 \\ &= \frac{1}{p}\sum_{k=1}^{p}(w(t_k))^2 - \left[\frac{1}{p}\sum_{k=1}^{p} w(t_k)\right]^2 \\ &= \frac{1}{p}\sum_{k=1}^{p}(w(t_k))^2 - \frac{1}{p^2} \end{aligned} \tag{2.108}$$

where $E(w(t_k)) = \frac{1}{p}\sum_{k=1}^{p} w(t_k)$ stands for the arithmetic mean of the weights $w(t_k)$ $(k = 1, 2, \cdots, p)$.

Yager (2007b) defined the concept of the average age of the data as

$$\bar{t} = \sum_{k=1}^{p}(p-k)w(t_k) \tag{2.109}$$

Then, similar to Yager (1988), here we can utilize (2.109) to introduce a measure of "orness" associated with the time series weight vector $\boldsymbol{w}(t) = (w(t_1), w(t_2), \cdots, w(t_p))^{\mathrm{T}}$:

$$\text{orness}(\boldsymbol{w}(t)) = \frac{\bar{t}}{p-1} = \frac{1}{p-1}\sum_{k=1}^{p}(p-k)w(t_k) \tag{2.110}$$

which characterizes the degree to which the aggregation is like an or operation.

In order to obtain the minimum variability time series weights under given level of orness, we need to solve the following constrained mathematical programming problem (Fullér and Majlender, 2003):

$$\min\ D^2\left(\boldsymbol{w}(t)\right)=\frac{1}{p}\sum_{k=1}^{p}\left(w(t_k)\right)^2-\frac{1}{p^2}$$

$$\text{s.t.}\ \operatorname{orness}\left(\boldsymbol{w}(t)\right)=\frac{1}{p-1}\sum_{k=1}^{p}(p-k)w(t_k)=\alpha,\quad 0\leqslant\alpha\leqslant 1$$

$$w(t_k)\geqslant 0,\ k=1,2,\cdots,p,\quad \sum_{k=1}^{p}w(t_k)=1$$

To solve this model, we construct the following Lagrange function:

$$\begin{aligned}l(w,\lambda_1,\lambda_2)=&\frac{1}{p}\sum_{k=1}^{p}\left(w(t_k)\right)^2-\frac{1}{p^2}-2\lambda_1\left[\frac{1}{p-1}\sum_{k=1}^{p}(p-k)w(t_k)-\alpha\right]\\&-2\lambda_2\left[\sum_{k=1}^{p}w(t_k)-1\right]\end{aligned}\tag{2.111}$$

Differentiating (2.111) with respect to $w(t_k)$ $(k=1,2,\cdots,p)$ and λ_i $(i=1,2)$, and setting these partial derivatives equal to zero, the following equations are obtained:

$$\frac{\partial l}{\partial w(t_k)}=\frac{2w(t_k)}{p}-2\lambda_1\frac{p-k}{p-1}-2\lambda_2=0,\quad k=1,2,\cdots,p\tag{2.112}$$

$$\frac{\partial l}{\partial\lambda_1}=-2\left[\frac{1}{p-1}\sum_{k=1}^{p}(p-k)w(t_k)-\alpha\right]=0\tag{2.113}$$

$$\frac{\partial l}{\partial\lambda_2}=-2\left(\sum_{k=1}^{p}w(t_k)-1\right)=0\tag{2.114}$$

Solving (2.112)~(2.114), we get

$$w(t_k)=\frac{(6p-12k+6)\alpha-2p+6k-2}{p(p+1)},\quad k=1,2,\cdots,p\tag{2.115}$$

Considering that $w(t_k)\geqslant 0$ for all k, we have

$$\frac{(6p-12k+6)\alpha-2p+6k-2}{p(p+1)}\geqslant 0,\quad k=1,2,\cdots,p\tag{2.116}$$

that is,

$$(6p-12k+6)\alpha\geqslant 2p-6k+2,\quad k=1,2,\cdots,p\tag{2.117}$$

Then we discuss the following three cases:

Case 1 If $6p-12k+6=0$, i.e., $k=\dfrac{p+1}{2}$, then

$$2p-6k+2=2p-6\times\frac{p+1}{2}+2=-(p+1)<0 \tag{2.118}$$

Thus, (2.117) holds for all $0\leqslant\alpha\leqslant1$.

Case 2 If $6p-12k+6>0$, i.e., $k<\dfrac{p+1}{2}$, then from (2.117), it follows that

$$\alpha\geqslant\frac{p-2}{3p-3} \tag{2.119}$$

i.e., (2.119) holds for $\dfrac{p-2}{3p-3}\leqslant\alpha\leqslant1$.

Case 3 If $6p-12k+6<0$, i.e., $k>\dfrac{p+1}{2}$, then from (2.117), it follows that

$$\alpha\leqslant\frac{2p-1}{3p-3} \tag{2.120}$$

i.e., (2.117) holds for $0\leqslant\alpha\leqslant\dfrac{2p-1}{3p-3}$.

Hence, from Cases 1~3, we know that (2.117) holds, with the condition:

$$\frac{p-2}{3p-3}\leqslant\alpha\leqslant\frac{2p-1}{3p-3} \tag{2.121}$$

Therefore, we can obtain the time series weights $w(t_k)$ $(k=1,2,\cdots,p)$ from (2.115) with the condition (2.121).

Xu (2009a) further studied the properties of the time series weights derived from (2.115):

Let

$$f(x)=\frac{(6p-12x+6)\alpha-2p+6x-2}{p(p+1)} \tag{2.122}$$

Then

$$f'(x)=\frac{-12\alpha+6}{p(p+1)} \tag{2.123}$$

Thus,

(1) If $\dfrac{p-2}{3p-3}\leqslant\alpha<\dfrac{1}{2}$, then $f'(x)>0$, i.e., $f(x)$ is a strictly monotonic increasing function.

(2) If $\alpha=\dfrac{1}{2}$, then $f'(x)=0$, i.e., $f(x)$ is a constant function.

(3) If $\dfrac{1}{2}<\alpha\leqslant\dfrac{2p-1}{3p-3}$, then $f'(x)<0$, i.e., $f(x)$ is a strictly monotonic decreasing function.

Therefore, by (2.115), we have

(1) If $\frac{p-2}{3p-3} \leqslant \alpha < \frac{1}{2}$, then $w(t_{k+1}) > w(t_k)$ $(k = 1, 2, \cdots, p-1)$, i.e., the sequence $\{w(t_k)\}$ is a monotonic increasing sequence, which indicates that the larger k, the greater $w(t_k)$. Furthermore,

$$\begin{aligned} w(t_{k+1}) - w(t_k) =& \frac{[6p - 12(k+1) + 6]\alpha - 2p + 6(k+1) - 2}{p(p+1)} \\ &- \frac{(6p - 12k + 6)\alpha - 2p + 6k - 2}{p(p+1)} \\ =& \frac{6 - 12\alpha}{p(p+1)} > 0, \quad k = 1, 2, \cdots, p-1 \end{aligned} \tag{2.124}$$

Thus, the sequence $\{w(t_k)\}$ is an increasing arithmetic sequence.

(2) If $\alpha = \frac{1}{2}$, then

$$w(t_k) = \frac{(6p - 12k + 6)\alpha - 2p + 6k - 2}{p(p+1)} = \frac{1}{p}, \quad k = 1, 2, \cdots, p \tag{2.125}$$

which indicates that all the weights $w(t_k)$ $(k = 1, 2, \cdots, p)$ are equal.

(3) If $\frac{1}{2} < \alpha \leqslant \frac{2p-1}{3p-3}$, then $w(t_{k+1}) < w(t_k)$ $(k = 1, 2, \cdots, p-1)$, i.e., the sequence $\{w(t_k)\}$ is a monotonic decreasing sequence, which indicates that the larger k, the smaller $w(t_k)$. Also since

$$w(t_{k+1}) - w(t_k) = \frac{6 - 12\alpha}{p(p+1)} < 0, \quad k = 1, 2, \cdots, p-1$$

then the sequence $\{w(t_k)\}$ is a decreasing arithmetic sequence.

From the above analysis, we know that the time series weights derived from the minimum variance model under given orness level α (or from (2.115)) are monotonic increasing with $\frac{p-2}{3p-3} \leqslant \alpha < \frac{1}{2}$ and monotonic decreasing with $\frac{1}{2} < \alpha \leqslant \frac{2p-1}{3p-3}$. In

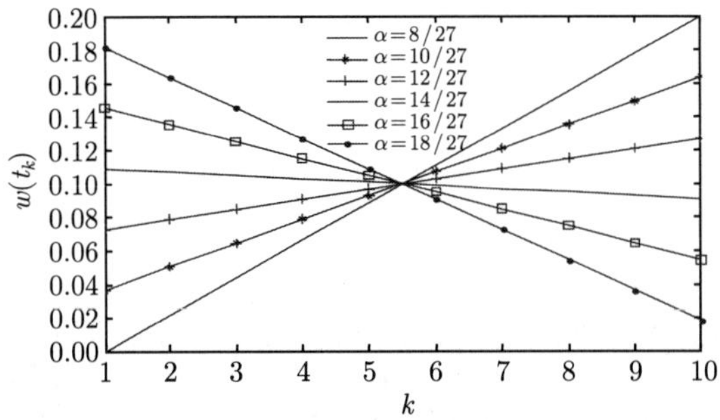

Figure 2.1 Some special time series weights derived from the minimum variance model (Xu, 2009a)

the case of $\alpha = \frac{1}{2}$, all the derived time series weights are equal.

Figure 2.1 depicts the distribution of the time series weights derived from the minimum variance model in the case of $p = 10$ under some special orness levels α.

2.4.2 Linguistic Geometric Operators

Similar to Subsection 2.4.1, in what follows, we introduce some linguistic geometric aggregation operators:

Let $S_4 = \{s_\alpha | \alpha = 1/\tau, \cdots, 1/2, 1, 2, \cdots, \tau\}$ be a multiplicative linguistic evaluation scale (see Section 1.2), and $\bar{S}_4 = \{s_\alpha | \alpha \in [1/q, q]\}$ be the extended multiplicative linguistic evaluation scale of S_4, where q $(q > \tau)$ is a sufficiently large positive integer. We also assume that s_{α_j} $(j = 1, 2, \cdots, n)$ are a collection of linguistic labels, and $s_{\alpha_j} \in S_4$ $(j = 1, 2, \cdots, n)$.

Based on the extended multiplicative linguistic evaluation scale $\bar{S}_4$, in the following, we introduce some operational laws:

Definition 2.35 (Xu, 2004e) Let $s_\alpha, s_\beta \in \bar{S}_4$ and $\lambda \in [0, 1]$. Then

(1) $s_\alpha \otimes s_\beta = s_{\alpha\beta}$;

(2) $(s_\alpha)^\lambda = s_{\alpha^\lambda}$.

Theorem 2.4 (Xu, 2004e) Let s_α, s_{α_1}, $s_{\alpha_2} \in \bar{S}_4$ and λ, $\lambda_1, \lambda_2 \geqslant 0$. Then

(1) $s_{\alpha_1} \otimes s_{\alpha_2} = s_{\alpha_2} \otimes s_{\alpha_1}$;

(2) $(s_{\alpha_1} \otimes s_{\alpha_2})^\lambda = s_{\alpha_1}^\lambda \otimes s_{\alpha_2}^\lambda$;

(3) $s_\alpha^{\lambda_1} \otimes s_\alpha^{\lambda_2} = s_\alpha^{\lambda_1 + \lambda_2}$.

Proof (1) Since

$$s_{\alpha_1} \otimes s_{\alpha_2} = s_{\alpha_1\alpha_2} = s_{\alpha_2\alpha_1} = s_{\alpha_2} \otimes s_{\alpha_1}$$

then (1) holds.

(2) Since

$$\begin{aligned}(s_{\alpha_1} \otimes s_{\alpha_2})^\lambda &= (s_{\alpha_1\alpha_2})^\lambda = s_{(\alpha_1\alpha_2)^\lambda} = s_{\alpha_1^\lambda \alpha_2^\lambda} \\ &= s_{\alpha_1^\lambda} \otimes s_{\alpha_2^\lambda} = s_{\alpha_1}^\lambda \otimes s_{\alpha_2}^\lambda\end{aligned}$$

then (2) holds.

(3) Since

$$s_\alpha^{\lambda_1} \otimes s_\alpha^{\lambda_2} = s_{\alpha^{\lambda_1}} \otimes s_{\alpha^{\lambda_2}} = s_{\alpha^{\lambda_1}\alpha^{\lambda_2}} = s_{\alpha^{\lambda_1+\lambda_2}} = s_\alpha^{\lambda_1+\lambda_2}$$

then (3) holds.

This completes the proof.

Definition 2.36 (Xu, 2006b) Let $\mathrm{LG} : (\bar{S}_4)^n \to \bar{S}_4$. Then

$$\mathrm{LG}(s_{\alpha_1}, s_{\alpha_2}, \cdots, s_{\alpha_n}) = (s_{\alpha_1} \otimes s_{\alpha_2} \otimes \cdots \otimes s_{\alpha_n})^{\frac{1}{n}} = s_{\bar{\alpha}} \tag{2.126}$$

is called a linguistic geometric (LG) operator, where

$$\bar{\alpha} = \left(\prod_{j=1}^{n} \alpha_j\right)^{\frac{1}{n}} \tag{2.127}$$

Example 2.32 Let $(s_{1/2}, s_1, s_2, s_{1/4})$ be a collection of linguistic labels in $\bar{S}_4$. Then

$$\text{LG}(s_{1/2}, s_1, s_2, s_{1/4}) = (s_{1/2} \otimes s_1 \otimes s_2 \otimes s_{1/4})^{\frac{1}{4}} = s_{0.71}$$

Definition 2.37 (Xu, 2006b) Let LWG : $\bar{S}_4^n \to \bar{S}_4$. If

$$\text{LWG}(s_{\alpha_1}, s_{\alpha_2}, \cdots, s_{\alpha_n}) = s_{\alpha_1}^{w_1} \otimes s_{\alpha_2}^{w_2} \otimes \cdots \otimes s_{\alpha_n}^{w_n} = s_{\dot{\alpha}} \tag{2.128}$$

then the function LWG is called a linguistic weighted geometric (LWG) operator, where

$$\dot{\alpha} = \prod_{j=1}^{n} \alpha_j^{w_j} \tag{2.129}$$

and $\boldsymbol{w} = (w_1, w_2, \cdots, w_n)^{\text{T}}$ is the weight vector of $s_{\alpha_i} (i = 1, 2, \cdots, n)$, $w_j \geqslant 0$ $(j = 1, 2, \cdots, n)$ and $\sum_{j=1}^{n} w_j = 1$. The LWG operator is the extension of the WG operator (Saaty, 1980; Xu and Da, 2003b).

In particular, if $\boldsymbol{w} = (1/n, 1/n, \cdots, 1/n)^{\text{T}}$, then the LWG operator reduces to the LG operator.

The characteristic of the LWG operator is that it first exponentially weights each linguistic argument, and then aggregates the weighted linguistic arguments multiplicatively.

Example 2.33 Let $(s_{1/3}, s_3, s_4, s_{1/2})$ be a collection of linguistic labels in $\bar{S}_4$, and $\boldsymbol{w} = (0.3, 0.2, 0.4, 0.1)^{\text{T}}$ be their weight vector. Then

$$\begin{aligned}\text{LWG}(s_{1/3}, s_3, s_4, s_{1/2}) &= (s_{1/3})^{0.3} \otimes (s_3)^{0.2} \otimes (s_4)^{0.4} \otimes (s_{1/2})^{0.1} \\ &= s_{1.46}\end{aligned}$$

Definition 2.38 (Xu, 2004e) Let LOWG : $(\bar{S}_4)^n \to \bar{S}_4$, and the exponential weighting vector associated with the function LOWG be $\boldsymbol{\omega} = (\omega_1, \omega_2, \cdots, \omega_n)^{\text{T}}$, $\omega_i \geqslant 0$ $(i = 1, 2, \cdots, n)$ and $\sum_{i=1}^{n} \omega_i = 1$, such that

$$\text{LOWG}(s_{\alpha_1}, s_{\alpha_2}, \cdots, s_{\alpha_n}) = s_{\beta_1}^{\omega_1} \otimes s_{\beta_2}^{\omega_2} \otimes \cdots \otimes s_{\beta_n}^{\omega_n} = s_{\bar{\beta}} \tag{2.130}$$

where

$$\bar{\beta} = \prod_{j=1}^{n} \beta_j^{\omega_j} \tag{2.131}$$

and s_{β_j} $(j = 1,2,\cdots,n)$ is the jth largest of the linguistic arguments s_{α_i} $(i = 1,2,\cdots,n)$. Then the function LOWG is called a linguistic OWG (LOWG) operator.

In particular, if $\boldsymbol{w} = (1,0,\cdots,0)^{\mathrm{T}}$, where $w_1 = 1$, $w_i = 0$ $(i = 2,\cdots,n)$, then the LOWG operator reduces to the LM_1 operator; if $\boldsymbol{\omega} = (0,\cdots,0,1)^{\mathrm{T}}$, where $w_i = 0$ $(i = 1,2,\cdots,n-1)$, $w_n = 1$, then the LOWG operator reduces to the LM_2 operator; if $\boldsymbol{\omega} = (1/n,1/n,\cdots,1/n)^{\mathrm{T}}$, then the LOWG operator reduces to the LG operator.

The characteristic of the LOWG operator is that it needs to reorder the linguistic arguments in descending order and exponentially weights the ordered position of each argument, and then aggregates the weighted linguistic arguments multiplicatively.

The LOWG operator is the extension of the ordered weighted geometric (OWG) operator (Chiclana et al., 2001; Herrera et al., 2003; 2001a; Xu, 2004d; Xu and Da, 2003b; 2002c; Yager and Xu, 2006) in linguistic environment. It has many desirable properties similar to the OWA and OWG operators, including monotonicity, idempotency and commutativity, and is a linguistic aggregation operator that lies between the linguistic max and linguistic min operators.

Example 2.34 Let $(s_1,s_4,s_{1/3},s_2,s_3)$ be a collection of linguistic labels in $\bar{S}_4$. Reordering them in descending order, we have

$$s_{\beta_1} = s_4,\quad s_{\beta_2} = s_3,\quad s_{\beta_3} = s_2,\quad s_{\beta_4} = s_1,\quad s_{\beta_5} = s_{1/3}$$

Suppose that the weighting vector associated with the LOWG operator is $\boldsymbol{\omega} = (0.1,0.3,0.3,0.2,0.1)^{\mathrm{T}}$. Then

$$\begin{aligned}\mathrm{LOWG}(s_1,s_4,s_{1/3},s_2,s_3) &= (s_4)^{0.1}\otimes(s_3)^{0.3}\otimes(s_2)^{0.3}\otimes(s_1)^{0.2}\otimes(s_{1/3})^{0.1}\\ &= s_{1.76}\end{aligned}$$

Based on the LWG and LOWG operators, Xu (2004f) defined a linguistic hybrid aggregation operator:

Definition 2.39 (Xu, 2004f) A linguistic hybrid geometric (LHG) operator is a mapping LHG : $\bar{S}_4^n \rightarrow \bar{S}_4$, which has an associated weighting vector $\boldsymbol{\omega} = (\omega_1, \omega_2,\cdots,\omega_n)^{\mathrm{T}}$, with $\omega_i \geqslant 0$ $(i = 1,2,\cdots,n)$ and $\sum_{i=1}^{n}\omega_i = 1$, such that

$$\mathrm{LHG}(s_{\alpha_1},s_{\alpha_2},\cdots,s_{\alpha_n}) = s_{\beta_1}^{\omega_1}\otimes s_{\beta_2}^{\omega_2}\otimes\cdots\otimes s_{\beta_n}^{\omega_n} \tag{2.132}$$

where s_{β_j} is the jth largest of the linguistic weighted arguments $\bar{s}_{\alpha_i}$ $(i = 1,2,\cdots,n)$ $(\bar{s}_{\alpha_i} = s_{\alpha_i}^{nw_i},\ i = 1,2,\cdots,n)$, $\boldsymbol{w} = (w_1,w_2,\cdots,w_n)^{\mathrm{T}}$ is the weight vector of s_{α_i} $(i = 1,2,\cdots,n)$, with $w_i \in [0,1]$ $(i = 1,2,\cdots,n)$, $\sum_{i=1}^{n} w_i = 1$, and n is the balancing coefficient (if $(w_1, w_2, \cdots, w_n)^{\mathrm{T}}$ approaches $(1/n, 1/n, \cdots, 1/n)^{\mathrm{T}}$, then the vector $(s_{\alpha_1}^{nw_1}, s_{\alpha_2}^{nw_2}, \cdots, s_{\alpha_n}^{nw_n})^{\mathrm{T}}$ approaches $(s_{\alpha_1}, s_{\alpha_2}, \cdots, s_{\alpha_n})^{\mathrm{T}}$).

In particular, if $\boldsymbol{\omega} = (1/n, 1/n, \cdots, 1/n)^{\mathrm{T}}$, then the LHG operator reduces to the LWG operator; if $\boldsymbol{w} = (1/n, 1/n, \cdots, 1/n)^{\mathrm{T}}$, then the LHG operator reduces to the LOWG operator.

Clearly, the LHG operator can not only consider the importance of a linguistic argument itself, but also reflect the ordered position of the argument, and thus, extends the LWG and LOWG operators.

Example 2.35 Let $(s_1, s_{1/2}, s_{1/3}, s_3, s_4)$ be a collection of linguistic labels in $\bar{S}_4$, and $\boldsymbol{w} = (0.30, 0.10, 0.20, 0.25, 0.15)^{\mathrm{T}}$ be their weight vector. We first calculate the linguistic weighted arguments:

$$\bar{s}_{\alpha_1} = (s_1)^{5\times 0.30} = s_1, \quad \bar{s}_{\alpha_2} = (s_{1/2})^{5\times 0.10} = s_{0.71}$$

$$\bar{s}_{\alpha_3} = (s_{1/3})^{5\times 0.20} = s_{0.33}, \quad \bar{s}_{\alpha_4} = (s_3)^{5\times 0.25} = s_{3.95}$$

$$\bar{s}_{\alpha_5} = (s_4)^{5\times 0.15} = s_{2.83}$$

Reordering $\bar{s}_{\alpha_i} (i = 1, 2, 3, 4, 5)$ in descending order, we have

$$s_{\beta_1} = \bar{s}_{\alpha_3} = s_{3.95}, \quad s_{\beta_2} = \bar{s}_{\alpha_4} = s_{2.83}, \quad s_{\beta_3} = \bar{s}_{\alpha_1} = s_1$$

$$s_{\beta_4} = \bar{s}_{\alpha_2} = s_{0.71}, \quad s_{\beta_5} = \bar{s}_{\alpha_3} = s_{0.33}$$

Suppose that $\boldsymbol{\omega} = (0.10, 0.15, 0.50, 0.15, 0.10)^{\mathrm{T}}$ is the weighting vector associated with the LHG operator. Then

$$\begin{aligned}\mathrm{LHG}(s_1, s_{1/2}, s_{1/3}, s_3, s_4) =& (s_{3.95})^{0.10} \otimes (s_{2.83})^{0.15} \otimes (s_1)^{0.50} \\ & \otimes (s_{0.71})^{0.15} \otimes (s_{0.33})^{0.10} \\ =& s_{1.64}\end{aligned}$$

Xu and Da (2003b) defined an induced OWG (IOWG) operator, which can only be used to aggregate the exact numberical values. Xu (2006b) gave an induced linguistic OWG (ILOWG) operator, which is used to aggregate the linguistic arguments:

Definition 2.40 (Xu, 2006b) A function ILOWG is called an induced linguistic OWG (ILOWG) operator, if

$$\mathrm{ILOWG}\left(\langle u_1, s_{\alpha_1}\rangle, \langle u_2, s_{\alpha_2}\rangle, \cdots, \langle u_n, s_{\alpha_n}\rangle\right) = s_{\gamma_1}^{\omega_1} \otimes s_{\gamma_2}^{\omega_2} \otimes \cdots \otimes s_{\gamma_n}^{\omega_n} = s_{\bar{\gamma}} \tag{2.133}$$

where

$$\bar{\gamma} = \prod_{i=1}^{n} \gamma_i^{\omega_i} \tag{2.134}$$

and $\boldsymbol{\omega} = (\omega_1, \omega_2, \cdots, \omega_n)^{\mathrm{T}}$ is the weighting vector associated with the ILOWG operator, such that $\omega_i \geqslant 0$ $(i = 1, 2, \cdots, n)$ and $\sum_{i=1}^{n} \omega_i = 1$. s_{γ_j} is the s_{α_i} value of the OWG pair $\langle u_i,\ s_{\alpha_i}\rangle$ having the jth largest u_i, and u_i in $\langle u_i,\ s_{\alpha_i}\rangle$ is referred to as the order inducing variable, and s_{α_i} as the linguistic label.

In particular, if $\boldsymbol{\omega} = (1/n,\ 1/n, \cdots,\ 1/n)^{\mathrm{T}}$, then the ILOWG operator reduces to the LG operator; if $u_i = s_{\alpha_i}$ for any i, then the LOWG operator reduces to the LOWG operator; if $u_i =$ No. i for any i, where No. i is the ordered position of s_i, then the ILOWG operator reduces to the LWG operator.

If there is a tie between $\langle u_i,\ s_{\alpha_i}\rangle$ and $\langle u_j,\ s_{\alpha_j}\rangle$ with respect to order inducing variables, then in the process of aggregation, we can average s_{α_i} and s_{α_j}, and get the corresponding pairs $\left\langle u_i,\ (s_{\alpha_i} \otimes s_{\alpha_j})^{1/2}\right\rangle$ and $\left\langle u_j,\ (s_{\alpha_i} \otimes s_{\alpha_j})^{1/2}\right\rangle$. If k items are tied, then we replace these by k replica's of their average.

Example 2.36 Given a collection of the OWG pairs $\langle u_i,\ s_{\alpha_i}\rangle$ $(i = 1, 2, 3, 4, 5)$:

$$\left(\langle 0.20,\ s_2\rangle, \langle 0.30,\ s_3\rangle, \langle 0.10,\ s_4\rangle, \left\langle 0.15,\ s_{1/3}\right\rangle, \left\langle 0.25,\ s_{1/2}\right\rangle\right)$$

We first use the values u_i $(i = 1, 2, 3, 4, 5)$ to rank the OWG pairs above. Since

$$u_2 > u_5 > u_1 > u_4 > u_3$$

we get the ordered OWG pairs:

$$\langle 0.30,\ s_3\rangle, \quad \left\langle 0.25,\ s_{1/2}\right\rangle, \quad \langle 0.20,\ s_2\rangle, \quad \left\langle 0.15,\ s_{1/3}\right\rangle, \quad \langle 0.10,\ s_4\rangle$$

i.e.,

$$s_{\gamma_1} = s_3, \quad s_{\gamma_2} = s_{1/2}, \quad s_{\gamma_3} = s_2, \quad s_{\gamma_4} = s_{1/3}, \quad s_{\gamma_5} = s_4$$

Suppose that the weighting vector associated with the ILOWG operator is $\boldsymbol{\omega} = (0.05, 0.20, 0.50, 0.20, 0.05)^{\mathrm{T}}$. Then

$$\begin{aligned}
&\mathrm{ILOWG}\left(\langle u_1, s_{\alpha_1}\rangle, \langle u_2, s_{\alpha_2}\rangle, \langle u_3, s_{\alpha_3}\rangle, \langle u_4, s_{\alpha_4}\rangle, \langle u_5, s_{\alpha_5}\rangle\right)\\
=&(s_3)^{0.05} \otimes \left(s_{1/2}\right)^{0.20} \otimes (s_2)^{0.50} \otimes \left(s_{1/3}\right)^{0.20} \otimes (s_4)^{0.05}\\
=&s_{1.12}
\end{aligned}$$

If we replace $\langle \mu_1,\ s_{\alpha_1}\rangle = \langle 0.20,\ s_2\rangle$ and $\langle \mu_2,\ s_{\alpha_2}\rangle = \langle 0.30,\ s_3\rangle$ with $\langle \mu_1,\ s_{\alpha_1}\rangle = \langle 0.10,\ s_2\rangle$ and $\langle \mu_2,\ s_{\alpha_2}\rangle = \langle 0.40,\ s_3\rangle$ respectively, then $u_1 = \mu_3$. Thus, in the process of aggregation, we substitute $s_{\alpha_1} = s_2$ and $s_{\alpha_3} = s_4$ with $(s_2 \otimes s_4)^{1/2} = s_{2.83}$, and get

$$\left\langle u_1,\ (s_{\alpha_1} \otimes s_{\alpha_3})^{1/2}\right\rangle = \left\langle u_3,\ (s_{\alpha_1} \otimes s_{\alpha_3})^{1/2}\right\rangle = \langle 0.1,\ s_{2.83}\rangle$$

i.e.,

$$s_{\gamma_1} = s_3, \quad s_{\gamma_2} = s_{1/2}, \quad s_{\gamma_3} = s_{1/3}, \quad s_{\gamma_4} = s_{\gamma_5} = s_{2.83}$$

Accordingly,

$$\begin{aligned}&\text{ILOWG}\left(\langle u_1, s_{\alpha_1}\rangle, \langle u_2, s_{\alpha_2}\rangle, \langle u_3, s_{\alpha_3}\rangle, \langle u_4, s_{\alpha_4}\rangle, \langle u_5, s_{\alpha_5}\rangle\right)\\ =&(s_3)^{0.05} \otimes (s_{1/2})^{0.20} \otimes (s_{1/3})^{0.50} \otimes (s_{2.83})^{0.20} \otimes (s_{2.83})^{0.05}\\ =&s_{0.69}\end{aligned}$$

Similar to the GILOWA operator (2.58), Xu (2006b) introduced a more general aggregation technique as below.

Definition 2.41 (Xu, 2006b) A generalized induced LOWG (GILOWG) operator is given as

$$\begin{aligned}&\text{GILOWG}\left(\langle v_1, u_1, s_{\alpha_1}\rangle, \langle v_2, u_2, s_{\alpha_2}\rangle, \cdots, \langle v_n, u_n, s_{\alpha_n}\rangle\right)\\ =&s_{\beta_1}^{\omega_1} \otimes s_{\beta_2}^{\omega_2} \otimes \cdots \otimes s_{\beta_n}^{\omega_n}\end{aligned} \tag{2.135}$$

where $\boldsymbol{\omega} = (\omega_1, \omega_2, \cdots, \omega_n)^{\mathrm{T}}$ is the associated weighting vector with $\omega_i \geqslant 0$ $(i = 1, 2, \cdots, n)$ and $\sum_{i=1}^{n} \omega_i = 1$. The object $\langle v_i, u_i, s_{\alpha_i}\rangle$ consists of three components, where the first component v_i represents the importance degree or character of the second component u_i, and the second component u_i is used to induce an ordering through the first component v_i over the third components s_{α_i}, which are then aggregated. Here, s_{β_j} is the s_{α_i} value of the object having the jth largest of v_i $(i = 1, 2, \cdots, n)$, v_i is a direct order inducing variable, u_i an indirect order inducing variable, and s_{α_i} an argument variable.

In particular, if $v_i = u_i$ for all i, then the GILOWG operator reduces to the ILOWG operator; if $v_i = s_{\alpha_i}$ for all i, then the GILOWG operator reduces to the LOWG operator; if $v_i =$ No. i for all i, where No. i is the ordered position of s_{α_i}, then the GILOWG operator reduces to the LWG operator; if $\boldsymbol{\omega} = (1/n,\ 1/n, \cdots, 1/n)^{\mathrm{T}}$, then the GILOWG operator reduces to the LG operator.

The GILOWG operator has some properties similar to the GILOWA operator.

Example 2.37 If we substitute the objects in Example 2.22 as $\langle v_i, u_i, s_{\alpha_i}\rangle$ $(i = 1, 2, 3, 4)$:

$$(\langle \text{No.3, Johnson, } s_1\rangle, \langle \text{No.2, Brown, } s_2\rangle, \langle \text{No.1, Smith, } s_4\rangle,$$
$$\langle \text{No.4, Anderson, } s_{1/2}\rangle)$$

and suppose that the weighting vector associated with the ILOWG operator is $\boldsymbol{\omega} = (0.15, 0.30, 0.40, 0.15)^{\mathrm{T}}$. Then we first use the values v_i $(i = 1, 2, 3, 4)$ to rank the objects $\langle v_i, u_i, s_{\alpha_i}\rangle$, and get

$$\langle \text{No.1, Smith, } s_4\rangle$$
$$\langle \text{No.2, Brown, } s_2\rangle$$
$$\langle \text{No.3, Johnson, } s_1\rangle$$

$$\langle \text{No.4, Anderson, } s_{1/2} \rangle$$

As a result,

$$s_{\gamma_1} = s_4, \quad s_{\gamma_2} = s_2, \quad s_{\gamma_3} = s_1, \quad s_{\gamma_4} = s_{1/2}$$

Consequently,

$$\begin{aligned}
&\text{GILOWG}\left(\langle v_1, u_1, s_{\alpha_1}\rangle, \langle v_2, u_2, s_{\alpha_2}\rangle, \langle v_3, u_3, s_{\alpha_3}\rangle, \langle v_4, u_4, s_{\alpha_4}\rangle\right) \\
=&(s_4)^{0.15} \otimes (s_2)^{0.30} \otimes (s_1)^{0.40} \otimes (s_{1/2})^{0.15} \\
=&s_{1.37}
\end{aligned}$$

Based on the multiplicative linguistic evaluation scale $\bar{S}_4$ and the Choquet integral (Choquet, 1953), Xu (2009b) gave an operator for aggregating multiplicative linguistic labels together with their correlative weights:

Definition 2.42 (Xu, 2009b) Let μ be a fuzzy measure on $\bar{S}_4$, and s_{α_i} $(i = 1, 2, \cdots, n)$ be n multiplicative linguistic labels in $\bar{S}_4$. Then

$$(C_3) \int s_\alpha \mathrm{d}\mu = \text{LCG}\left(s_{\alpha_1}, s_{\alpha_2}, \cdots, s_{\alpha_n}\right) = \bigotimes_{i=1}^{n} \left(s_{\alpha_{\sigma(i)}}\right)^{\mu(A_{\sigma(i)}) - \mu(A_{\sigma(i-1)})} \tag{2.136}$$

is called a linguistic correlated geometric (LCG) operator, where $(C_3) \displaystyle\int s_\alpha \mathrm{d}\mu$ is a notation of Choquet integral, $(\sigma(1), \sigma(2), \cdots, \sigma(n))$ is a permutation of $(1, 2, \cdots, n)$ such that $s_{\alpha_{\sigma(1)}} \geqslant s_{\alpha_{\sigma(2)}} \geqslant \cdots \geqslant s_{\alpha_{\sigma(n)}}$, $A_{\sigma(k)} = \{s_{\alpha_{\sigma(j)}} | j \leqslant k\}$, when $k \geqslant 1$ and $A_{\sigma(0)} = \varnothing$.

Now we discuss some special cases of the LCG operator (Xu, 2009b):

(1) If (2.60) and (2.62) hold, then the LCG operator reduces to the LWG operator. Especially, if $w_i = \dfrac{1}{n}$ for all $i = 1, 2, \cdots, n$, then the LWG operator reduces to the LG operator.

(2) If (2.63) and (2.64) hold, then the LCG operator reduces to the LOWG operator. Especially, if $\mu(A) = \dfrac{|A|}{n}$ for all $A \subseteq \bar{S}_4$, then both the LCG and LOWG operators reduce to the LG operator.

(3) If (2.65) and (2.66) hold, then the LCG operator reduces to the following form:

$$\text{LWOWG}\left(s_{\alpha_1}, s_{\alpha_2}, \cdots, s_{\alpha_n}\right) = \bigotimes_{i=1}^{n} \left(s_{\alpha_{\sigma(i)}}\right)^{\omega_i} \tag{2.137}$$

which is called a linguistic weighted OWG (LWOWG) operator (Xu, 2009b). Especially, if $\mu\left(\{s_{\alpha_i}\}\right) = \dfrac{1}{n}$ for all $i = 1, 2, \cdots, n$, then the LWOWG operator reduces to the LOWG operator.

In what follows, we further consider situations where the linguistic information is expressed in uncertain linguistic variables:

For convenience, let $\tilde{S}_4 = \{\tilde{s}|\tilde{s} = [s_\alpha, s_\beta],\ s_\alpha, s_\beta \in \bar{S}_4\}$ be the set of all the uncertain linguistic variables, which is on the basis of the extended multiplicative linguistic evaluation scale $\bar{S}_4$.

Definition 2.43 (Xu, 2006e) Let $\tilde{s}=[s_\alpha, s_\beta]$, $\tilde{s}_1=[s_{\alpha_1}, s_{\beta_1}]$ and $\tilde{s}_2=[s_{\alpha_2}, s_{\beta_2}] \in \tilde{S}_4$. Then

(1) $\tilde{s}_1 \otimes \tilde{s}_2 = [s_{\alpha_1}, s_{\beta_1}] \otimes [s_{\alpha_2}, s_{\beta_2}] = [s_{\alpha_1} \otimes s_{\alpha_2},\ s_{\beta_1} \otimes s_{\beta_2}]$;

(2) $\tilde{s}^\lambda = [s_\alpha^\lambda, s_\beta^\lambda]\ (\lambda \in [0,1])$.

Based on Definition 2.43, Xu (2006b; 2006e) developed some uncertain linguistic aggregation operators:

Definition 2.44 (Xu, 2006e) Let ULG : $\tilde{S}_4^n \to \tilde{S}_4$. Then

$$\mathrm{ULG}(\tilde{s}_1, \tilde{s}_2, \cdots, \tilde{s}_n) = (\tilde{s}_1 \otimes \tilde{s}_2 \otimes \cdots \otimes \tilde{s}_n)^{\frac{1}{n}} \tag{2.138}$$

is called an uncertain linguistic geometric (ULG) operator.

Example 2.38 Let $([s_{1/2}, s_1], [s_1, s_2], [s_2, s_3], [s_{1/4}, s_{1/3}])$ be a collection of uncertain linguistic variables in $\tilde{S}_4$. Then

$$\begin{aligned}
&\mathrm{ULG}([s_{1/2}, s_1], [s_1, s_2], [s_2, s_3], [s_{1/4}, s_{1/3}])\\
=&([s_{1/2}, s_1] \otimes [s_1, s_2] \otimes [s_2, s_3] \otimes [s_{1/4}, s_{1/3}])^{\frac{1}{4}}\\
=&[s_{0.71}, s_{1.19}]
\end{aligned}$$

Definition 2.45 (Xu, 2006e) Let ULWG : $\tilde{S}_4^n \to \tilde{S}_4$. If

$$\mathrm{ULWG}(\tilde{s}_1, \tilde{s}_2, \cdots, \tilde{s}_n) = \tilde{s}_1^{w_1} \otimes \tilde{s}_2^{w_2} \otimes \cdots \otimes \tilde{s}_n^{w_n} \tag{2.139}$$

where $\boldsymbol{w} = (w_1, w_2, \cdots, w_n)^{\mathrm{T}}$ is the weight vector of $\tilde{s}_i$ $(i = 1, 2, \cdots, n)$, with $w_i \geqslant 0$ $(i = 1, 2, \cdots, n)$ and $\sum_{i=1}^{n} w_i = 1$, then the function ULWG is called an uncertain linguistic weighted geometric (ULWG) operator.

In particular, if $\boldsymbol{w} = (1/n, 1/n, \cdots, 1/n)^{\mathrm{T}}$, then the ULWG operator reduces to the ULG operator.

Example 2.39 Let $([s_{1/2}, s_1], [s_1, s_2], [s_2, s_3], [s_{1/4}, s_{1/3}])$ be a collection of uncertain linguistic variables in $\tilde{S}_4$, and $\boldsymbol{w} = (0.3, 0.2, 0.1, 0.4)^{\mathrm{T}}$ be their weight vector. Then

$$\begin{aligned}
&\mathrm{ULWG}([s_{1/2}, s_1], [s_1, s_2], [s_2, s_3], [s_{1/4}, s_{1/3}])\\
=&[s_{1/2}, s_1]^{0.3} \otimes [s_1, s_2]^{0.2} \otimes [s_2, s_3]^{0.1} \otimes [s_{1/4}, s_{1/3}]^{0.4}\\
=&[s_{0.50}, s_{0.83}]
\end{aligned}$$

Definition 2.46 (Xu, 2006e) Let a mapping ULOWG : $\tilde{S}_4^n \to \tilde{S}_4$. If

$$\text{ULOWG}(\tilde{s}_1, \tilde{s}_2, \cdots, \tilde{s}_n) = \tilde{s}_{\sigma_1}^{\omega_1} \otimes \tilde{s}_{\sigma_2}^{\omega_2} \otimes \cdots \otimes \tilde{s}_{\sigma_n}^{\omega_n} \tag{2.140}$$

then ULOWG is called an uncertain linguistic ordered weighted geometric (ULOWG) operator, which has an associated weighting vector $\boldsymbol{\omega} = (\omega_1, \omega_2, \cdots, \omega_n)^{\mathrm{T}}$, with $\omega_i \geqslant 0$ $(i = 1, 2, \cdots, n)$, $\sum_{i=1}^{n} \omega_i = 1$, and $\tilde{s}_{\sigma_j}$ is the jth largest of the uncertain linguistic variables $\tilde{s}_i$ $(i = 1, 2, \cdots, n)$.

In particular, if $\boldsymbol{\omega} = (1/n, 1/n, \cdots, 1/n)^{\mathrm{T}}$, then the ULOWG operator reduces to the ULG operator.

Example 2.40 Let $\tilde{s}_1 = [s_{1/2}, s_2]$, $\tilde{s}_2 = [s_{1/2}, s_1]$, $\tilde{s}_3 = [s_{1/3}, s_1]$ and $\tilde{s}_4 = [s_1, s_2]$ be a collection of uncertain linguistic variables in $\tilde{S}_4$. In order to rank $\tilde{s}_i$ $(i = 1, 2, 3, 4)$, we first utilize (2.68) to compare $\tilde{s}_i$ and $\tilde{s}_j$, and construct the possibility degree matrix:

$$\boldsymbol{P} = \begin{pmatrix} 0.5 & 0.750 & 0.769 & 0.400 \\ 0.250 & 0.5 & 0.572 & 0.000 \\ 0.231 & 0.428 & 0.5 & 0.000 \\ 0.600 & 1.000 & 1.000 & 0.5 \end{pmatrix}$$

Then we utilize (2.82) to derive the priority vector of $\boldsymbol{P}$:

$$\boldsymbol{\zeta} = (0.285, 0.194, 0.180, 0.342)^{\mathrm{T}}$$

and employ the priority vector $\boldsymbol{\zeta}$ to rank the uncertain linguistic variables $\tilde{s}_i$ $(i = 1, 2, 3, 4)$:

$$\tilde{s}_{\sigma_1} = \tilde{s}_4 = [s_1, s_2], \quad \tilde{s}_{\sigma_2} = \tilde{s}_1 = [s_{1/2}, s_2], \quad \tilde{s}_{\sigma_3} = \tilde{s}_2 = [s_{1/2}, s_1], \quad \tilde{s}_{\sigma_4} = \tilde{s}_3 = [s_{1/3}, s_1]$$

Suppose that $\boldsymbol{\omega} = (0.2, 0.3, 0.3, 0.2)^{\mathrm{T}}$ is the weighting vector associated with the ULOWG operator. Then

$$\begin{aligned} \text{ULOWG}(\tilde{s}_1, \tilde{s}_2, \tilde{s}_3, \tilde{s}_4) &= [s_1, s_2]^{0.2} \otimes [s_{1/2}, s_2]^{0.3} \otimes [s_{1/2}, s_1]^{0.3} \otimes [s_{1/3}, s_1]^{0.2} \\ &= [s_{0.53}, s_{1.41}] \end{aligned}$$

Definition 2.47 (Xu, 2006e) Let a mapping ULHG : $\tilde{S}_4^n \to \tilde{S}_4$. If

$$\text{ULHG}(\tilde{s}_1, \tilde{s}_2, \cdots, \tilde{s}_n) = \tilde{s}_{\beta_1}^{\omega_1} \otimes \tilde{s}_{\beta_2}^{\omega_2} \otimes \cdots \otimes \tilde{s}_{\beta_n}^{\omega_n} \tag{2.141}$$

then ULHG is called an uncertain linguistic hybrid geometric (ULHG) operator, which has an associated weighting vector $\boldsymbol{\omega} = (\omega_1, \omega_2, \cdots, \omega_n)^{\mathrm{T}}$, with $\omega_i \geqslant 0$ $(i = 1, 2, \cdots, n)$ and $\sum_{j=1}^{n} \omega_j = 1$. $\tilde{s}_{\beta_j}$ is the jth largest of the uncertain linguistic variables $\tilde{s}'_i$ ($\tilde{s}'_i = \tilde{s}_i^{nw_i}$, $i = 1, 2, \cdots, n$), $\boldsymbol{w} = (w_1, w_2, \cdots, w_n)^{\mathrm{T}}$ is the weight vector of $\tilde{s}_i$ ($i = 1, 2, \cdots, n$), with $w_i \geqslant 0$ ($i = 1, 2, \cdots, n$), $\sum_{i=1}^{n} w_i = 1$, and n is the balancing coefficient.

Especially, if $\boldsymbol{\omega} = (1/n, 1/n, \cdots, 1/n)^{\mathrm{T}}$, then the ULHG reduces to the ULWG operator; if $\boldsymbol{w} = (1/n, 1/n, \cdots, 1/n)^{\mathrm{T}}$, then the ULHG reduces to the ULOWG operator:

Example 2.41 Let $\tilde{s}_1 = [s_{1/3}, s_{1/2}]$, $\tilde{s}_2 = [s_{1/2}, s_1]$, $\tilde{s}_3 = [s_2, s_3]$, $\tilde{s}_4 = [s_2, s_4]$ and $\tilde{s}_5 = [s_1, s_3]$ be a collection of uncertain linguistic variables in $\tilde{S}_4$, and let $\boldsymbol{w} = (0.20, 0.15, 0.25, 0.30, 0.10)^{\mathrm{T}}$ be their weight vector. We first calculate the weighted uncertain linguistic variables:

$$\tilde{s}'_1 = (\tilde{s}_1)^{5w_1} = [s_{1/3}, s_{1/2}]^{5\times 0.20} = [s_{1/3}, s_{1/2}]$$

$$\tilde{s}'_2 = (\tilde{s}_2)^{5w_2} = [s_{1/2}, s_1]^{5\times 0.15} = [s_{0.59}, s_1]$$

$$\tilde{s}'_3 = (\tilde{s}_3)^{5w_3} = [s_2, s_3]^{5\times 0.25} = [s_{2.38}, s_{3.95}]$$

$$\tilde{s}'_4 = (\tilde{s}_4)^{5w_4} = [s_2, s_4]^{5\times 0.30} = [s_{2.83}, s_8]$$

$$\tilde{s}'_5 = (\tilde{s}_5)^{5w_5} = [s_1, s_3]^{5\times 0.10} = [s_1, s_{1.73}]$$

In order to rank the uncertain linguistic variables $\tilde{s}'_i$ ($i = 1, 2, \cdots, 5$), we employ (2.68) to compare $\tilde{s}'_i$ and $\tilde{s}'_j$, and construct the possibility degree matrix:

$$\boldsymbol{P} = \begin{pmatrix} 0.5 & 0 & 0 & 0 & 0 \\ 1 & 0.5 & 0 & 0 & 0 \\ 1 & 1 & 0.5 & 0.166 & 1 \\ 1 & 1 & 0.834 & 0.5 & 1 \\ 1 & 1 & 0 & 0 & 0.5 \end{pmatrix}$$

Then we utilize (2.82) to derive the priority vector of $\boldsymbol{P}$:

$$\boldsymbol{\zeta} = (0.100, 0.150, 0.258, 0.292, 0.200)^{\mathrm{T}}$$

and utilize the priority vector $\boldsymbol{\zeta}$ to rank $\tilde{s}'_i$ ($i = 1, 2, \cdots, 5$):

$$\tilde{s}_{\sigma_1} = \tilde{s}'_4 = [s_{2.83}, s_8], \quad \tilde{s}_{\sigma_2} = \tilde{s}'_3 = [s_{2.38}, s_{3.95}], \quad \tilde{s}_{\sigma_3} = \tilde{s}'_5 = [s_1, s_{1.73}]$$

$$\tilde{s}_{\sigma_4} = \tilde{s}'_2 = [s_{0.59}, s_1], \quad \tilde{s}_{\sigma_5} = \tilde{s}'_1 = [s_{1/3}, s_{1/2}]$$

Assume that $\boldsymbol{\omega} = (0.15, 0.2, 0.3, 0.2, 0.15)^{\mathrm{T}}$ is the weighting vector associated with the ULHG operator. Then

$$\mathrm{ULHG}(\tilde{s}_1, \tilde{s}_2, \tilde{s}_3, \tilde{s}_4, \tilde{s}_5) = [s_{2.83}, s_8]^{0.15} \otimes [s_{2.38}, s_{3.95}]^{0.20} \otimes [s_1, s_{1.73}]^{0.30}$$

$$\otimes \left[s_{0.59}, s_1\right]^{0.20} \otimes \left[s_{1/3}, s_{1/2}\right]^{0.15}$$
$$= \left[s_{1.06}, s_{1.91}\right]$$

Definition 2.48 (Xu, 2006e) An induced uncertain LOWG (IULOWG) operator is of the form:

$$\text{IULOWG}\left(\langle u_1, \tilde{s}_1\rangle, \langle u_2, \tilde{s}_2\rangle, \cdots, \langle u_n, \tilde{s}_n\rangle\right) = \tilde{s}_{\sigma_1}^{\omega_1} \otimes \tilde{s}_{\sigma_2}^{\omega_2} \otimes \cdots \otimes \tilde{s}_{\sigma_n}^{\omega_n} \qquad (2.142)$$

where $\boldsymbol{\omega} = (\omega_1, \omega_2, \cdots, \omega_n)^{\mathrm{T}}$ is the weighting vector, such that $\omega_i \geqslant 0\ (i = 1, 2, \cdots, n)$, $\sum_{j=1}^{n} \omega_j = 1$, and $\tilde{s}_{\sigma_j}$ is the $\tilde{s}_i$ value of the OWG pairs $\langle u_i,\ \tilde{s}_i\rangle\ (i = 1, 2, \cdots, n)$ having the jth largest u_i value. The term u_i is referred as the order inducing variable, and $\tilde{s}_i$ is referred as the argument variable.

In particular, if $u_i = \tilde{s}_i$ for all i, then the IULOWG operator reduces to the ULOWG operator; if $u_i =$ No. i for all i, where No. i is the ordered position of $\tilde{s}_i$, then the IULOWG operator reduces to the ULWG operator; if $\boldsymbol{\omega} = (1/n, 1/n, \cdots, 1/n)^{\mathrm{T}}$, then the IULOWG operator reduces to the ULG operator.

If there is a tie between $\langle u_i,\ \tilde{s}_i\rangle$ and $\langle u_j,\ \tilde{s}_j\rangle$ with respect to order inducing variables, then in the process of aggregation, we can replace $\tilde{s}_i$ and $\tilde{s}_j$ with their geometric mean $(\tilde{s}_i \otimes \tilde{s}_j)^{1/2}$, and get the corresponding pairs $\left\langle u_i, (\tilde{s}_i \oplus \tilde{s}_j)^{1/2}\right\rangle$ and $\left\langle u_j, (\tilde{s}_i \oplus \tilde{s}_j)^{1/2}\right\rangle$. If k items are tied, we replace these by k replica's of their geometric mean.

Example 2.42 Given a collection of uncertain LOWG pairs $\langle u_i,\ \tilde{s}_i\rangle\ (i = 1, 2, 3, 4)$:

$$\left(\langle -1,\ [s_2, s_3]\rangle, \left\langle -3,\ [s_{1/2}, s_2]\right\rangle, \langle -2,\ [s_1, s_3]\rangle, \left\langle -4,\ [s_{1/3}, s_1]\right\rangle\right)$$

We first utilize the values $u_i\ (i = 1, 2, 3, 4)$ to rank $\langle u_i,\ \tilde{s}_i\rangle\ (i = 1, 2, 3, 4)$, and then we get the ordered uncertain OWG pairs:

$$\left(\langle -1,\ [s_2, s_3]\rangle, \langle -2,\ [s_1, s_3]\rangle, \left\langle -3,\ [s_{1/2}, s_2]\right\rangle, \left\langle -4,\ [s_{1/3}, s_1]\right\rangle\right)$$

i.e.,

$$\tilde{s}_{\sigma_1} = [s_2, s_3], \quad \tilde{s}_{\sigma_2} = [s_1, s_3], \quad \tilde{s}_{\sigma_3} = [s_{1/2}, s_2], \quad \tilde{s}_{\sigma_4} = [s_{1/3}, s_1]$$

Assume that the weighting vector associated with the IULOWG operator is $\boldsymbol{\omega} = (0.10, 0.40, 0.40, 0.10)^{\mathrm{T}}$. Then

$$\text{IULOWG}\left(\langle u_1, \tilde{s}_1\rangle, \langle u_2, \tilde{s}_2\rangle, \langle u_3, \tilde{s}_3\rangle, \langle u_4, \tilde{s}_4\rangle\right)$$
$$= [s_2, s_3]^{0.10} \otimes [s_1, s_3]^{0.40} \otimes \left[s_{1/2}, s_2\right]^{0.40} \otimes \left[s_{1/3}, s_1\right]^{0.10}$$

If we revise $\langle u_2, \tilde{s}_2\rangle = \left\langle -3, \left[s_{1/2}, s_2\right]\right\rangle$ as $\langle u_2, \tilde{s}_2\rangle = \left\langle -2, \left[s_{1/2}, s_2\right]\right\rangle$, then $u_2 = u_3$. In the process of aggregation, we replace $\tilde{s}_2$ and $\tilde{s}_3$ with their geometric mean, and get

$$\left\langle u_2, (\tilde{s}_2 \otimes \tilde{s}_3)^{1/2}\right\rangle = \left\langle u_3, (\tilde{s}_2 \otimes \tilde{s}_3)^{1/2}\right\rangle = \langle 0.2, [s_{0.71}, s_{2.45}]\rangle$$

Thus

$$\tilde{s}_{\sigma_1} = [s_2, s_3], \quad \tilde{s}_{\sigma_2} = \tilde{s}_{\sigma_3} = [s_{0.71}, s_{2.45}], \quad \tilde{s}_{\sigma_4} = [s_{1/3}, s_1]$$

Accordingly,

$$\begin{aligned} &\text{IULOWG}\left(\langle u_1, \tilde{s}_1\rangle, \langle u_2, \tilde{s}_2\rangle, \langle u_3, \tilde{s}_3\rangle, \langle u_4, \tilde{s}_4\rangle\right) \\ =&[s_2, s_3]^{0.10} \otimes [s_{0.71}, s_{2.45}]^{0.40} \otimes [s_{0.71}, s_{2.45}]^{0.40} \otimes [s_{1.3}, s_1]^{0.10} \\ =&[s_{0.73}, s_{2.29}] \end{aligned}$$

If we substitute (2.85) with

$$\begin{aligned} &\text{GIULOWG}\left(\langle v_1, u_1, \tilde{s}_1\rangle, \langle v_2, u_2, \tilde{s}_2\rangle, \cdots, \langle v_n, u_n, \tilde{s}_n\rangle\right) \\ =&\tilde{s}_{\sigma_1}^{\omega_1} \otimes \tilde{s}_{\sigma_2}^{\omega_2} \otimes \cdots \otimes \tilde{s}_{\sigma_n}^{\omega_n} \end{aligned} \tag{2.143}$$

which is called a generalized induced uncertain LOWG (GIULOWG) operator.

In particular, if $v_i = u_i$ for all i, then the GIULOWG operator reduces to the IULOWG operator; if $v_i = \tilde{s}_i$ for all i, then the GIULOWG operator reduces to the ULOWG operator; if $v_i =$ No. i for all i, where No. i is the ordered position of $\tilde{s}_i$, then the GIULOWG operator reduces to the ULWG operator; if $\boldsymbol{\omega} = (1/n, 1/n, \cdots, 1/n)^{\mathrm{T}}$, then the GIULOWG operator reduces to the ULG operator.

Example 2.43 Given a collection of uncertain objects $\langle v_i, u_i, \tilde{s}_i\rangle$ $(i = 1, 2, 3, 4)$:

$$\left(\langle 0.2, \text{Johnson}, [s_1, s_3]\rangle, \langle 0.4, \text{Brown}, [s_2, s_3]\rangle, \left\langle 0.3, \text{Smith}, [s_{1/3}, s_1]\right\rangle\right.$$

$$\left.\left\langle 0.1, \text{Anderson}, [s_{1/2}, s_1]\right\rangle\right)$$

We first rank $\langle v_i, u_i, s_i\rangle$ $(i = 1, 2, 3, 4)$ according to v_i $(i = 1, 2, 3, 4)$:

$$\langle 0.4, \text{Brown}, [s_2, s_3]\rangle$$

$$\left\langle 0.3, \text{Smith}, [s_{1/3}, s_1]\right\rangle$$

$$\langle 0.2, \text{Johnson}, [s_1, s_3]\rangle$$

$$\left.\left\langle 0.1, \text{Anderson}, [s_{1/2}, s_1]\right\rangle\right)$$

i.e.,

$$\tilde{s}_{\sigma_1} = [s_2, s_3], \ . \ \tilde{s}_{\sigma_2} = [s_{1/3}, s_1], \quad \tilde{s}_{\sigma_3} = [s_1, s_3], \quad \tilde{s}_{\sigma_4} = [s_{1/2}, s_1]$$

Suppose that the weighting vector associated with the GIULOWG operator is $\omega = (0.20, 0.40, 0.30, 0.10)^{\mathrm{T}}$. Then

$$\begin{aligned}&\mathrm{GILOWG}\left(\langle v_1, u_1, \tilde{s}_1\rangle, \langle v_2, u_2, \tilde{s}_2\rangle, \langle v_3, u_3, \tilde{s}_3\rangle, \langle v_4, u_4, \tilde{s}_4\rangle\right)\\ =&[s_2, s_3]^{0.20} \otimes [s_{1/3}, s_1]^{0.40} \otimes [s_1, s_3]^{0.3} \otimes [s_{1/2}, s_1]^{0.10}\\ =&[s_{0.69}, s_{1.73}]\end{aligned}$$

Yager and Xu (2006) developed a continuous ordered weighted geometric (C-OWG) operator. Zhang and Xu (2005) extended the C-OWG operator to uncertain linguistic situations:

Definition 2.49 (Yager and Xu, 2006) Let $[a, b]$ be an interval number, and

$$g_{\varphi}([a,b]) = b\left(\frac{a}{b}\right)^{\int_0^1 \frac{\mathrm{d}\varphi(y)}{\mathrm{d}y} y \mathrm{d}y} = a^{\left(1-\int_0^1 \varphi(y)\mathrm{d}y\right)} b^{\int_0^1 \varphi(y)\mathrm{d}y} \tag{2.144}$$

where φ is a BUM function. Then g is called a continuous OWG (C-OWG) operator.

Definition 2.50 (Zhang and Xu, 2005) Let $\tilde{s} = [s_\alpha, s_\beta] \in \tilde{S}_4$ be an uncertain multiplicative linguistic variable, and $l\, g_{\varphi}([s_\alpha, s_\beta]) = s_\gamma$, where

$$\gamma = g_{\varphi}([I(s_\alpha), I(s_\beta)]) = g_{\varphi}([\alpha, \beta]) \tag{2.145}$$

and φ is a BUM function. Then $l\, g$ is called an uncertain linguistic C-OWG (ULC-OWG) operator.

Example 2.44 Let $\tilde{s} = [s_{1/2}, s_2]$ be an uncertain multiplicative linguistic variable, and the BUM function $\varphi(x) = x$. Then

$$g_{\varphi}([1/2, 2]) = \left(\frac{1}{2}\right)^{\left(1-\int_0^1 x\mathrm{d}x\right)} 2^{\int_0^1 x\mathrm{d}x} = 1$$

and hence, $l\, g_{\varphi}([s_{1/2}, s_2]) = s_1$.

Xu (2009b) further extended the LCG operator (2.136) to uncertain linguistic situations:

Definition 2.51 (Xu, 2009b) Let μ be a fuzzy measure on $\tilde{S}_4$, and $\tilde{s}_i$ $(i = 1, 2, \cdots, n)$ be n uncertain multiplicative linguistic variables in $\tilde{S}_4$. Then

$$(C_4)\int \tilde{s}\, \mathrm{d}\mu = \mathrm{ULCG}\left(\tilde{s}_1, \tilde{s}_2, \cdots, \tilde{s}_n\right) = \bigotimes_{i=1}^{n} \left(\tilde{s}_{\sigma(i)}\right)^{\mu(\tilde{A}_{\sigma(i)})-\mu(\tilde{A}_{\sigma(i-1)})} \tag{2.146}$$

is called an uncertain linguistic correlated geometric (ULCG) operator, where $(C_4)\int \tilde{s}\, \mathrm{d}\mu$ is a notation of Choquet integral, $(\sigma(1), \sigma(2), \cdots, \sigma(n))$ is a permutation of $(1, 2, \cdots, n)$ such that $\tilde{s}_{\sigma(1)} \geqslant \tilde{s}_{\sigma(2)} \geqslant \cdots \geqslant \tilde{s}_{\sigma(n)}$, $\tilde{A}_{\sigma(k)} = \{\tilde{s}_{\sigma(j)} | j \leqslant k\}$, when $k \geqslant 1$ and $\tilde{A}_{\sigma(0)} = \varnothing$.

Below we discuss some special cases of the ULCG operator (Xu, 2009b):

(1) If (2.89) and (2.90) hold, then the ULCG operator (2.146) reduces to the ULWG operator. Especially, if $w_i = \frac{1}{n}$ for all $i = 1, 2, \cdots, n$, then the ULWG operator reduces to the ULG operator.

(2) If (2.91) and (2.92) hold, then the ULCG operator reduces to the ULOWG operator. Especially, if $\mu(\tilde{A}) = \frac{|\tilde{A}|}{n}$ for all $\tilde{A} \subseteq \tilde{S}_4$, then both the ULCG and ULOWG operators reduce to the ULG operator.

(3) If (2.93) and (2.94) hold, then the ULCG operator reduces to the following form:

$$\text{ULWOWG}\left(\tilde{s}_1, \tilde{s}_2, \cdots, \tilde{s}_n\right) = \bigotimes_{i=1}^{n} \left(\tilde{s}_{\sigma(i)}\right)^{w_i} \tag{2.147}$$

which we call an uncertain linguistic weighted ordered weighted geometric (ULWOWG) operator. Especially, if $\mu\left(\{\tilde{s}_i\}\right) = \frac{1}{n}$ for all $i = 1, 2, \cdots, n$, then the ULWOWG operator reduces to the ULOWG operator.

In cases where the linguistic information provided by using the multiplicative linguistic labels in $\bar{S}_4$ is time dependent, Xu (2009a) gave the following definition:

Definition 2.52 (Xu, 2009a) Let t be a time variable. Then $s_\alpha(t)$ is called a multiplicative linguistic label at the time t, where $s_\alpha(t) \in \bar{S}_4$ for any t. If $t = t_1, t_2, \cdots, t_p$, then $s_\alpha(t_k)\,(k = 1, 2, \cdots, p)$ denote p multiplicative linguistic labels collected from p different periods.

Based on Definition 2.52, Xu (2009a) defined the concept of dynamic linguistic weighted geometric operator:

Definition 2.53 (Xu, 2009a) Let $s_\alpha(t_k)$ $(k = 1, 2, \cdots, p)$ be p multiplicative linguistic labels collected from p different periods t_k $(k = 1, 2, \cdots, p)$, where $s_\alpha(t_k) \in \bar{S}_4\,(k = 1, 2, \cdots, p)$, and $\boldsymbol{w}(t) = (w(t_1), w(t_2), \cdots, w(t_p))^{\mathrm{T}}$ is the weight vector of the periods t_k $(k = 1, 2, \cdots, p)$. Then

$$\begin{aligned} &\text{DLWG}_{\boldsymbol{w}(t)}(s_\alpha(t_1), s_\alpha(t_2), \cdots, s_\alpha(t_p)) \\ =&(s_\alpha(t_1))^{w(t_1)} \otimes (s_\alpha(t_2))^{w(t_2)} \otimes \cdots \otimes (s_\alpha(t_p))^{w(t_p)} \end{aligned} \tag{2.148}$$

is called a dynamic linguistic weighted geometric (DLWG) operator. Especially, if $\boldsymbol{w}(t) = (1/p, 1/p, \cdots, 1/p)^{\mathrm{T}}$, then (2.148) reduces to the dynamic linguistic geometric (DLG) operator:

$$\text{DLG}\left(s_\alpha(t_1), s_\alpha(t_2), \cdots, s_\alpha(t_p)\right) = \left(s_\alpha(t_1) \otimes s_\alpha(t_2) \otimes \cdots \otimes s_\alpha(t_p)\right)^{\frac{1}{p}} \tag{2.149}$$

The weight vector $\boldsymbol{w}(t) = (w(t_1), w(t_2), \cdots, w(t_p))^{\mathrm{T}}$ of the periods t_k $(k=1, 2, \cdots, p)$ can be derived by the methods introduced in Subsection 2.4.1.

References

Adamopoulos G I, Pappis G P. 1996. A fuzzy linguistic approach to a multicriteria-sequencing problem. European Journal of Operational Research, 92: 628-636.

Becker H. 2001. Computing with words and machine learning in medical diagnostics. Information Sciences, 134: 53-69.

Bonissone P P, Decker K S. 1986. Selecting uncertainty calculi and granularity: an experiment in trading-off precision and complexity. *In*: Kanal L H, Lemmer J F. eds. Uncertainty in Artificial Intelligence. Amsterdam: North-Holland: 217-247.

Bordogna G, Fedrizzi M, Passi G. 1997. A linguistic modeling of consensus in group decision making based on OWA operator. IEEE Transactions on Systems, Man, and Cybernetics, 27: 126-132.

Bordogna G, Passi G. 1993. A fuzzy linguistic approach generalizing boolean information retrieval: a model and its evaluation. Journal of the American Society for Information Science and Technology, 44: 70-82.

Chang P, Chen Y. 1994. A fuzzy multicriteria decision making method for technology transfer strategy selection in biotechnology. Fuzzy Sets and Systems, 63: 131-139.

Chen S J, Hwang C L. 1992. Fuzzy Multiple Attribute Decision Making: Methods and Applications. Berlin: Springer-Verlag.

Chen S M. 1997. A new method for tool steel materials selection under fuzzy environment. Fuzzy Sets and Systems, 92: 265-274.

Chiclana F, Herrera F, Herrera-Viedma E. 1998. Integrating three representation models in fuzzy multipurpose decision making based on fuzzy preference relations. Fuzzy Sets and Systems, 97: 33-48.

Chiclana F, Herrera F, Herrera-Viedma E. 2001. Integrating multiplicative preference relations in a multipurpose decision-making model based on fuzzy preference relations. Fuzzy Sets and Systems, 122: 277-291.

Choquet G. 1953. Theory of capacities. Annales de l'institut Fourier, 5: 131-296.

Degani R, Bortolan G. 1988. The problem of linguistic approximation in clinical decision making. Internatioanl Journal of Approximate Reasoning, 2: 143-162.

Delgado M, Herrera F, Herrera-Viedma E, et al. 1998. Combining numerical and linguistic information in group decision making. Information Sciences, 107: 177-194.

Delgado M, Herrera F, Herrera-Viedma E, et al. 2002. A communication model based on the 2-tuple fuzzy linguistic representation for a distributed intelligent agent system on internet. Soft Computing, 6: 320-328.

Delgado M, Verdegay J L, Vila M A. 1993a. Linguistic decision making models. International Journal of Intelligent Systems, 7: 479-492.

Delgado M, Verdegay J L, Vila M A. 1993b. On aggregation operators of linguistic labels. International Journal of Intelligent Systems, 8: 351-370.

Delgado M, Verdegay J L, Vila M A. 1994. A model for incomplete and vague in information in decision making problems. International Journal of Intelligent Systems, 9: 365-378.

Dubois D, Prade H. 1980. Fuzzy Sets and Systems: Theory and Applications. New York: Academic Press.

Fan Z P, Ma J, Zhang Q. 2002. An approach to multiple attribute decision making based on fuzzy preference information on alternatives. Fuzzy Sets and Systems, 131: 101-106.

Fullér R, Majlender P. 2003. On obtaining minimal variability OWA operator weights. Fuzzy Sets and Systems, 136: 203-215.

Grabisch M. 1995. Fuzzy integral in multicriteria decision making. Fuzzy Sets and Systems, 69: 279-298.

Harsanyi J C. 1955. Cardinal welfare, individualistic ethics, and interpersonal comparisons of utility. Journal of Political Economy, 63: 309-321.

Herrera F, Herrera-Viedma E, Chiclana F. 2001a. Multiperson decision making based on multiplicative preference relations. European Journal of Operational Research, 129: 372-385.

Herrera F, Herrera-Viedma E, Chiclana F. 2003. A study of the origin and uses of the ordered weighted geometric operator in multicriteria decision making. International Journal of Intelligent Systems, 18: 689-707.

Herrera F, Herrera-Viedma E, Martínez L. 2000. A fusion approach for managing multi-granularity linguistic term sets in decision making. Fuzzy Sets and Systems, 114: 43-58.

Herrera F, Herrera-Viedma E, Verdegay J L. 1995. A sequential selection process in group decision making with a linguistic assessment approach. Information Sciences, 85: 223-239.

Herrera F, Herrera-Viedma E, Verdegay J L. 1996a. A model of consensus in group decision making under linguistic assessments. Fuzzy Sets and Systems, 78: 73-87.

Herrera F, Herrera-Viedma E, Verdegay J L. 1996b. Direct approach processes in group decision making using linguistic OWA operators. Fuzzy Sets and Systems, 79: 175-190.

Herrera F, Herrera-Viedma E, Verdegay J L. 1997. A rational consensus model in group decision making using linguistic assessments. Fuzzy Sets and Systems, 88: 31-49.

Herrera F, Herrera-Viedma E. 1997. Aggregation operators for linguistic weighted information. IEEE Transactions on Systems, Man, and Cybernetics-Part A, 27: 646-656.

Herrera F, Herrera-Viedma E. 2000a. Choice functions and mechanisms for linguistic preference relations. European Journal of Operational Research, 120: 144-161.

Herrera F, Herrera-Viedma E. 2000b. Linguistic decision analysis: steps for solving decision problems under linguistic information. Fuzzy Sets and Systems, 115: 67-82.

Herrera F, Herrera-Viedma E. 2003. Linguistic preference modeling: foundation models and new trends. *In*: Proceedings of 4th International Workshop on Preferences and Decisions, Trento, Italy: 47-51.

Herrera F, López E, Mendaña, et al. 2001b. A linguistic decision model for personnel management solved with a linguistic biobjective genetic algorithm. Fuzzy Sets and Systems, 118: 47-64.

Herrera F, López E, Rodríguez M A. 2002. A linguistic decision model for promotion mix management solved with genetic algorithms. Fuzzy Sets and Systems, 131: 47-61.

Herrera F, Martínez L. 2000a. A 2-tuple fuzzy linguistic representation model for computing with words. IEEE Transactions on Fuzzy Systems, 8: 746-752.

Herrera F, Martínez L. 2000b. An approach for combining linguistic and numerical information based on the 2-tuple fuzzy linguistic representation model in decision-making. International Journal of Uncertainty, Fuzziness and Knowledge-Based Systems, 8: 539-562.

Herrera F, Martínez L. 2001a. The 2-tuple linguistic computational model: advantages of its linguistic description, accuracy and consistency. Internal Journal of Uncertainty, Fuzziness and Knowledge-Based Systems, 9 (Suppl.): 33-48.

Herrera F, Verdegay J L. 1993. Linguistic assessments in group decision. *In*: Proceedings of the First European Congress on Fuzzy and Intelligent Technologies, Aachen: 941-948.

Herrera-Viedma E, Cordon O, Luque M, et al. 2003. A model of fuzzy linguistic IRS based on multi-granular linguistic information. International Journal of Approximate Reasoning, 34: 221-239.

Herrera-Viedma E, Peis E. 2003. Evaluating the informative quality of documents in SGML format from judgments by means of fuzzy linguistic techniques based on computing with words. Information Processing and Management, 39: 233-249.

Herrera-Viedma E. 2001. Modeling the retrieval process of an information retrieval system using an ordinal fuzzy linguistic approach. Journal of the American Society for Information Science and Technology, 52: 460-475.

Huynh V N, Nakamori Y. 2005. A satisfactory-oriented approach to multiexpert decision-making with linguistic assessments. IEEE Transactions on Systems, Man, and Cybernetics-Part B, 35: 184-196.

Kostek B. 2003. Computing with words" concept applied to musical information retrieval. Electronic Notes in Theoretical Computer Science, 82(4): 1-12.

Law C K. 1996. Using fuzzy numbers in educational grading systems. Fuzzy Sets and Systems, 83: 311-323.

Lee H M. 1996. Applying fuzzy set theory to evaluate the rate of aggregative risk in software development. Fuzzy Sets and Systems, 80: 323-336.

Levrat E, Voisin A, Bombardier S, et al. 1997. Subjective evaluation of car seat comfort with fuzzy set techniques. International Journal of Intelligent Systems, 12: 891-913.

Ma J, Fan Z P, Jiang Y P, et al. 2006. A method for repairing the inconsistency of fuzzy preference relations. Fuzzy Sets and Systems, 157: 20-33.

Ovchinnikov S. 1999. Invariance properties of ordinal OWA operators. International Journal of Intelligent Systems, 14: 413-418.

Saaty T L. 1980. The Analytic Hierarchy Process. New York: McGraw-Hill.

Tong R M, Bonissone P P. 1980. A linguistic approach to decision making with fuzzy sets. IEEE Transactions on Systems, Man, and Cybernetics, 10: 716-723.

Torra V. 1997. The weighted OWA operator. International Journal of Intelligent Systems, 12: 153-166.

Torra V. 2003. Information Fusion in Data Mining. Berlin: Springer.

Wang Z, Klir G. 1992. Fuzzy Measure Theory. New York: Plenum Press.

Xu R N, Zhai X Y. 1992. Extensions of the analytic hierarchy process in fuzzy environment. Fuzzy Sets and Systems, 52: 251-257.

Xu Z S, Chen J. 2008. Group decision-making procedures based on incomplete fuzzy preference relations. Soft Computing, 12: 515-521.

Xu Z S, Da Q L. 2002a. Multi-attribute decision making based on fuzzy linguistic assessments. Journal of Southeast University, 32: 656–658.

Xu Z S, Da Q L. 2002b. The uncertain OWA operator. International Journal of Intelligent Systems, 17: 569-575.

Xu Z S, Da Q L. 2002c. The ordered weighted geometric averaging operators. International Journal of Intelligent Systems, 17: 709-716.

Xu Z S, Da Q L. 2003a. Method based on fuzzy linguistic scale and fuzzy induced ordered weighted geometric averaging (FIOWGA) operator for decision making problems. Journal of Southeast University, 19: 88-91.

Xu Z S, Da Q L. 2003b. An overview of operators for aggregating information. International Journal of Intelligent Systems, 18: 953-969.

Xu Z S, Da Q L. 2005. A least deviation method to obtain a priority vector of a fuzzy preference relation. European Journal of Operational Research, 164: 206-216.

Xu Z S. 2000. A simulation-based evaluation of several scales in the analytic hierarchy process. Systems Engineering-Theory & Practice, 20(7): 58-62.

Xu Z S. 2001a. The least variance priority method (LVM) for fuzzy complementary judgment matrix. Systems Engineering-Theory & Practice, 21(1): 93-96.

Xu Z S. 2001b. Algorithm for priority of fuzzy complementary judgment matrix. Journal of Systems Engineering, 16(4): 311-314.

Xu Z S. 2004a. Uncertain linguistic aggregation operators based approach to multiple attribute group decision making under uncertain linguistic environment. Information Sciences, 168: 171-184.

Xu Z S. 2004b. On method of multi-attribute group decision making under pure linguistic information. Control and Decision, 19(7): 778-781.

Xu Z S. 2004c. Method based on fuzzy linguistic assessments and GIOWA operator in multi-attribute group decision making. Journal of Systems Science and Mathematical Sciences, 24: 218-224.

Xu Z S. 2004d. Uncertain Multiple Attribute Decision Making: Methods and Applications. Beijing: Tsinghua University Press.

Xu Z S. 2004e. EOWA and EOWG operators for aggregating linguistic labels based on linguistic preference relations. International Journal of Uncertainty, Fuzziness and Knowledge-Based Systems, 12: 791-810.

Xu Z S. 2004f. A method based on linguistic aggregation operators for group decision making with linguistic preference relations. Information Sciences, 166: 19-30.

Xu Z S. 2004g. Goal programming models for obtaining the priority vector of incomplete fuzzy preference relation. International Journal of Approximate Reasoning, 36: 261-270.

Xu Z S. 2004h. Incomplete complementary judgment matrix. Systems Engineering-Theory & Practice, 24(6): 91-97.

Xu Z S. 2005a. An approach to pure linguistic multiple attribute decision making under uncertainty. International Journal of Information Technology and Decision Making, 4: 197-206.

Xu Z S. 2005b. An approach to group decision making based on incomplete linguistic preference relations. International Journal of Information Technology and Decision Making, 4: 153-160.

Xu Z S. 2005c. Deviation measures of linguistic preference relations in group decision making. Omega, 33: 249-254.

Xu Z S. 2005d. An overview of methods for determining OWA weights. International Journal of Intelligent Systems, 20: 843-865.

Xu Z S. 2006a. A note on linguistic hybrid arithmetic averaging operator in group decision making with linguistic information. Group Decision and Negotiation, 15: 581-591.

Xu Z S. 2006b. On generalized induced linguistic aggregation operators. International Journal of General Systems, 35: 17-28.

Xu Z S. 2006c. Induced uncertain linguistic OWA operators applied to group decision making. Information Fusion, 7: 231-238.

Xu Z S. 2006d. A direct approach to group decision making with uncertain additive linguistic preference relations. Fuzzy Optimization and Decision Making, 5: 23-35.

Xu Z S. 2006e. An approach based on the uncertain LOWG and the induced uncertain LOWG operators to group decision making with uncertain multiplicative linguistic preference relations. Decision Support Systems, 41: 488-499.

Xu Z S. 2006f. Goal programming models for multiple attribute decision making under linguistic setting. Journal of Management Sciences in China, 9: 9-17.

Xu Z S. 2006g. Incomplete linguistic preference relations and their fusion. Information Fusion, 7: 331-337.

Xu Z S. 2006h. A practical procedure for group decision making under incomplete multiplicative linguistic preference relations. Group Decision and Negotiation, 15: 593-604.

Xu Z S. 2006i. Multiple attribute decision making based on different types of linguistic information. Journal of Southeast University, 22(1): 134-136.

Xu Z S. 2007a. A method for multiple attribute decision making with incomplete weight information in linguistic setting. Knowledge-Based Systems, 20: 719-725.

Xu Z S. 2007b. Linguistic aggregation operators: an overview. *In*: Bustince H, Herrera F, Montero J. eds. Fuzzy Sets and Their Extensions: Representation, Aggregation and Models. Heidelberg: Springer. 163-181.

Xu Z S. 2007c. A survey of preference relations. International Journal of General Systems, 36: 179-203.

Xu Z S. 2007d. Dynamic linguistic preference relations and their use in multi-period decision making. *In*: International Conference on Management Science and Engineering. Proceedings of 2007 International Conference on Management Science and Engineering. Harbin, China, 345-350.

Xu Z S. 2009a. Multi-period multi-attribute group decision making under linguistic assessments. International Journal of General Systems, 38: 823-850.

Xu Z S. 2009b. Correlated linguistic information aggregation. International Journal of Uncertainty, Fuzziness and Knowledge-Based System, 17: 633-647.

Yager R R, Filev D P. 1999. Induced ordered weighted averaging operators. IEEE Transactions on Systems, Man, and Cybernetics-Part B, 29: 141-150.

Yager R R, Goldstein L S, Mendels. 1994. Fuzmar: an approach to aggregating market research data based on fuzzy reasoning. Fuzzy Sets and Systems, 68: 1-11.

Yager R R, Kacprzyk J. 1997. The Ordered Weighted Averaging Operators: Theory and Applications. Boston: Kluwer.

Yager R R, Rybalov A. 1997. Understanding the median as a fusion operator. International Journal of General Systems, 26: 239-263.

Yager R R, Xu Z S. 2006. The continuous ordered weighted geometric operator and its application to decision making. Fuzzy Sets and Systems, 157: 1393-1402.

Yager R R. 1988. On ordered weighted averaging aggregation operators in multi-criteria decision making. IEEE Transactions on Systems, Man, and Cybernetics, 18: 183-190.

Yager R R. 1992. Applications and extensions of OWA aggregations. International Journal of Man-Machine Studied, 37: 103-132.

Yager R R. 1993. Families of OWA operators. Fuzzy Sets and Systems, 59: 125-145.

Yager R R. 1995. An approach to ordinal decision making. International Journal of Approximate Reasoning, 12: 237-261.

Yager R R. 1996. Quantifier guided aggregation using OWA operators. International Journal of Intelligent Systems, 11: 49-73.

Yager R R. 1998. Fusion of ordinal information using weighted median aggregation. International Journal of Approximate Reasoning, 18: 35-52.

Yager R R. 2004. OWA aggregation over a continuous interval argument with applications to decision Making. IEEE Transaction on Systems, Man, and Cybernetics-Part B, 34: 1952-1963.

Yager R R. 2007a. Centered OWA operators. Soft Computing, 11: 631-639.

Yager R R. 2007b. Time series smoothing and OWA aggregation. Technical Report# MII-2701, Machine Intelligence Institute, Iona College, New Rochelle, NY.

Zadeh L A, Kacprzyk J, eds. 1999a. Computing with Words in Information/Intelligent Systems 1: Foundations. Heidelberg: Physica-Verlag.

Zadeh L A, Kacprzyk J, eds. 1999b. Computing with Words in Information/Intelligent Systems 2: Applications. Heidelberg: Physica-Verlag.

Zadeh L A. 1983. A computational approach to fuzzy quantifiers in natural languages. Computers & Mathematics with Applications, 9: 149-184.

Zhang H M, Xu Z S. 2005. Uncertain linguistic information based C-OWA and C-OWG operators and their applications. Journal of PLA University of Science and Technology, 6(6): 604-608.

Chapter 3

Linguistic Preference Relations

Preference relations (or called pairwise comparison matrices, judgment matrices) are very useful in expressing decision maker's preference information over objects by comparing each pair of them in decision making problems of various fields, including politics, social psychology, engineering, management, business and economics, etc. During the past years, the use of preference relations is receiving increasing attention, and a number of studies have focused on this issue. In some situations, such as personnel appraisal, performance evaluation of weapon equipments, the partner selection of supply chain management, etc., a decision maker usually provides his/her preference information by using linguistic labels and constructs linguistic preference relations (Chen and Fan, 2005; Delgado et al., 1998; Fan and Xiao, 2002; Herrera and Herrera-Viedma, 2003; 2000; 1997; Herrera et al., 2005; 1997; 1996a; 1996b; Xu, 2008; 2007b; 2006a; 2006b; 2006c; 2005b; 2004a; 2004b; Xu and Wu, 2004). In this chapter, we shall introduce the concept of linguistic preference relation, uncertain linguistic preference relation, incomplete linguistic preference relation, consistent linguistic preference relation, and acceptable linguistic preference relation, etc., and their desirable properties. We also introduce in detail a series of approaches for decision making based on these linguistic preference relations.

3.1 Additive Linguistic Preference Relations

For a decision making problem, let $X = \{x_1, x_2, \cdots, x_n\}$ be a set of n alternatives. A decision maker compares each pair of alternatives in X by means of the additive linguistic evaluation scale $S_2 = \{s_\alpha | \alpha = -\tau, \cdots, -1, 0, 1, \cdots, \tau\}$, and constructs an additive linguistic preference relation (Xu, 2007a) $\boldsymbol{A} = (a_{ij})_{n\times n}$, where

$$s_{-\tau} \leqslant a_{ij} \leqslant s_\tau, \quad a_{ij} \oplus a_{ji} = s_0, \quad a_{ii} = s_0, \quad i, j = 1, 2, \cdots, n \tag{3.1}$$

and a_{ij} denotes the preference degree that the alternative x_i over x_j. Especially, $a_{ij} = s_0$ indicates indifference between x_i and x_j; $a_{ij} > s_0$ means that x_i is preferred to x_j, the greater a_{ij}, the stronger the preference intensity of the alternative x_i over x_j, especially, $a_{ij} = s_\tau$ indicates that x_i is absolutely preferred to x_j; $a_{ij} < s_0$ means that x_j is preferred to x_i, the smaller a_{ij}, the greater the preference intensity of the alternative x_j over x_i, especially, $a_{ij} = s_{-\tau}$ indicates that x_j is absolutely preferred to x_i.

Based on the concept of additive linguistic preference relation, we have the following theorems:

Theorem 3.1 Let $\boldsymbol{A} = (a_{ij})_{n\times n}$ be an additive linguistic preference relation. If we remove the ith line and ith column from $\boldsymbol{A}$, then the preference relation composed by the remainder $(n-1)$ lines and $(n-1)$ columns of $\boldsymbol{A}$ is also an additive linguistic preference relation.

Theorem 3.2 Let $\boldsymbol{A} = (a_{ij})_{n\times n}$ be an additive linguistic preference relation. Then

$$\bigoplus_{i,j=1}^{n} a_{ij} = s_0 \tag{3.2}$$

Proof By (3.1), we have

$$\bigoplus_{i,j=1}^{n} a_{ij} = \left[\bigoplus_{i<j}(a_{ij} \oplus a_{ji})\right] \oplus \left(\bigoplus_{i=1}^{n} a_{ii}\right) = \left(\bigoplus_{i<j} s_0\right) \oplus \left(\bigoplus_{i=1}^{n} s_0\right) = s_0$$

which completes the proof.

Definition 3.1 Let $\boldsymbol{A} = (a_{ij})_{n\times n}$ be an additive linguistic preference relation. If $a_{ij}^{\mathrm{T}} = a_{ji}\ (i,j = 1,2,\cdots,n)$, then $\boldsymbol{A}^{\mathrm{T}} = (a_{ij}^{\mathrm{T}})_{n\times n}$ is called the transpose of $\boldsymbol{A}$.

Theorem 3.3 Let $\boldsymbol{A} = (a_{ij})_{n\times n}$ be an additive linguistic preference relation. Then the transpose $\boldsymbol{A}^{\mathrm{T}} = (a_{ij}^{\mathrm{T}})_{n\times n}$ is also an additive linguistic preference relation.

Proof Since

$$s_{-\tau} \leqslant a_{ij}^{\mathrm{T}} = a_{ji} \leqslant s_{\tau}, \quad a_{ij}^{\mathrm{T}} \oplus a_{ji}^{\mathrm{T}} = a_{ji} \oplus a_{ij} = s_0, \quad a_{ii}^{\mathrm{T}} = a_{ii} = s_0$$

then it should be given as in Theorem 3.3 is also an additive linguistic preference relation. This completes the proof.

Based on the concept of additive linguistic preference relation, we can directly derive the following result:

Theorem 3.4 Let $\boldsymbol{A} = (a_{ij})_{n\times n}$ be an additive linguistic preference relation, and $\boldsymbol{A}^{\mathrm{T}}$ be the transpose of $\boldsymbol{A}$. Then $\boldsymbol{A}^{\mathrm{T}} = -\boldsymbol{A}$, i.e., $\boldsymbol{A}$ is an anti-symmetric matrix.

Definition 3.2 (Xu, 2006c) Let $\boldsymbol{A} = (a_{ij})_{n\times n}$ be an additive linguistic preference relation. If

$$a_{ij} = a_{ik} \oplus a_{kj}, \quad i,j,k = 1,2,\cdots,n \tag{3.3}$$

then $\boldsymbol{A}$ is called a consistent additive linguistic preference relation.

(3.3) is a type of additive transitivity. It represents the idea as follows: Interpreting the value a_{ik} as the intensity of preference of the alternative x_i over x_k, and a_{kj} as the intensity of preference of the alternative x_k over x_j, it is reasonable to suppose that the intensity of preference of x_i over x_j should be equal to the sum of the intensities of preferences when using an intermediate alternative x_k. In particular, if the decision maker provides his/her linguistic preferences over these three alternatives in an indifference situation, that is $x_i \sim x_k \sim x_j$, then it follows that $a_{ik} = a_{kj} = a_{ij} = s_0$, which coincides with (3.3).

Additively, the additive linguistic preference relation $\boldsymbol{A} = (a_{ij})_{n\times n}$ has the following properties:

(1) If $a_{ik} \oplus a_{kj} \geqslant a_{ij}$ for all $i, j, k = 1, 2, \cdots, n$, then we say $\boldsymbol{A}$ satisfies the triangle condition.

The condition can be explained geometrically: If we take the alternatives x_i, x_k and x_j for the vertices of a triangle with length sides a_{ik}, a_{kj} and a_{ij}, then the length corresponding to the vertices x_i, x_j should not exceed the sum of the lengths corresponding to the vertices x_i, x_k and x_k, x_j.

(2) If

$$a_{ik} \geqslant s_0, \quad a_{kj} \geqslant s_0 \Rightarrow a_{ij} \geqslant s_0 \quad \text{for all} \quad i, j, k = 1, 2, \cdots, n$$

then we say $\boldsymbol{A}$ satisfies the weak transitivity property.

This property can be interpreted as follows: If x_i is preferred to x_k, and x_k is preferred to x_j, then x_i should be preferred to x_j.

(3) If $a_{ij} \geqslant \min\{a_{ik}, a_{kj}\}$ for all $i, j, k = 1, 2, \cdots, n$, then we say $\boldsymbol{A}$ satisfies the max-min transitivity property.

The max-min transitivity property is that the linguistic preference value obtained by a direct comparison between two alternatives should be equal to or greater than the minimum partial values obtained by comparing both alternatives with an intermediate one.

(4) If $a_{ij} \geqslant \max\{a_{ik}, a_{kj}\}$ for all $i, j, k = 1, 2, \cdots, n$, then we say $\boldsymbol{A}$ satisfies the max-max transitivity property.

This property can be described as follows: The linguistic preference value obtained by a direct comparison between two alternatives should be equal to or greater than the maximum partial values obtained by comparing both alternatives with an intermediate one.

(5) If

$$a_{ik} \geqslant s_0, \quad a_{kj} \geqslant s_0 \Rightarrow a_{ij} \geqslant \min\{a_{ik}, a_{kj}\} \quad \text{for all} \quad i, j, k = 1, 2, \cdots, n$$

then we say $\boldsymbol{A}$ satisfies the restricted max-min transitivity property.

Wc can interpret the restricted max-min transitivity property in the following way: When an alternative x_i is preferred to x_k with a value a_{ik} and x_k is preferred to x_j with a value a_{kj}, then x_i should be preferred to x_j with at least an intensity of preference a_{ij} equal to the minimum of the above values. With the equality only when there is indifference between at least two of the three alternatives.

(6) If

$$a_{ik} \geqslant s_0, \quad a_{kj} \geqslant s_0 \Rightarrow a_{ij} \geqslant \max\{a_{ik}, a_{kj}\} \quad \text{for all} \quad i,j,k = 1,2,\cdots,n$$

then we say $\boldsymbol{A}$ satisfies the restricted max-max transitivity property.

The restricted max-max transitivity property implies that when an alternative x_i is preferred to x_k with a value a_{ik}, and x_k is preferred to x_j with a value a_{kj}, then x_i should be preferred to x_j with at least an intensity of preference a_{ij} equal to the maximum of the above values. The equality holds only when there is indifference between at least two of the three alternatives.

Theorem 3.5 (Xu, 2005b) Let $\boldsymbol{A}_k = (a_{ij}^{(k)})_{n\times n}$ $(k = 1,2,\cdots,l)$ be the additive linguistic preference relations provided by l decision makers e_k $(k = 1,2,\cdots,l)$, where

$$s_{-\tau} \leqslant a_{ij}^{(k)} \leqslant s_\tau, \quad a_{ij}^{(k)} \oplus a_{ji}^{(k)} = s_0, \quad a_{ii}^{(k)} = s_0, \quad k = 1,2,\cdots,l, \quad i,j = 1,2,\cdots,n$$

$\boldsymbol{\upsilon} = (\upsilon_1, \upsilon_2, \cdots, \upsilon_l)^{\mathrm{T}}$ is the weight vector of the decision makers e_k $(k = 1,2,\cdots,l)$, with $\upsilon_k \geqslant 0\,(k = 1,2,\cdots,l)$ and $\sum\limits_{k=1}^{l} \upsilon_k = 1$. Then the weighted combination $\boldsymbol{A} = (a_{ij})_{n\times n}$ of $\boldsymbol{A}_k = (a_{ij}^{(k)})_{n\times n}$ $(k = 1,2,\cdots,l)$ is also an additive linguistic preference relation, where

$$a_{ij} = \mathrm{LWA}_2(a_{ij}^{(1)}, a_{ij}^{(2)}, \cdots, a_{ij}^{(l)}) = \upsilon_1 a_{ij}^{(1)} \oplus \upsilon_2 a_{ij}^{(2)} \oplus \cdots \oplus \upsilon_l a_{ij}^{(l)}$$

$$s_{-\tau} \leqslant a_{ij} \leqslant s_\tau, \quad a_{ij} \oplus a_{ji} = s_0, \quad a_{ii} = s_0, \quad i,j = 1,2,\cdots,n \tag{3.4}$$

Especially, if all $\boldsymbol{A}_k = (a_{ij}^{(k)})_{n\times n}$ $(k = 1,2,\cdots,l)$ are consistent additive linguistic preference relations, then the weighted combination $\boldsymbol{A} = (a_{ij})_{n\times n}$ is also a consistent additive linguistic preference relation.

Proof Since all $\boldsymbol{A}_k = (a_{ij}^{(k)})_{n\times n}$ $(k = 1,2,\cdots,l)$ are the additive linguistic preference relations, we have

$$\begin{aligned} a_{ij} &= \upsilon_1 a_{ij}^{(1)} \oplus \upsilon_2 a_{ij}^{(2)} \oplus \cdots \oplus \upsilon_l a_{ij}^{(l)} \\ &\geqslant \upsilon_1 s_{-\tau} \oplus \upsilon_2 s_{-\tau} \oplus \cdots \oplus \upsilon_l s_{-\tau} \\ &= (\upsilon_1 + \upsilon_2 + \cdots + \upsilon_l) s_{-\tau} \\ &= s_{-\tau}, \quad i,j = 1,2,\cdots,n \\ a_{ij} &= \upsilon_1 a_{ij}^{(1)} \oplus \upsilon_2 a_{ij}^{(2)} \oplus \cdots \oplus \upsilon_l a_{ij}^{(l)} \\ &\leqslant \upsilon_1 s_\tau \oplus \upsilon_2 s_\tau \oplus \cdots \oplus \upsilon_l s_\tau \\ &= (\upsilon_1 + \upsilon_2 + \cdots + \upsilon_l) s_\tau \\ &= s_\tau, \quad i,j = 1,2,\cdots,n \end{aligned}$$

$$a_{ij} \oplus a_{ji} = [\upsilon_1 a_{ij}^{(1)} \oplus \upsilon_2 a_{ij}^{(2)} \oplus \cdots \oplus \upsilon_l a_{ij}^{(l)}] \oplus [\upsilon_1 a_{ji}^{(1)} \oplus \upsilon_2 a_{ji}^{(2)} \oplus \cdots \oplus \upsilon_l a_{ji}^{(l)}]$$

$$
\begin{aligned}
&= v_1[a_{ij}^{(1)} \oplus a_{ji}^{(1)}] \oplus v_2[a_{ij}^{(2)} \oplus a_{ji}^{(2)}] \oplus \cdots \oplus v_l[a_{ij}^{(l)} \oplus a_{ji}^{(l)}] \\
&= v_1 s_0 \oplus v_2 s_0 \oplus \cdots \oplus v_l s_0 \\
&= (v_1 + v_2 + \cdots + v_l) s_0 \\
&= s_0, \quad i, j = 1, 2, \cdots, n \\
a_{ii} &= v_1 a_{ii}^{(1)} \oplus v_2 a_{ii}^{(2)} \oplus \cdots \oplus v_l a_{ii}^{(l)} \\
&= v_1 s_0 \oplus v_2 s_0 \oplus \cdots \oplus v_l s_0 \\
&= (v_1 + v_2 + \cdots + v_l) s_0 \\
&= s_0, \quad i = 1, 2, \cdots, n
\end{aligned}
$$

Thus, $\boldsymbol{A} = (a_{ij})_{n\times n}$ is an additive linguistic preference relation.

If all $\boldsymbol{A}_k = (a_{ij}^{(k)})_{n\times n}$ $(k = 1, 2, \cdots, l)$ are the consistent additive linguistic preference relations, i.e., $a_{ij}^{(k)} = a_{is}^{(k)} \oplus a_{sj}^{(k)}$ $(i, j, s = 1, 2, \cdots, n,\ k = 1, 2, \cdots, l)$, then we get

$$
\begin{aligned}
a_{ij} &= v_1 a_{ij}^{(1)} \oplus v_2 a_{ij}^{(2)} \oplus \cdots \oplus v_l a_{ij}^{(l)} \\
&= v_1[a_{is}^{(1)} \oplus a_{sj}^{(1)}] \oplus v_2[a_{is}^{(2)} \oplus a_{sj}^{(2)}] \oplus \cdots \oplus v_l[a_{is}^{(n)} \oplus a_{sj}^{(n)}] \\
&= [v_1 a_{is}^{(1)} \oplus v_2 a_{is}^{(2)} \oplus \cdots \oplus v_l a_{is}^{(n)}] \oplus [v_1 a_{sj}^{(1)} \oplus v_2 a_{sj}^{(2)} \oplus \cdots \oplus v_l a_{sj}^{(n)}] \\
&= a_{is} \oplus a_{sj}, \quad i, j, s = 1, 2, \cdots, n
\end{aligned}
$$

Accordingly, the weighted combination $\boldsymbol{A} = (a_{ij})_{n\times n}$ is a consistent additive linguistic preference relation. This completes the proof.

Similar to Chiclana et al. (2003), we can prove the following theorem:

Theorem 3.6 (Xu, 2004a) Let $\boldsymbol{A}_k = (a_{ij}^{(k)})_{n\times n}$ $(k = 1, 2, \cdots, l)$ be the additive linguistic preference relations provided by l decision makers e_k $(k = 1, 2, \cdots, l)$, where

$$
s_{-\tau} \leqslant a_{ij}^{(k)} \leqslant s_\tau, \quad a_{ij}^{(k)} \oplus a_{ji}^{(k)} = s_0, \quad a_{ii}^{(k)} = s_0, \quad k = 1, 2, \cdots, l,\ i, j = 1, 2, \cdots, n
$$

Then the weighted combination $\boldsymbol{A} = (a_{ij})_{n\times n}$ of $\boldsymbol{A}_k = (a_{ij}^{(k)})_{n\times n}$ $(k = 1, 2, \cdots, l)$ is also an additive linguistic preference relation, where

$$
a_{ij} = \mathrm{LOWA}_2(a_{ij}^{(1)}, a_{ij}^{(2)}, \cdots, a_{ij}^{(l)}) = \omega_1 a_{ij}^{\sigma(1)} \oplus \omega_2 a_{ij}^{\sigma(2)} \oplus \cdots \oplus \omega_l a_{ij}^{\sigma(l)}
$$

$$
s_{-\tau} \leqslant a_{ij} \leqslant s_\tau, \quad a_{ji} \oplus a_{ji} = s_0, \quad a_{ii} = s_0, \quad i, j = 1, 2, \cdots, n \tag{3.5}
$$

$(\sigma(1), \sigma(2), \cdots, \sigma(n))$ is any permutation of $(1, 2, \cdots, n)$, such that $a_{ij}^{\sigma(k)} \leqslant a_{ij}^{\sigma(k-1)}$ $(k = 2, \cdots, l)$, $\boldsymbol{\omega} = (\omega_1, \omega_2, \cdots, \omega_l)^{\mathrm{T}}$ can be obtained by

$$
\omega_k = Q\left(\frac{k}{l}\right) - Q\left(\frac{k-1}{l}\right), \quad k = 1, 2, \cdots, l \tag{3.6}
$$

where

$$Q(r)=\begin{cases}0, & 0\leqslant r<a\\ \dfrac{r-a}{b-a}, & a\leqslant r\leqslant b\\ 1, & b<r\leqslant 1\end{cases} \tag{3.7}$$

is a nondecreasing proportional fuzzy linguistic quantifier, $a,b,r\in[0,1]$, $a+b=1$. If all $\boldsymbol{A}_k=(a_{ij}^{(k)})_{n\times n}$ $(k=1,2,\cdots,l)$ are the consistent additive linguistic preference relations, then the weighted combination $\boldsymbol{A}=(a_{ij})_{n\times n}$ is also a consistent additive linguistic preference relation.

Xu (2004a) utilized the LOWA$_2$ and LA operators to develop a group decision making approach based on additive linguistic preference relations, i.e., he first employed the LOWA$_2$ operator to aggregate the additive linguistic preference relations provided by all the decision makers into the collective additive linguistic preference relation based on fuzzy majority, and then used the LA operator to aggregate all the elements of each line in the collective additive linguistic preference relation into the overall preference degree of this alternative over all the alternatives, by which the ranking and selection of the alternatives can be achieved. The approach involves the following steps:

Step 1 For a group decision making problem with additive linguistic preference relations, let $X=\{x_1,x_2,\cdots,x_n\}$ be a finite set of alternatives, $E=\{e_1,e_2,\cdots,e_l\}$ be the set of decision makers. Each decision maker $e_k\in E$ utilizes the additive linguistic evaluation scale $S_2=\{s_\alpha|\alpha=-\tau,\cdots,-1,0,1,\cdots,\tau\}$ to compare each pair alternatives in X, and constructs an additive linguistic preference relation $\boldsymbol{A}_k=(a_{ij}^{(k)})_{n\times n}$, where

$$s_{-\tau}\leqslant a_{ij}^{(k)}\leqslant s_\tau,\quad a_{ij}^{(k)}\oplus a_{ji}^{(k)}=s_0,\quad a_{ii}^{(k)}=s_0,\quad i,j=1,2,\cdots,n,\quad k=1,2,\cdots,l$$

Step 2 Use the LOWA$_2$ operator (3.5) to aggregate the additive linguistic preference relations $\boldsymbol{A}_k=(a_{ij}^{(k)})_{n\times n}$ $(k=1,2,\cdots,l)$ provided by all the decision makers e_k $(k=1,2,\cdots,l)$ into the collective additive linguistic preference relation $\boldsymbol{A}=(a_{ij})_{n\times n}$.

Step 3 Utilize the LA operator:

$$a_i=\mathrm{LA}(a_{i1},a_{i2},\cdots,a_{in}) \tag{3.8}$$

to aggregate all the elements in the ith line of the collective additive linguistic preference relation $\boldsymbol{A}=(a_{ij})_{n\times n}$ into the overall preference degree a_i of the alternative x_i over all the alternatives.

Step 4 Rank and select the alternatives x_i $(i=1,2,\cdots,n)$ according the values a_i $(i=1,2,\cdots,n)$.

Example 3.1 (Xu, 2004e) Consider a group decision making problem that involves the evaluation of five schools (alternatives) x_i $(i=1,2,\cdots,5)$ of a university. One main criterion used is "research". Three decision makers e_k $(k=1,2,3)$ compare each pair of these five alternatives x_i $(i = 1, 2, \cdots, 5)$ with respect to the criterion "research" by using the additive linguistic evaluation scale:

$$\begin{aligned} S_2 = \{ & s_{-4} = \text{extremely poor}, \quad s_{-3} = \text{very poor}, \quad s_{-2} = \text{poor}, \quad s_{-1} = \text{slightly poor}, \\ & s_0 = \text{fair}, \quad s_1 = \text{slightly good}, \quad s_2 = \text{good}, \quad s_3 = \text{very good}, \\ & s_4 = \text{extremely good}\} \end{aligned}$$

and construct the additive linguistic preference relations $\boldsymbol{A}_k$ $(k = 1, 2, 3)$ respectively:

$$\boldsymbol{A}_1 = \begin{pmatrix} s_0 & s_1 & s_{-2} & s_3 & s_{-1} \\ s_{-1} & s_0 & s_4 & s_1 & s_{-2} \\ s_2 & s_{-4} & s_0 & s_2 & s_3 \\ s_{-3} & s_{-1} & s_{-2} & s_0 & s_4 \\ s_1 & s_2 & s_{-3} & s_{-4} & s_0 \end{pmatrix}$$

$$\boldsymbol{A}_2 = \begin{pmatrix} s_0 & s_2 & s_{-1} & s_2 & s_{-2} \\ s_{-2} & s_0 & s_4 & s_2 & s_{-1} \\ s_1 & s_{-4} & s_0 & s_3 & s_2 \\ s_{-2} & s_{-2} & s_{-3} & s_0 & s_4 \\ s_2 & s_1 & s_{-2} & s_{-4} & s_0 \end{pmatrix}$$

$$\boldsymbol{A}_3 = \begin{pmatrix} s_0 & s_3 & s_{-3} & s_1 & s_{-3} \\ s_{-3} & s_0 & s_3 & s_2 & s_{-2} \\ s_3 & s_{-3} & s_0 & s_3 & s_1 \\ s_{-1} & s_{-2} & s_{-3} & s_0 & s_4 \\ s_3 & s_2 & s_{-1} & s_{-4} & s_0 \end{pmatrix}$$

To get the optimal alternative, the following procedure is given:

We first use the LOWA$_2$ operator $a_{ij} = \text{LOWA}_2(a_{ij}^{(1)}, a_{ij}^{(2)}, a_{ij}^{(3)})$ $(i, j = 1, 2, 3, 4, 5)$ (whose weighting vector can be obtained by (3.6) and (3.7). Without loss of generality, we take $a = 0.25$ and $b = 0.75$, and then $\boldsymbol{\omega} = (1/6, 2/3, 1/6)^{\mathrm{T}}$) to aggregate the additive linguistic preference relations $\boldsymbol{A}_k = (a_{ij}^{(k)})_{5\times 5}$ $(k = 1, 2, 3)$ into the collective additive linguistic preference relation $\boldsymbol{A} = (a_{ij})_{5\times 5}$:

$$a_{11} = \frac{1}{6} \times s_0 \oplus \frac{2}{3} \times s_0 \oplus \frac{1}{6} \times s_0 = s_0$$

$$a_{12} = \frac{1}{6} \times s_3 \oplus \frac{2}{3} \times s_2 \oplus \frac{1}{6} \times s_1 = s_2$$

$$a_{13} = \frac{1}{6} \times s_{-1} \oplus \frac{2}{3} \times s_{-2} \oplus \frac{1}{6} \times s_{-3} = s_{-2}$$

$$a_{14} = \frac{1}{6} \times s_3 \oplus \frac{2}{3} \times s_2 \oplus \frac{1}{6} \times s_1 = s_2$$

$$a_{15} = \frac{1}{6} \times s_{-1} \oplus \frac{2}{3} \times s_{-2} \oplus \frac{1}{6} \times s_{-3} = s_{-2}$$

$$a_{21} = \frac{1}{6} \times s_{-1} \oplus \frac{2}{3} \times s_{-2} \oplus \frac{1}{6} \times s_{-3} = s_{-2}$$

$$a_{22} = \frac{1}{6} \times s_0 \oplus \frac{2}{3} \times s_0 \oplus \frac{1}{6} \times s_0 = s_0$$
$$a_{23} = \frac{1}{6} \times s_4 \oplus \frac{2}{3} \times s_4 \oplus \frac{1}{6} \times s_3 = s_{3.83}$$
$$a_{24} = \frac{1}{6} \times s_2 \oplus \frac{2}{3} \times s_2 \oplus \frac{1}{6} \times s_1 = s_{1.83}$$
$$a_{25} = \frac{1}{6} \times s_{-1} \oplus \frac{2}{3} \times s_{-2} \oplus \frac{1}{6} \times s_{-2} = s_{-1.83}$$
$$a_{31} = \frac{1}{6} \times s_3 \oplus \frac{2}{3} \times s_2 \oplus \frac{1}{6} \times s_1 = s_2$$
$$a_{32} = \frac{1}{6} \times s_{-3} \oplus \frac{2}{3} \times s_{-4} \oplus \frac{1}{6} \times s_{-4} = s_{-3.83}$$
$$a_{33} = \frac{1}{6} \times s_0 \oplus \frac{2}{3} \times s_0 \oplus \frac{1}{6} \times s_0 = s_0$$
$$a_{34} = \frac{1}{6} \times s_3 \oplus \frac{2}{3} \times s_2 \oplus \frac{1}{6} \times s_2 = s_{2.83}$$
$$a_{35} = \frac{1}{6} \times s_3 \oplus \frac{2}{3} \times s_2 \oplus \frac{1}{6} \times s_2 = s_{2.83}$$
$$a_{35} = \frac{1}{6} \times s_3 \oplus \frac{2}{3} \times s_2 \oplus \frac{1}{6} \times s_1 = s_2$$
$$a_{41} = \frac{1}{6} \times s_{-1} \oplus \frac{2}{3} \times s_{-2} \oplus \frac{1}{6} \times s_{-3} = s_{-2}$$
$$a_{42} = \frac{1}{6} \times s_{-1} \oplus \frac{2}{3} \times s_{-2} \oplus \frac{1}{6} \times s_{-2} = s_{-1.83}$$
$$a_{43} = \frac{1}{6} \times s_{-2} \oplus \frac{2}{3} \times s_{-3} \oplus \frac{1}{6} \times s_{-3} = s_{-2.83}$$
$$a_{44} = \frac{1}{6} \times s_0 \oplus \frac{2}{3} \times s_0 \oplus \frac{1}{6} \times s_0 = s_0$$
$$a_{45} = \frac{1}{6} \times s_4 \oplus \frac{2}{3} \times s_4 \oplus \frac{1}{6} \times s_4 = s_4$$
$$a_{51} = \frac{1}{6} \times s_3 \oplus \frac{2}{3} \times s_2 \oplus \frac{1}{6} \times s_1 = s_2$$
$$a_{52} = \frac{1}{6} \times s_2 \oplus \frac{2}{3} \times s_2 \oplus \frac{1}{6} \times s_1 = s_{1.83}$$
$$a_{53} = \frac{1}{6} \times s_{-1} \oplus \frac{2}{3} \times s_{-2} \oplus \frac{1}{6} \times s_{-3} = s_{-2}$$
$$a_{54} = \frac{1}{6} \times s_{-4} \oplus \frac{2}{3} \times s_{-4} \oplus \frac{1}{6} \times s_{-4} = s_{-4}$$
$$a_{55} = \frac{1}{6} \times s_0 \oplus \frac{2}{3} \times s_0 \oplus \frac{1}{6} \times s_0 = s_0$$

$$\boldsymbol{A} = \begin{pmatrix} s_0 & s_2 & s_{-2} & s_2 & s_{-2} \\ s_{-2} & s_0 & s_{3.83} & s_{1.83} & s_{-1.83} \\ s_2 & s_{-3.83} & s_0 & s_{2.83} & s_2 \\ s_{-2} & s_{-1.83} & s_{-2.83} & s_0 & s_4 \\ s_2 & s_{1.83} & s_{-2} & s_{-4} & s_0 \end{pmatrix}$$

Then we utilize the LA operator $a_i = \text{LA}(a_{i1}, a_{i2}, a_{i3}, a_{i4}, a_{i5})$ to aggregate all the elements in the ith line of the collective additive linguistic preference relation $\boldsymbol{A} = (a_{ij})_{n\times n}$ into the overall preference degree a_i of the alternative x_i over all the alternatives:

$$a_1 = s_0, \quad a_2 = s_{0.37}, \quad a_3 = s_{0.6}, \quad a_4 = s_{-0.53}, \quad a_5 = s_{-0.43}$$

Consequently,

$$a_3 > a_2 > a_1 > a_5 > a_4$$

from which we rank the alternatives x_i $(i = 1, 2, 3, 4, 5)$:

$$x_3 \succ x_2 \succ x_1 \succ x_5 \succ x_4$$

Hence, the optimal alternative is x_3.

In Chapter 2, we have introduced the LOWA$_2$ operator, whose characteristic is that it first reorders all linguistic arguments in descending order, and then aggregates these ordered liguistic arguments together with the weights of the ordered positions. Based on the LOWA$_2$ operator, Zhang and Xu (2006) gave a linguistic ascending ordered weighted averaging operator:

Definition 3.3 (Zhang and Xu, 2006) Let a mapping LOWA$_3$: $(\bar{S}_2)^n \to \bar{S}_2$. If

$$\text{LOWA}_3(s_{\alpha_1}, s_{\alpha_2}, \cdots, s_{\alpha_n}) = \omega_1 s_{\beta'_1} \oplus \omega_2 s_{\beta'_2} \oplus \cdots \oplus \omega_n s_{\beta'_n} = s_{\bar{\beta}'} \tag{3.9}$$

then LOWA$_3$ is called a linguistic ascending ordered weighted averaging (LOWA$_3$) operator, where $\boldsymbol{\omega} = (\omega_1, \omega_2, \cdots, \omega_n)^{\mathrm{T}}$, and

$$\bar{\beta}' = \sum_{j=1}^{n} \omega_j \beta'_j \tag{3.10}$$

is the weighting vector associated with the LOWA$_3$ operator, $\omega_j \geqslant 0\, (j = 1, 2, \cdots, n)$, $\sum_{j=1}^{n} \omega_j = 1$, and $s_{\beta'_j}$ is the jth smallest of the linguistic arguments $(s_{\alpha_1}, s_{\alpha_2}, \cdots, s_{\alpha_n})$.

Definition 3.4 (Zhang and Xu, 2006) Let $\boldsymbol{\omega} = (\omega_1, \omega_2, \cdots, \omega_n)^{\mathrm{T}}$ be a weighting vector, if $\omega_j = \omega_{n+1-j}$ for all j. Then $\boldsymbol{\omega}$ is called symmetric.

Theorem 3.7 (Zhang and Xu, 2006) For any symmetric weighting vector $\boldsymbol{\omega} = (\omega_1, \omega_2, \cdots, \omega_n)^{\mathrm{T}}$, the LOWA$_2$ and LOWA$_3$ operators are equivalent, i.e.,

$$\text{LOWA}_2(s_{\alpha_1}, s_{\alpha_2}, \cdots, s_{\alpha_n}) = \text{LOWA}_3(s_{\alpha_1}, s_{\alpha_2}, \cdots, s_{\alpha_n}) \tag{3.11}$$

Proof Since the weighting vector $\boldsymbol{\omega} = (\omega_1, \omega_2, \cdots, \omega_n)^{\mathrm{T}}$ is symmetric, i.e., $\omega_j = \omega_{n+1-j}$ for all j, we have

(1) If n is even, then

$$\text{LOWA}_2(s_{\alpha_1}, s_{\alpha_2}, \cdots, s_{\alpha_n}) = \omega_1 s_{\beta'_1} \oplus \omega_2 s_{\beta'_2} \oplus \cdots \oplus \omega_n s_{\beta'_n}$$

$$\begin{aligned}
&= \bigoplus_{j=1}^{\frac{n}{2}} (\omega_j s_{\beta'_j} \oplus \omega_{n+1-j} s_{\beta'_{n+1-j}}) \\
&= \bigoplus_{j=1}^{\frac{n}{2}} (\omega_j s_{\beta'_{n+1-j}} \oplus \omega_{n+1-j} s_{\beta'_j}) \\
&= \omega_1 s_{\beta'_n} \oplus \omega_2 s_{\beta'_{n-1}} \oplus \cdots \oplus \omega_n s_{\beta'_1} \\
&= \mathrm{LOWA}_3(s_{\alpha_1}, s_{\alpha_2}, \cdots, s_{\alpha_n})
\end{aligned}$$

(2) If n is odd, then

$$\begin{aligned}
\mathrm{LOWA}_2(s_{\alpha_1}, s_{\alpha_2}, \cdots, s_{\alpha_n}) &= \omega_1 s_{\beta'_1} \oplus \omega_2 s_{\beta'_2} \oplus \cdots \oplus \omega_n s_{\beta'_n} \\
&= \bigoplus_{j=1}^{\frac{n-1}{2}} (\omega_j s_{\beta'_j} \oplus \omega_{n+1-j} s_{\beta'_{n+1-j}}) \oplus \omega_{(n+1)/2} s_{\beta_{(n+1)/2}} \\
&= \bigoplus_{j=1}^{\frac{n-1}{2}} (\omega_j s_{\beta'_{n+1-j}} \oplus \omega_{n+1-j} s_{\beta'_j}) \oplus \omega_{(n+1)/2} s_{\beta_{(n+1)/2}} \\
&= \omega_1 s_{\beta'_n} \oplus \omega_2 s_{\beta'_{n-1}} \oplus \cdots \oplus \omega_n s_{\beta'_1} \\
&= \mathrm{LOWA}_3(s_{\alpha_1}, s_{\alpha_2}, \cdots, s_{\alpha_n})
\end{aligned}$$

Thus, (3.11) holds for any symmetric weighting vector $\boldsymbol{\omega} = (\omega_1, \omega_2, \cdots, \omega_n)^{\mathrm{T}}$.

Zhang and Xu (2006) applied the LOWA_2 operator with the symmetric weighting vector to group decision making:

Theorem 3.8 (Zhang and Xu, 2006) Assume that the decision maker $e_k \in E$ compares each pair of alternatives in X by using the additive linguistic evaluation scale S_2, and constructs the additive linguistic preference relations $\boldsymbol{A}_k = (a_{ij}^{(k)})_{n\times n}$ $(k = 1, 2, \cdots, l)$, where $s_{-\tau} \leqslant a_{ij}^{(k)} \leqslant s_\tau$, $a_{ij}^{(k)} \oplus a_{ji}^{(k)} = s_0$, $a_{ii}^{(k)} = s_0$ $(i, j = 1, 2, \cdots, n)$. Then the weighted combination $\boldsymbol{A} = (a_{ij})_{n\times n}$ derived from $\boldsymbol{A}_k = (a_{ij}^{(k)})_{n\times n}$ $(k = 1, 2, \cdots, l)$ by using the LOWA_2 operator with the symmetric weighting vector is also an additive linguistic preference relation.

Proof Let $\boldsymbol{\omega} = (\omega_1, \omega_2, \cdots, \omega_l)^{\mathrm{T}}$ be the weighting vector associated with the LOWA_2 operator. Using the OWA_2 operator to aggregate the additive linguistic preference relation $\boldsymbol{A}_k = (a_{ij}^{(k)})_{n\times n}$ $(k = 1, 2, \cdots, l)$, we get $\boldsymbol{A} = (a_{ij})_{n\times n}$, where

$$a_{ij} = \mathrm{LOWA}_2(a_{ij}^{(1)}, a_{ij}^{(2)}, \cdots, a_{ij}^{(l)}) = \omega_1 b_{ij}^{(1)} \oplus \omega_2 b_{ij}^{(2)} \oplus \cdots \oplus \omega_l b_{ij}^{(l)}, \quad i, j = 1, 2, \cdots, n$$

$b_{ij}^{(k)}$ is the kth largest of the linguistic arguments $(a_{ij}^{(1)}, a_{ij}^{(2)}, \cdots, a_{ij}^{(l)})$. Since $b_{ij}^{(k)} \oplus b_{ji}^{(k)} = s_0$, then $b_{ji}^{(k)}$ is the k smallest of the linguistic arguments $(a_{ji}^{(1)}, a_{ji}^{(2)}, \cdots, a_{ji}^{(l)})$. Also since the weighting vector $\boldsymbol{\omega} = (\omega_1, \omega_2, \cdots, \omega_l)^{\mathrm{T}}$ is symmetric, i.e., $\omega_k = \omega_{n+1-k}$ for any k, it yields

$$a_{ij} \oplus a_{ji} = \mathrm{LOWA}_2(a_{ij}^{(1)}, a_{ij}^{(2)}, \cdots, a_{ij}^{(l)}) \oplus \mathrm{LOWA}_2(a_{ji}^{(1)}, a_{ji}^{(2)}, \cdots, a_{ji}^{(l)})$$

$$
\begin{aligned}
&= [\omega_1 b_{ij}^{(1)} \oplus \omega_2 b_{ij}^{(2)} \oplus \cdots \oplus \omega_l b_{ij}^{(l)}] \oplus [\omega_l b_{ji}^{(1)} \oplus \omega_{l-1} b_{ji}^{(2)} \oplus \cdots \oplus \omega_1 b_{ji}^{(l)}] \\
&= [\omega_1 b_{ij}^{(1)} \oplus \omega_2 b_{ij}^{(2)} \oplus \cdots \oplus \omega_l b_{ij}^{(l)}] \oplus [\omega_1 b_{ji}^{(1)} \oplus \omega_2 b_{ji}^{(2)} \oplus \cdots \oplus \omega_l b_{ji}^{(l)}] \\
&= \omega_1 [b_{ij}^{(1)} \oplus b_{ji}^{(1)}] \oplus \omega_2 [b_{ij}^{(2)} \oplus b_{ji}^{(2)}] \oplus \cdots \oplus \omega_l [b_{ij}^{(l)} \oplus b_{ji}^{(l)}] \\
&= \omega_1 s_0 \oplus \omega_2 s_0 \oplus \cdots \oplus \omega_l s_0 \\
&= (\omega_1 + \omega_2 + \cdots + \omega_l) s_0 \\
&= s_0, \quad i, j = 1, 2, \cdots, n
\end{aligned}
$$

Considering that $s_{-\tau} \leqslant a_{ij}^{(k)} \leqslant s_\tau$ for all i, j, and $a_{ii}^{(k)} = s_0$, we have

$$
\begin{aligned}
a_{ij} &= \mathrm{LOWA}_2(a_{ij}^{(1)}, a_{ij}^{(2)}, \cdots, a_{ij}^{(l)}) = \omega_1 b_{ij}^{(1)} \oplus \omega_2 b_{ij}^{(2)} \oplus \cdots \oplus \omega_l b_{ij}^{(l)} \\
&\geqslant \omega_1 s_{-\tau} \oplus \omega_2 s_{-\tau} \oplus \cdots \oplus \omega_l s_{-\tau} \\
&= (\omega_1 + \omega_2 + \cdots + \omega_l) s_{-\tau} = s_{-\tau}, \quad i, j = 1, 2, \cdots, n \\
a_{ij} &= \mathrm{LOWA}_2(a_{ij}^{(1)}, a_{ij}^{(2)}, \cdots, a_{ij}^{(l)}) = \omega_1 b_{ij}^{(1)} \oplus \omega_2 b_{ij}^{(2)} \oplus \cdots \oplus \omega_l b_{ij}^{(l)} \\
&\leqslant \omega_1 s_{\tau} \oplus \omega_2 s_{\tau} \oplus \cdots \oplus \omega_l s_{\tau} \\
&= (\omega_1 + \omega_2 + \cdots + \omega_l) s_{\tau} = s_{\tau}, \quad i, j = 1, 2, \cdots, n \\
a_{ii} &= \mathrm{LOWA}_2(a_{ii}^{(1)}, a_{ii}^{(2)}, \cdots, a_{ii}^{(l)}) = \omega_1 b_{ii}^{(1)} \oplus \omega_2 b_{ii}^{(2)} \oplus \cdots \oplus \omega_l b_{ii}^{(l)} \\
&= \omega_1 s_0 \oplus \omega_2 s_0 \oplus \cdots \oplus \omega_l s_0 \\
&= (\omega_1 + \omega_2 + \cdots + \omega_l) s_0 = s_0, \quad i, j = 1, 2, \cdots, n
\end{aligned}
$$

and then the weighted combination $\boldsymbol{A} = (a_{ij})_{n\times n}$ is an additive linguistic preference relation. This completes the proof.

In what follows, we introduce some common symmetric weighting vectors (Zhang and Xu, 2006) for the convenience of practical applications:

(1) Let $\boldsymbol{\omega} = (\omega_1, \omega_2, \cdots, \omega_n)^{\mathrm{T}}$ be the weighting vector, where (Xu, 2005c)

$$
\omega_i = \frac{\mathrm{e}^{-\frac{\left(i-\frac{1+n}{2}\right)^2}{2\sigma_n^2}}}{\sum_{j=1}^{n} \mathrm{e}^{-\frac{\left(j-\frac{1+n}{2}\right)^2}{2\sigma_n^2}}}, \quad \sigma_n = \sqrt{\frac{1}{n}\sum_{i=1}^{n}\left(i - \frac{1+n}{2}\right)^2}, \quad i = 1, 2, \cdots, n \tag{3.12}
$$

then $\boldsymbol{\omega}$ is symmetric.

Proof By (3.12), we have

$$
\omega_{n+1-i} = \frac{\mathrm{e}^{-\frac{\left(n+1-i-\frac{1+n}{2}\right)^2}{2\sigma_n^2}}}{\sum_{j=1}^{n} \mathrm{e}^{-\frac{\left(n+1-j-\frac{1+n}{2}\right)^2}{2\sigma_n^2}}} = \frac{\mathrm{e}^{-\frac{\left[-\left(i-\frac{1+n}{2}\right)\right]^2}{2\sigma_n^2}}}{\sum_{j=1}^{n} \mathrm{e}^{-\frac{\left[-\left(j-\frac{1+n}{2}\right)\right]^2}{2\sigma_n^2}}}
$$

$$= \frac{\mathrm{e}^{-\frac{\left(i-\frac{1+n}{2}\right)^2}{2\sigma_n^2}}}{\sum_{j=1}^{n} \mathrm{e}^{-\frac{\left(j-\frac{1+n}{2}\right)^2}{2\sigma_n^2}}} = \omega_i, \quad i = 1, 2, \cdots, n$$

which completes the proof.

(2) Let $\boldsymbol{\omega} = (\omega_1, \omega_2, \cdots, \omega_n)^{\mathrm{T}}$ be the weighting vector, where ω_k satisfies (3.6) and (3.7). Then $\boldsymbol{\omega}$ is symmetric if and only if $a + b = 1$ see Chiclana et al. (2003) for details. Therefore, Theorem 3.6 is a special of Theorem 3.8.

(3) Let $\boldsymbol{\omega} = (\omega_1, \omega_2, \cdots, \omega_n)^{\mathrm{T}}$ be the weighting vector, where (Yager, 1993)

$$\omega_i = \begin{cases} 0, & i < k \\ \frac{1}{m}, & k \leqslant i \leqslant k + m \\ 0, & i \geqslant k + m \end{cases} \tag{3.13}$$

where k and m are the positive integers, and $k + m \leqslant n + 1$. Then $\boldsymbol{\omega}$ is symmetric if and only if $2k + m - 1 = n + 1$.

Proof **Sufficiency** (i) Since $\omega_i = 0$ for $i < k$, and

$$n + 1 - i = 2k + m - i > 2k + m - 1 - k = k + m - 1$$

i.e., $\omega_{n+1-i} = 0$ for $n + 1 - i \geqslant k + m$, we get $\omega_i = \omega_{n+1-i}$.

(ii) Since $\omega_i = \dfrac{1}{m}$ for $k \leqslant i \leqslant k + m$, we have

$$n + 1 - i = 2k + m - 1 - i > 2k + m - 1 - (k + m) = k - 1$$

i.e., $n + 1 - i \geqslant k$. Also since

$$n + 1 - i = 2k + m - 1 - i \leqslant 2k + m - 1 - k = k + m - 1$$

i.e., $n + 1 - i < k + m$, we have $\omega_{n+1-i} = \dfrac{1}{m}$ for $k \leqslant n + 1 - i < k + m$, i.e., $\omega_i = \omega_{n+1-i}$.

(iii) Since $\omega_i = 0$ for $k \geqslant k + m$, then

$$n + 1 - i = 2k + m - 1 - i \leqslant 2k + m - 1 - (k + m) = k - 1$$

i.e., for $n + 1 - i < k$, $\omega_{n+1-i} = 0$. Accordingly, $\omega_i = \omega_{n+1-i}$.

Necessity Since $\omega_i = \omega_{n+1-i}$, then

(i) If $k = 1$, then we have

$$\omega_i = \begin{cases} 0, & i < 1 \\ \frac{1}{m}, & 1 \leqslant i \leqslant 1 + m \\ 0, & i \geqslant 1 + m \end{cases}$$

where m is a positive integer, and $1+m \leqslant n+1$. Since $\omega_1 = \frac{1}{m}$, then $\omega_n = \omega_1 = \frac{1}{m}$, i.e.,

$$m = n, \quad 2k + m - 1 = 2 + m - 1 = 2 + n - 1 = n + 1$$

(ii) If $k \geqslant 2$, then we have

$$\omega_1 = \omega_2 = \cdots = \omega_{k-1} = 0, \quad \omega_k = \omega_{k+1} = \cdots = \omega_{k+m-1} = \frac{1}{m}$$

$$\omega_{k+m} = \omega_{k+m+1} = \cdots = \omega_n = 0$$

Since $\omega_i = \omega_{n+1-i}$, then $k-1 = n-(k+m-1)$ and $k+(k+m-1) = n+1$, i.e.,

$$2k + m - 1 = n + 1$$

(4) Let $\boldsymbol{\omega} = (\omega_1, \omega_2, \cdots, \omega_m)^{\mathrm{T}}$ be the weighting vector, where (Yager, 1993)

$$\omega_i = \begin{cases} 0, & i < k - m \\ \dfrac{1}{2m+1}, & k - m \leqslant i \leqslant k + m \\ 0, & i \geqslant k + m \end{cases} \tag{3.14}$$

where k and m are the positive integers, and $k + m \leqslant n + 1, k \geqslant m + 1$. Then $\boldsymbol{\omega}$ is symmetric if and only if n is odd and $k = \frac{n+1}{2}$.

Proof **Sufficiency** If n is odd and $k = \frac{n+1}{2}$, then

$$(k - m - 1) + (k + m + 1) = 2k = 2 \cdot \frac{n+1}{2} = n + 1$$

and $k - m + (k + m) = 2k = n + 1$. Thus $\boldsymbol{\omega}$ is symmetric.

Necessity If $\omega_i = \omega_{n+1-i}$, then

$$(k - m - 1) + (k + m + 1) = n + 1 \quad \text{and} \quad k - m + (k + m) = n + 1$$

i.e., $k = \frac{n+1}{2}$. Thus, if n is even, then $n+1$ is odd, and $\frac{n+1}{2}$ is not an integer, which contradicts the condition that k is an integer. Therefore, the weighting vector $\boldsymbol{\omega}$ is symmetric only if n is odd and $k = \frac{n+1}{2}$.

(5) Let $\boldsymbol{\omega} = (\omega_1, \omega_2, \cdots, \omega_n)^{\mathrm{T}}$ be the weighting vector, where (Yager, 1993)

$$\omega_i = \begin{cases} \dfrac{1}{k}, & i \leqslant k \\ 0, & i > k \end{cases} \tag{3.15}$$

Then $\boldsymbol{\omega}$ is symmetric if and only if $k = n$.

Proof **Sufficiency** If $k = n$, then $\omega_i = \dfrac{1}{n}$ for any i. Thus

$$\omega_i = \omega_{n+1-i} = \frac{1}{n} \quad \text{for any} \quad i$$

Consequently, $\boldsymbol{\omega}$ is symmetric.

Necessity If $\boldsymbol{\omega}$ is symmetric, i.e., $\omega_i = \omega_{n+1-i}$, then $\omega_n = \omega_1 = \dfrac{1}{k}$. Thus $k = n$.

Similar to (5), we can prove the following result:

(6) Let $\boldsymbol{\omega} = (\omega_1, \omega_2, \cdots, \omega_n)^{\mathrm{T}}$ be the weighting vector, where (Yager, 1993)

$$\omega_i = \begin{cases} 0, & i < k \\ \dfrac{1}{(n+1)-k}, & i \geqslant k \end{cases} \tag{3.16}$$

Then $\boldsymbol{\omega}$ is symmetric if and only if $k = 1$.

Xu (2005b) studied the similarity property of additive linguistic preference relations:

Definition 3.5 (Xu, 2005b) Let s_α, $s_\beta \in S_2$. Then

$$d(s_\alpha, s_\beta) = \frac{|\alpha - \beta|}{2\tau} \tag{3.17}$$

is called the deviation degree between s_α and s_β, where 2τ is the number of linguistic labels in the additive linguistic evaluation scale S_2.

Clearly, the smaller $d(s_\alpha, s_\beta)$, the smaller the deviation degree between s_α and s_β. In particular, if $d(s_\alpha, s_\beta) = 0$, then $s_\alpha = s_\beta$.

Theorem 3.9 (Xu, 2005b) Let s_α, $s_\beta \in S_2$. Then

$$0 \leqslant d(s_\alpha, s_\beta) \leqslant 1 \tag{3.18}$$

Proof Since $-\tau \leqslant \alpha, \beta \leqslant \tau$, then

$$0 \leqslant \frac{|\alpha - \beta|}{2\tau} \leqslant 1 \tag{3.19}$$

i.e., (3.18) holds, which completes the proof.

Definition 3.6 (Xu, 2005b) Let s_α, $s_\beta \in S_2$. Then

$$\rho(s_\alpha, s_\beta) = 1 - d(s_\alpha, s_\beta) = 1 - \frac{|\alpha - \beta|}{2\tau} \tag{3.20}$$

is called the similarity degree between s_α and s_β.

Theorem 3.10 (Xu, 2005b) Let s_α, $s_\beta \in S_2$. Then

$$0 \leqslant \rho(s_\alpha, s_\beta) \leqslant 1 \tag{3.21}$$

Proof Since $0 \leqslant d(s_\alpha, s_\beta) \leqslant 1$, we have

$$0 \leqslant 1 - d(s_\alpha, s_\beta) \leqslant 1 \tag{3.22}$$

i.e., (3.21) holds. This completes the proof.

Definition 3.7 (Xu, 2005b) Let $\boldsymbol{A} = (a_{ij})_{n\times n}$ and $\boldsymbol{B} = (b_{ij})_{n\times n}$ be two additive linguistic preference relations. Then

$$d(\boldsymbol{A}, \boldsymbol{B}) = \frac{1}{n^2}\sum_{i=1}^{n}\sum_{j=1}^{n} d(a_{ij}, b_{ij}) \tag{3.23}$$

is called the deviation degree between $\boldsymbol{A}$ and $\boldsymbol{B}$.

If we denote the subscripts of the linguistic labels a_{ij} and b_{ij} as $I(a_{ij})$ and $I(b_{ij})$ respectively, then (3.23) can be expressed as

$$d(\boldsymbol{A}, \boldsymbol{B}) = \frac{1}{n^2}\sum_{i=1}^{n}\sum_{j=1}^{n} \frac{|I(a_{ij}) - I(b_{ij})|}{2\tau} \tag{3.24}$$

For example, if $a_{23} = s_2$, then $I(a_{23}) = 2$.

Theorem 3.11 (Xu, 2005b) Let $\boldsymbol{A}$, $\boldsymbol{B}$ and $\boldsymbol{C}$ be three additive linguistic preference relations. Then

(1) $0 \leqslant d(\boldsymbol{A}, \boldsymbol{B}) \leqslant 1$. In particular, $d(\boldsymbol{A}, \boldsymbol{B}) = 0 \Leftrightarrow \boldsymbol{A} = \boldsymbol{B}$;

(2) $d(\boldsymbol{A}, \boldsymbol{B}) = d(\boldsymbol{B}, \boldsymbol{A})$;

(3) $d(\boldsymbol{A}, \boldsymbol{C}) \leqslant d(\boldsymbol{A}, \boldsymbol{B}) + d(\boldsymbol{B}, \boldsymbol{C})$.

Proof (1) Since $0 \leqslant d(a_{ij}, b_{ij}) \leqslant 1$ for all i, j, then

$$0 = \frac{0}{n^2} = \frac{1}{n^2}\sum_{i=1}^{n}\sum_{j=1}^{n} d(a_{ij}, b_{ij}) \leqslant \frac{n^2}{n^2} = 1$$

i.e.,

$$0 \leqslant d(\boldsymbol{A}, \boldsymbol{B}) \leqslant 1$$

In particular,

$$\begin{aligned} d(\boldsymbol{A}, \boldsymbol{B}) = 0 &\Leftrightarrow d(a_{ij}, b_{ij}) = 0\, (i, j = 1, 2, \cdots, n) \\ &\Leftrightarrow a_{ij} = b_{ij}\, (i, j = 1, 2, \cdots, n) \\ &\Leftrightarrow \boldsymbol{A} = \boldsymbol{B} \end{aligned}$$

(2) It follows from (3.24) that

$$d(\boldsymbol{A}, \boldsymbol{B}) = \frac{1}{n^2}\sum_{i=1}^{n}\sum_{j=1}^{n} \frac{|I(a_{ij}) - I(b_{ij})|}{2\tau}$$

$$= \frac{1}{n^2}\sum_{i=1}^{n}\sum_{j=1}^{n}\frac{|I(b_{ij})-I(a_{ij})|}{2\tau}$$
$$= d(\boldsymbol{B},\boldsymbol{A})$$

(3) Since

$$\begin{aligned}
d(a_{ij},c_{ij}) &= \frac{|I(a_{ij})-I(c_{ij})|}{2\tau}\\
&= \frac{|(I(a_{ij})-I(b_{ij}))+(I(b_{ij})-I(c_{ij}))|}{2\tau}\\
&\leqslant \frac{|I(a_{ij})-I(b_{ij})|+|I(b_{ij})-I(c_{ij})|}{2\tau}\\
&= \frac{|I(a_{ij})-I(b_{ij})|}{2\tau}+\frac{|I(b_{ij})-I(c_{ij})|}{2\tau}\\
&= d(a_{ij},b_{ij})+d(b_{ij},c_{ij})
\end{aligned}$$

we get

$$\begin{aligned}
d(\boldsymbol{A},\boldsymbol{C}) &= \frac{1}{n^2}\sum_{i=1}^{n}\sum_{j=1}^{n}d(a_{ij},c_{ij})\\
&\leqslant \frac{1}{n^2}\sum_{i=1}^{n}\sum_{j=1}^{n}[d(a_{ij},b_{ij})+d(b_{ij},c_{ij})]\\
&= \frac{1}{n^2}\sum_{i=1}^{n}\sum_{j=1}^{n}d(a_{ij},b_{ij})+\frac{1}{n^2}\sum_{i=1}^{n}\sum_{j=1}^{n}d(b_{ij},c_{ij})\\
&= d(\boldsymbol{A},\boldsymbol{B})+d(\boldsymbol{B},\boldsymbol{C})
\end{aligned}$$

which completes the proof.

Definition 3.8 (Xu, 2005b) Let $\boldsymbol{A}=(a_{ij})_{n\times n}$ and $\boldsymbol{B}=(b_{ij})_{n\times n}$ be two additive linguistic preference relations. Then

$$\rho(\boldsymbol{A},\boldsymbol{B}) = 1-d(\boldsymbol{A},\boldsymbol{B}) = 1-\frac{1}{n^2}\sum_{i=1}^{n}\sum_{j=1}^{n}\frac{|I(a_{ij})-I(b_{ij})|}{2\tau} \tag{3.25}$$

is called the similarity degree between $\boldsymbol{A}$ and $\boldsymbol{B}$.

Theorem 3.12 (Xu, 2005b) Let $\boldsymbol{A}$, $\boldsymbol{B}$ and $\boldsymbol{C}$ be three additive linguistic preference relations. Then

(1) $0\leqslant\rho(\boldsymbol{A},\boldsymbol{B})\leqslant 1$. Especially, $\rho(\boldsymbol{A},\boldsymbol{B})=1\Leftrightarrow\boldsymbol{A}=\boldsymbol{B}$;

(2) $\rho(\boldsymbol{A},\boldsymbol{B})=\rho(\boldsymbol{B},\boldsymbol{A})$.

Proof (1) It follows from Theorem 3.11 that $0\leqslant d(\boldsymbol{A},\boldsymbol{B})\leqslant 1$. Thus, $0\leqslant 1-d(\boldsymbol{A},\boldsymbol{B})\leqslant 1$, i.e., $0\leqslant\rho(\boldsymbol{A},\boldsymbol{B})\leqslant 1$. Also since $d(\boldsymbol{A},\boldsymbol{B})=0\Leftrightarrow\boldsymbol{A}=\boldsymbol{B}$, we have

$$\rho(\boldsymbol{A},\boldsymbol{B}) = 1-d(\boldsymbol{A},\boldsymbol{B}) = 1\Leftrightarrow\boldsymbol{A}=\boldsymbol{B}$$

(2) By Theorem 3.11, we have $d(\boldsymbol{A}, \boldsymbol{B}) = d(\boldsymbol{B}, \boldsymbol{A})$, and then

$$\rho(\boldsymbol{A}, \boldsymbol{B}) = 1 - d(\boldsymbol{A}, \boldsymbol{B}) = 1 - d(\boldsymbol{B}, \boldsymbol{A}) = \rho(\boldsymbol{B}, \boldsymbol{A})$$

which completes the proof.

Theorem 3.13 (Xu, 2005b) Let $\boldsymbol{A}_k = (a_{ij}^{(k)})_{n\times n}$ $(k = 1, 2, \cdots, l)$ be l additive linguistic preference relations provided by the decision makers e_k $(k = 1, 2, \cdots, l)$, and $\boldsymbol{v} = (v_1, v_2, \cdots, v_l)^{\mathrm{T}}$ be the weight vector of the decision makers e_k $(k = 1, 2, \cdots, l)$, and let $\boldsymbol{A} = (a_{ij})_{n\times n}$ be the weighted combination derived by (3.4). Then

$$d(\boldsymbol{A}, \boldsymbol{A}_k) \leqslant \max_s d(\boldsymbol{A}_s, \boldsymbol{A}_k), \quad k = 1, 2, \cdots, l \tag{3.26}$$

Proof

$$\begin{aligned}
d(\boldsymbol{A}, \boldsymbol{A}_k) &= d(v_1\boldsymbol{A}_1 \oplus v_2\boldsymbol{A}_2 \oplus \cdots \oplus v_l\boldsymbol{A}_l, \boldsymbol{A}_k)\\
&= \frac{1}{n^2}\sum_{i=1}^{n}\sum_{j=1}^{n} d\left(\sum_{s=1}^{l} v_s a_{ij}^{(s)}, a_{ij}^{(k)}\right)\\
&= \frac{1}{n^2}\sum_{i=1}^{n}\sum_{j=1}^{n} d\left(\sum_{s=1}^{l} v_s a_{ij}^{(s)}, \sum_{s=1}^{l} v_s a_{ij}^{(k)}\right)\\
&= \frac{1}{n^2}\sum_{i=1}^{n}\sum_{j=1}^{n} \frac{\left|\sum_{s=1}^{l} v_s I(a_{ij}^{(s)}), \sum_{s=1}^{l} v_s I(a_{ij}^{(k)})\right|}{2\tau}\\
&= \frac{1}{n^2}\sum_{i=1}^{n}\sum_{j=1}^{n} \frac{\left|\sum_{s=1}^{l} v_s [I(a_{ij}^{(s)}) - I(a_{ij}^{(k)})]\right|}{2\tau}\\
&\leqslant \frac{1}{n^2}\sum_{i=1}^{n}\sum_{j=1}^{n}\sum_{s=1}^{l} v_s \frac{\left|I(a_{ij}^{(s)}) - I(a_{ij}^{(k)})\right|}{2\tau}\\
&= \sum_{s=1}^{l} v_s \left[\frac{1}{n^2}\sum_{i=1}^{n}\sum_{j=1}^{n} \frac{\left|I(a_{ij}^{(s)}) - I(a_{ij}^{(k)})\right|}{2\tau}\right]\\
&= \sum_{s=1}^{l} v_s d(\boldsymbol{A}_s, \boldsymbol{A}_k)\\
&\leqslant \sum_{s=1}^{l} v_s \max_s d(\boldsymbol{A}_s, \boldsymbol{A}_k)\\
&= \max_s d(\boldsymbol{A}_s, \boldsymbol{A}_k) \sum_{s=1}^{l} v_s\\
&= \max_s d(\boldsymbol{A}_s, \boldsymbol{A}_k)
\end{aligned}$$

which completes the proof.

Definition 3.9 (Xu, 2005b) Let $\boldsymbol{A} = (a_{ij})_{n\times n}$ and $\boldsymbol{B} = (b_{ij})_{n\times n}$ be two additive linguistic preference relations. If

$$d(\boldsymbol{A}, \boldsymbol{B}) \leqslant \alpha_0 \tag{3.27}$$

then $\boldsymbol{A}$ and $\boldsymbol{B}$ are called of acceptable deviation degree. In actual applications, we generally take $\alpha_0 = 0.15$.

By Theorem 3.13, we get the following two corollaries:

Corollary 3.1 Let $\boldsymbol{A}_k = (a_{ij}^{(k)})_{n\times n}$ $(k = 1, 2, \cdots, l)$ and $\boldsymbol{B} = (b_{ij})_{n\times n}$ be $l+1$ additive linguistic preference relations provided by the decision makers e_k $(k = 1, 2, \cdots, l)$ and e. Let $\boldsymbol{v} = (v_1, v_2, \cdots, v_l)^{\mathrm{T}}$ be the weight vector of the decision makers e_k $(k = 1, 2, \cdots, l)$, $v_k \geqslant 0\,(k = 1, 2, \cdots, l)$, $\sum_{k=1}^{l} v_k = 1$, and $\boldsymbol{A} = (a_{ij})_{n\times n}$ be the weighted combination derived by (3.4). If $d(\boldsymbol{A}_k, \boldsymbol{B}) \leqslant \alpha_0\,(k = 1, 2, \cdots, l)$, then $d(\boldsymbol{A}, \boldsymbol{B}) \leqslant \alpha_0$.

Corollary 3.2 Let $\boldsymbol{A}_k = (a_{ij}^{(k)})_{n\times n}$ $(k = 1, 2, \cdots, l)$ and $\boldsymbol{B} = (b_{ij})_{n\times n}$ be $l+1$ additive linguistic preference relations provided by the decision makers e_k $(k = 1, 2, \cdots, l)$ and e. Let $\boldsymbol{v} = (v_1, v_2, \cdots, v_l)^{\mathrm{T}}$ be the weight vector of the decision makers e_k $(k = 1, 2, \cdots, l)$, $v_k \geqslant 0\,(k = 1, 2, \cdots, l)$, $\sum_{k=1}^{l} v_k = 1$, and $\boldsymbol{A} = (a_{ij})_{n\times n}$ be the weighted combination derived by (3.4). If $d(\boldsymbol{A}_k, \boldsymbol{B}) = 0\,(k = 1, 2, \cdots, l)$, then $d(\boldsymbol{A}, \boldsymbol{B}) = 0$.

Example 3.2 (Xu, 2005b) A group decision making problem with linguistic preference relations involves the investment of an investment company. Let us suppose an investment company, which wants to invest a sum of money in the best option (adapted from Herrera et al. (1995)). There is a panel with five possible alternatives in which to invest the money:

(1) x_1 is a car industry;

(2) x_2 is a food industry;

(3) x_3 is a computer industry;

(4) x_4 is an arms industry;

(5) x_5 is a TV company.

One main criterion used is "growth analysis". There are five decision makers e_k $(k = 1, 2, \cdots, 5)$. The decision makers compare these five companies with respect to the criterion "growth analysis" by using the additive linguistic evaluation scale:

$$S_2 = \{s_{-4} = \text{extremely poor},\ \ s_{-3} = \text{very poor},\ \ s_{-2} = \text{poor},\ \ s_{-1} = \text{slightly poor},$$

$$s_0 = \text{fair},\ \ s_1 = \text{slightly good},\ \ s_2 = \text{good},\ \ s_3 = \text{very good},\ \ s_4 = \text{extremely good}\}$$

and construct, respectively, the additive linguistic preference relations. Suppose that the additive linguistic preference relation $\boldsymbol{B}$ is given by a leading decision maker e_1, and the linguistic preference relations $\boldsymbol{A}_k$ $(k = 1, 2, 3, 4)$ are given by the other decision makers e_k $(k = 2, 3, 4, 5)$ respectively:

$$\boldsymbol{B} = \begin{pmatrix} s_0 & s_0 & s_2 & s_{-1} & s_4 \\ s_0 & s_0 & s_{-1} & s_0 & s_3 \\ s_{-2} & s_1 & s_0 & s_2 & s_1 \\ s_1 & s_0 & s_{-2} & s_0 & s_2 \\ s_{-4} & s_{-3} & s_{-1} & s_{-2} & s_0 \end{pmatrix}$$

$$\boldsymbol{A}_1 = \begin{pmatrix} s_0 & s_{-1} & s_3 & s_{-1} & s_3 \\ s_1 & s_0 & s_1 & s_0 & s_2 \\ s_{-3} & s_{-1} & s_0 & s_{-1} & s_2 \\ s_1 & s_0 & s_1 & s_0 & s_0 \\ s_{-3} & s_{-2} & s_{-2} & s_0 & s_0 \end{pmatrix}$$

$$\boldsymbol{A}_2 = \begin{pmatrix} s_0 & s_1 & s_2 & s_0 & s_4 \\ s_{-1} & s_0 & s_{-1} & s_0 & s_0 \\ s_{-2} & s_1 & s_0 & s_{-1} & s_3 \\ s_0 & s_0 & s_1 & s_0 & s_1 \\ s_{-4} & s_0 & s_{-3} & s_{-1} & s_0 \end{pmatrix}$$

$$\boldsymbol{A}_3 = \begin{pmatrix} s_0 & s_0 & s_3 & s_1 & s_3 \\ s_0 & s_0 & s_{-2} & s_2 & s_2 \\ s_{-3} & s_2 & s_0 & s_1 & s_1 \\ s_{-1} & s_{-2} & s_{-1} & s_0 & s_{-1} \\ s_{-3} & s_2 & s_{-1} & s_1 & s_0 \end{pmatrix}$$

$$\boldsymbol{A}_4 = \begin{pmatrix} s_0 & s_2 & s_0 & s_{-1} & s_2 \\ s_{-2} & s_0 & s_{-1} & s_1 & s_0 \\ s_0 & s_1 & s_0 & s_{-1} & s_2 \\ s_1 & s_0 & s_1 & s_0 & s_1 \\ s_{-2} & s_0 & s_{-2} & s_{-1} & s_0 \end{pmatrix}$$

$$d(\boldsymbol{A}_1, \boldsymbol{B}) = 0.12 < 0.15$$
$$d(\boldsymbol{A}_2, \boldsymbol{B}) = 0.11 < 0.15$$
$$d(\boldsymbol{A}_3, \boldsymbol{B}) = 0.12 < 0.15$$
$$d(\boldsymbol{A}_4, \boldsymbol{B}) = 0.14 < 0.15$$

Assume that the weight vector of the decision makers e_k $(k = 2, 3, 4, 5)$ is $\boldsymbol{v} = (1/4, 1/4, 1/4, 1/4)^{\mathrm{T}}$. Then the weighted combination of the additive linguistic preference relations is

$$
\boldsymbol{A} = \begin{pmatrix} s_0 & s_{-5} & s_2 & s_{-1.75} & s_3 \\ s_{-0.5} & s_0 & s_{-1.25} & s_{0.5} & s_1 \\ s_{-2} & s_{-0.75} & s_0 & s_{-1.5} & s_2 \\ s_{0.25} & s_{-1.5} & s_{1.5} & s_0 & s_{0.25} \\ s_{-3} & s_{-1} & s_{-2} & s_{-0.25} & s_0 \end{pmatrix}
$$

Accordingly,

$$d(\boldsymbol{A}, \boldsymbol{B}) = 0.10 < 0.15$$

3.2 Incomplete Additive Linguistic Preference Relations

Considering that each linguistic preference relation necessitates the completion of all $n(n-1)/2$ judgments in its entire top triangular portion, sometimes, it is difficult to obtain such a linguistic preference relation, especially for the linguistic preference relation with high order (Xu, 2004c), because that ① a decision should be made under time pressure and lack of data; ② the decision makers' limited expertise related with problem domain; ③ the decision makers have limited attention and information processing capabilities; or ④ some of decision alternatives are incomparable, the decision makers may develop incomplete linguistic preference relations in which some of the elements cannot be provided. Therefore, it is worth paying attention to this issue (Xu, 2006c).

Definition 3.10 (Xu, 2006c) Let $\boldsymbol{A} = (a_{ij})_{n\times n}$ be an additive linguistic preference relation. Then $\boldsymbol{A}$ is called an incomplete additive linguistic preference relation if some of its elements cannot be given by the decision maker, which we denote by the unknown variable "x", and the others can be provided by the decision maker, which satisfy

$$s_{-\tau} \leqslant a_{ij} \leqslant s_{\tau}, \quad a_{ij} \oplus a_{ji} = s_0, \quad a_{ii} = s_0 \tag{3.28}$$

In this case, the decision maker is unable to provide preference values for all pairs of alternatives, and thus, some of them are missing. For the sake of convenience, let Ω be the set of all known elements in $\boldsymbol{A}$.

Definition 3.11 Let $\boldsymbol{A} = (a_{ij})_{n\times n}$ be an additive linguistic preference relation. If the elements of the main diagonal are s_0, and all the other elements in $\boldsymbol{A}$ are missing, then $\boldsymbol{A}$ is called a completely incomplete additive linguistic preference relation.

Definition 3.12 (Xu, 2006c) Let $\boldsymbol{A} = (a_{ij})_{n\times n}$ be an incomplete additive linguistic preference relation. If

$$a_{ij} = a_{ik} \oplus a_{kj}, \quad a_{ik}, a_{jk}, a_{ij} \in \Omega \tag{3.29}$$

then $\boldsymbol{A}$ is called a consistent incomplete additive linguistic preference relation.

Definition 3.13 (Xu, 2006c) Let $\boldsymbol{A} = (a_{ij})_{n\times n}$ be an incomplete additive linguistic preference relation. If $(i,j) \cap (k,l) \neq \varnothing$, then the elements a_{ij} and a_{kl} are called adjoining.

Definition 3.14 (Xu, 2006c) Let $\boldsymbol{A} = (a_{ij})_{n\times n}$ be an incomplete additive linguistic preference relation, and let $a_{i_0j_0}$ be an unknown element in $\boldsymbol{A}$. If there exist the adjoining known elements $a_{i_0j_1}, a_{j_1j_2}, \cdots, a_{j_kj_0}$, then $a_{i_0j_0}$ is called available indirectly.

Definition 3.15 (Xu, 2006c) Let $\boldsymbol{A} = (a_{ij})_{n\times n}$ be an incomplete additive linguistic preference relation. If each unknown element can be obtained indirectly by the known elements in $\boldsymbol{A}$, then $\boldsymbol{A}$ is called acceptable; otherwise, $\boldsymbol{A}$ is called unacceptable.

Definition 3.16 (Xu, 2006c) Let $\Phi(\boldsymbol{A})$ be the digraph of the incomplete additive linguistic preference relation $\boldsymbol{A} = (a_{ij})_{n\times n}$. If every two nodes are mutually reachable, then $\Phi(\boldsymbol{A})$ is called strong connected.

Similar to Xu (2004d), the following theorems can be easily proven:

Theorem 3.14 The incomplete additive linguistic preference relation $\boldsymbol{A} = (a_{ij})_{n\times n}$ is acceptable if and only if the digraph $\Phi(\boldsymbol{A})$ is strong connected.

According to the concept of incomplete additive linguistic preference relation, we have the following theorems:

Theorem 3.15 (Xu, 2006c) Let $\boldsymbol{A} = (a_{ij})_{n\times n}$ be an incomplete additive linguistic preference relation. If we remove the ith line and ith column from $\boldsymbol{A}$, then the preference relation composed by the remainder $(n-1)$ lines and $(n-1)$ columns of $\boldsymbol{A}$ is also an additive linguistic preference relation or incomplete additive linguistic preference relation.

Theorem 3.16 (Xu, 2006c) Let $\boldsymbol{A} = (a_{ij})_{n\times n}$ be an incomplete additive linguistic preference relation. If $\boldsymbol{A}$ is acceptable, then there exists at least one known element (except diagonal elements) in each line or each column of $\boldsymbol{A}$, i.e., there exist at least $n-1$ judgments provided by the decision maker.

Proof Let $a_{i_0j_0}$ be an arbitrary unknown element in the incomplete linguistic preference relation $\boldsymbol{A}$. If $\boldsymbol{A}$ is acceptable, then by Definition 3.15, we know that there must exist at least one sequence of the known elements $a_{i_0j_1}, a_{j_1j_2}, \cdots, a_{j_kj_0}$. Therefore, there is at least an element $a_{i_0j_1}$ $(i_0 \neq j_1)$ in the line i_0, and there is at least an element $a_{j_kj_0} (j_k \neq j_0)$ in the line j_0. By the arbitrariness property of i_0 and j_0, there exist at least $n-1$ judgments provided by the decision maker, i.e., there exists at least one known element (except diagonal elements) in each line or each column of $\boldsymbol{A}$. This completes the proof.

Similar to Section 3.1, the incomplete additive linguistic preference relation $\boldsymbol{A} = (a_{ij})_{n\times n}$ has the following properties (Xu, 2006c):

(1) If $a_{ik} \oplus a_{kj} \geqslant a_{ij}$ for all $a_{ik}, a_{kj}, a_{ij} \in \Omega$, then we say $\boldsymbol{A}$ satisfies the triangle condition;

(2) If $a_{ik} \geqslant s_0$, $a_{kj} \geqslant s_0 \Rightarrow a_{ij} \geqslant s_0$ for all $a_{ik}, a_{kj}, a_{ij} \in \Omega$, then we say $\boldsymbol{A}$ satisfies

the weak transitivity property;

(3) If $a_{ij} \geqslant \min\{a_{ik}, a_{kj}\}$ for all $a_{ik}, a_{kj}, a_{ij} \in \Omega$, then we say $\boldsymbol{A}$ satisfies the max-min transitivity property;

(4) If $a_{ij} \geqslant \max\{a_{ik}, a_{kj}\}$ for all $a_{ik}, a_{kj}, a_{ij} \in \Omega$, then we say $\boldsymbol{A}$ satisfies the max-max transitivity property;

(5) If $a_{ik} \geqslant s_0$, $a_{kj} \geqslant s_0 \Rightarrow a_{ij} \geqslant \min\{a_{ik}, a_{kj}\}$ for all $a_{ik}, a_{kj}, a_{ij} \in \Omega$, then we say $\boldsymbol{A}$ satisfies the restricted max-min transitivity property;

(6) If $a_{ik} \geqslant s_0$, $a_{kj} \geqslant s_0 \Rightarrow a_{ij} \geqslant \max\{ a_{ik}, a_{kj}\}$ for all $a_{ik}, a_{kj}, a_{ij} \in \Omega$, then we say $\boldsymbol{A}$ satisfies the restricted max-max transitivity property.

In the following, based on the acceptable incomplete additive linguistic preference relation with the least judgments (i.e., $n-1$ judgments), we introduce a simple and practical method for constructing a consistent complete additive linguistic preference relation as follows (Xu, 2006c):

Step 1 For a decision making problem, let $X = \{x_1, x_2, \cdots, x_n\}$ be a finite set of alternatives. The decision maker compares each pair of alternatives in X, and constructs an acceptable additive linguistic preference relation with only $n-1$ juegments $a_{ij_1}, a_{ij_2}, \ldots, a_{ij_{n-1}}$.

Step 2 Utilize the known elements in $\boldsymbol{A}$ and the following formula to determine all the unknown elements in $\boldsymbol{A}$:

$$\dot{a}_{ij} = \max\{s_{-\tau},\ \min\{s_\tau,\ a_{ik} \oplus a_{kj}\}\}, \quad a_{ik}, a_{kj} \in \Omega \tag{3.30}$$

until get a complete additive linguistic preference relation $\dot{\boldsymbol{A}} = (\dot{a}_{ij})_{n\times n}$, where

$$\dot{a}_{ij} = \begin{cases} \max\{s_{-\tau},\ \min\{s_\tau,\ a_{ik} \oplus a_{kj}\}\}, & a_{ij} \notin \Omega, a_{ik}, a_{kj} \in \Omega \\ a_{ij}, & a_{ij} \in \Omega \end{cases} \tag{3.31}$$

In the real world, however, many decision making processes take place in group settings because the increasing complexity and uncertainty of the socio-economic environment makes it less and less possible for single decision maker to consider all relevant aspects of a decision making problem. Therefore, it is interesting to consider the group decision making problem with incomplete linguistic preference relations.

Xu (2006c) developed an approach to group decision making based on the acceptable incomplete additive linguistic preference relations with the least judgments. The approach consists of the following steps:

Step 1 For a group decision making problem, let $E = \{e_1, e_2, \cdots, e_l\}$ be the set of decision makers, $\boldsymbol{\upsilon} = (\upsilon_1, \upsilon_2, \cdots, \upsilon_l)^{\mathrm{T}}$ be the weight vector of the decision makers e_k $(k = 1, 2, \cdots, l)$, $\upsilon_k \geqslant 0\,(k = 1, 2, \cdots, l)$ and $\sum_{k=1}^{l} \upsilon_k = 1$. The decision maker $e_k \in E$ uses the additive linguistic evaluation scale $S_2 = \{s_\alpha | \alpha = -\tau, \cdots, -1, 0, 1, \cdots, \tau\}$

to compare each pair of alternatives in X, and constructs an acceptable incomplete additive linguistic preference relation $\boldsymbol{A}_k = (a_{ij}^{(k)})_{n\times n}$ with only $n-1$ judgments $a_{ij_1}^{(k)}, a_{ij_2}^{(k)}, \cdots, a_{ij_{n-1}}^{(k)}$.

Step 2 Use the known elements in $\boldsymbol{A}_k$ $(k=1,2,\cdots,l)$ and (3.30) to determine all the missing elements in $\boldsymbol{A}_k$ $(k=1,2,\cdots,l)$, and thus, get the corresponding complete additive linguistic preference relations $\dot{\boldsymbol{A}}_k = (\dot{a}_{ij}^{(k)})_{n\times n}$ $(k=1,2,\cdots,l)$.

Step 3 Utilize the LWA$_2$ operator:

$$\dot{a}_{ij} = v_1\dot{a}_{ij}^{(1)} \oplus v_2\dot{a}_{ij}^{(2)} \oplus \cdots \oplus v_l\dot{a}_{ij}^{(l)}, \quad i,j=1,2,\cdots,n \tag{3.32}$$

to aggregate all the additive linguistic preference relations $\dot{\boldsymbol{A}}_k = (\dot{a}_{ij}^{(k)})_{n\times n}$ $(k = 1,2,\cdots,l)$ into the collective additive linguistic preference relation $\dot{\boldsymbol{A}} = (\dot{a}_{ij})_{n\times n}$.

Step 4 Utilize the LA operator:

$$\dot{a}_i = \frac{1}{n}(\dot{a}_{i1} \oplus \dot{a}_{i2} \oplus \cdots \oplus \dot{a}_{in}), \quad i=1,2,\cdots,n \tag{3.33}$$

to aggregate all the elements $\dot{a}_{ij}$ $(j=1,2,\cdots,n)$ in ith lime of $\dot{\boldsymbol{A}}$ into the averaged preference degree $\dot{a}_i$ of the alternative x_i over all the alternatives.

Step 5 Utilize $\dot{a}_i$ $(i=1,2,\cdots,n)$ to rank all the alternatives x_i $(i=1,2,\cdots,n)$, and then select the best one.

Example 3.3 (Xu, 2006c) Consider a decision making problem that involves the evaluation of eight schools (alternatives) x_i $(i=1,2,\cdots,8)$ of a university. One main criterion used is "research". Three decision makers e_k $(k=1,2,3)$ (whose weight vector $\boldsymbol{v} = (0.5,0.3,0.2)^{\mathrm{T}}$) compare each pair of these eight alternatives x_i $(i = 1,2,\cdots,8)$ with respect to the criterion "research" by using the additive linguistic evaluation scale:

$$\begin{aligned} S_2 = \{&s_{-4} = \text{extremely poor}, \quad s_{-3} = \text{very poor}, \quad s_{-2} = \text{poor}, \\ &s_{-1} = \text{slightly poor}, \quad s_0 = \text{fair}, \quad s_1 = \text{slightly good}, \\ &s_2 = \text{good}, \quad s_3 = \text{very good}, \quad s_4 = \text{extremely good}\} \end{aligned}$$

and offer the following judgments:

(1) e_1: $a_{12}^{(1)} = s_{-2}$, $a_{13}^{(1)} = s_{-1}$, $a_{14}^{(1)} = s_{-3}$, $a_{15}^{(1)} = s_1$, $a_{16}^{(1)} = s_0$, $a_{17}^{(1)} = s_{-2}$, $a_{18}^{(1)} = s_1$;

(2) e_2: $a_{41}^{(2)} = s_2$, $a_{42}^{(2)} = s_1$, $a_{43}^{(2)} = s_2$, $a_{45}^{(2)} = s_3$, $a_{46}^{(2)} = s_4$, $a_{47}^{(2)} = s_2$, $a_{48}^{(2)} = s_3$;

(3) e_3: $a_{81}^{(3)} = s_{-1}$, $a_{82}^{(3)} = s_{-3}$, $a_{83}^{(3)} = s_{-1}$, $a_{84}^{(3)} = s_{-2}$, $a_{85}^{(3)} = s_1$, $a_{86}^{(3)} = s_0$, $a_{87}^{(3)} = s_{-1}$.

To get the best alternative, the following steps are involved:

Step 1 According to Definition 3.10 and the preference information provided by the decision makers e_k $(k=1,2,3)$, we construct the following acceptable incomplete additive linguistic preference relations $\boldsymbol{A}_k$ $(k=1,2,3)$:

$$A_1 = (a_{ij}^{(1)})_{8\times 8} = \begin{pmatrix} s_0 & s_{-2} & s_{-1} & s_{-3} & s_1 & s_0 & s_{-2} & s_1 \\ s_2 & s_0 & x & x & x & x & x & x \\ s_1 & x & s_0 & x & x & x & x & x \\ s_3 & x & x & s_0 & x & x & x & x \\ s_{-1} & x & x & x & s_0 & x & x & x \\ s_0 & x & x & x & x & s_0 & x & x \\ s_2 & x & x & x & x & x & s_0 & x \\ s_{-1} & x & x & x & x & x & x & s_0 \end{pmatrix}$$

$$A_2 = (a_{ij}^{(2)})_{8\times 8} = \begin{pmatrix} s_0 & x & x & s_{-2} & x & x & x & x \\ x & s_0 & x & s_{-1} & x & x & x & x \\ x & x & s_0 & s_{-2} & x & x & x & x \\ s_2 & s_1 & s_2 & s_0 & s_3 & s_4 & s_2 & s_3 \\ x & x & x & s_{-3} & s_0 & x & x & x \\ x & x & x & s_{-4} & x & s_0 & x & x \\ x & x & x & s_{-2} & x & x & s_0 & x \\ x & x & x & s_{-3} & x & x & x & s_0 \end{pmatrix}$$

$$A_3 = (a_{ij}^{(3)})_{8\times 8} = \begin{pmatrix} s_0 & x & x & x & x & x & x & s_1 \\ x & s_0 & x & x & x & x & x & s_3 \\ x & x & s_0 & x & x & x & x & s_1 \\ x & x & x & s_0 & x & x & x & s_2 \\ x & x & x & x & s_0 & x & x & s_{-1} \\ x & x & x & x & x & s_0 & x & s_0 \\ x & x & x & x & x & x & s_0 & s_1 \\ s_{-1} & s_{-3} & s_{-1} & s_{-2} & s_1 & s_0 & s_{-1} & s_0 \end{pmatrix}$$

Step 2 Use the known elements in A_k $(k = 1, 2, 3)$ and (3.30) to determine all the missing elements of A_k $(k = 1, 2, 3)$:

$$\begin{aligned} \dot{a}_{23}^{(1)} &= \max\{s_{-4}, \min\{s_4, a_{21}^{(1)} \oplus a_{13}^{(1)}\}\} = \max\{s_{-4}, \min\{s_4, s_2 \oplus s_{-1}\}\} \\ &= \max\{s_{-4}, \min\{s_4, s_1\}\} = \max\{s_{-4}, s_1\} = s_1 \end{aligned}$$

Similarly, we have

$$\dot{a}_{24}^{(1)} = \max\{s_{-4}, \min\{s_4, a_{21}^{(1)} \oplus a_{14}^{(1)}\}\} = s_{-1}$$
$$\dot{a}_{25}^{(1)} = \max\{s_{-4}, \min\{s_4, a_{21}^{(1)} \oplus a_{15}^{(1)}\}\} = s_3$$
$$\dot{a}_{26}^{(1)} = \max\{s_{-4}, \min\{s_4, a_{21}^{(1)} \oplus a_{16}^{(1)}\}\} = s_2$$
$$\dot{a}_{27}^{(1)} = \max\{s_{-4}, \min\{s_4, a_{21}^{(1)} \oplus a_{17}^{(1)}\}\} = s_0$$
$$\dot{a}_{28}^{(1)} = \max\{s_{-4}, \min\{s_4, a_{21}^{(1)} \oplus a_{18}^{(1)}\}\} = s_3$$
$$\dot{a}_{34}^{(1)} = \max\{s_{-4}, \min\{s_4, a_{31}^{(1)} \oplus a_{14}^{(1)}\}\} = s_{-2}$$
$$\dot{a}_{35}^{(1)} = \max\{s_{-4}, \min\{s_4, a_{31}^{(1)} \oplus a_{15}^{(1)}\}\} = s_2$$
$$\dot{a}_{36}^{(1)} = \max\{s_{-4}, \min\{s_4, a_{31}^{(1)} \oplus a_{16}^{(1)}\}\} = s_1$$
$$\dot{a}_{37}^{(1)} = \max\{s_{-4}, \min\{s_4, a_{31}^{(1)} \oplus a_{17}^{(1)}\}\} = s_{-1}$$
$$\dot{a}_{38}^{(1)} = \max\{s_{-4}, \min\{s_4, a_{31}^{(1)} \oplus a_{18}^{(1)}\}\} = s_2$$

$$\dot{a}_{45}^{(1)} = \max\{s_{-4}, \min\{s_4, a_{41}^{(1)} \oplus a_{15}^{(1)}\}\} = s_4$$
$$\dot{a}_{46}^{(1)} = \max\{s_{-4}, \min\{s_4, a_{41}^{(1)} \oplus a_{16}^{(1)}\}\} = s_3$$
$$\dot{a}_{47}^{(1)} = \max\{s_{-4}, \min\{s_4, a_{41}^{(1)} \oplus a_{17}^{(1)}\}\} = s_1$$
$$\dot{a}_{48}^{(1)} = \max\{s_{-4}, \min\{s_4, a_{41}^{(1)} \oplus a_{18}^{(1)}\}\} = s_4$$
$$\dot{a}_{56}^{(1)} = \max\{s_{-4}, \min\{s_4, a_{51}^{(1)} \oplus a_{16}^{(1)}\}\} = s_{-1}$$
$$\dot{a}_{57}^{(1)} = \max\{s_{-4}, \min\{s_4, a_{51}^{(1)} \oplus a_{17}^{(1)}\}\} = s_{-3}$$
$$\dot{a}_{58}^{(1)} = \max\{s_{-4}, \min\{s_4, a_{51}^{(1)} \oplus a_{18}^{(1)}\}\} = s_0$$
$$\dot{a}_{67}^{(1)} = \max\{s_{-4}, \min\{s_4, a_{61}^{(1)} \oplus a_{17}^{(1)}\}\} = s_{-2}$$
$$\dot{a}_{68}^{(1)} = \max\{s_{-4}, \min\{s_4, a_{61}^{(1)} \oplus a_{18}^{(1)}\}\} = s_1$$
$$\dot{a}_{78}^{(1)} = \max\{s_{-4}, \min\{s_4, a_{71}^{(1)} \oplus a_{18}^{(1)}\}\} = s_3$$
$$\dot{a}_{12}^{(2)} = \max\{s_{-4}, \min\{s_4, a_{14}^{(2)} \oplus a_{42}^{(2)}\}\} = s_{-1}$$
$$\dot{a}_{13}^{(2)} = \max\{s_{-4}, \min\{s_4, a_{14}^{(2)} \oplus a_{43}^{(2)}\}\} = s_0$$
$$\dot{a}_{15}^{(2)} = \max\{s_{-4}, \min\{s_4, a_{14}^{(2)} \oplus a_{45}^{(2)}\}\} = s_1$$
$$\dot{a}_{16}^{(2)} = \max\{s_{-4}, \min\{s_4, a_{14}^{(2)} \oplus a_{46}^{(2)}\}\} = s_2$$
$$\dot{a}_{17}^{(2)} = \max\{s_{-4}, \min\{s_4, a_{14}^{(2)} \oplus a_{47}^{(2)}\}\} = s_0$$
$$\dot{a}_{18}^{(2)} = \max\{s_{-4}, \min\{s_4, a_{14}^{(2)} \oplus a_{48}^{(2)}\}\} = s_1$$
$$\dot{a}_{23}^{(2)} = \max\{s_{-4}, \min\{s_4, a_{24}^{(2)} \oplus a_{43}^{(2)}\}\} = s_1$$
$$\dot{a}_{25}^{(2)} = \max\{s_{-4}, \min\{s_4, a_{24}^{(2)} \oplus a_{45}^{(2)}\}\} = s_2$$
$$\dot{a}_{26}^{(2)} = \max\{s_{-4}, \min\{s_4, a_{24}^{(2)} \oplus a_{46}^{(2)}\}\} = s_3$$
$$\dot{a}_{27}^{(2)} = \max\{s_{-4}, \min\{s_4, a_{24}^{(2)} \oplus a_{47}^{(2)}\}\} = s_1$$
$$\dot{a}_{28}^{(2)} = \max\{s_{-4}, \min\{s_4, a_{24}^{(2)} \oplus a_{48}^{(2)}\}\} = s_2$$
$$\dot{a}_{35}^{(2)} = \max\{s_{-4}, \min\{s_4, a_{34}^{(2)} \oplus a_{45}^{(2)}\}\} = s_1$$
$$\dot{a}_{36}^{(2)} = \max\{s_{-4}, \min\{s_4, a_{34}^{(2)} \oplus a_{46}^{(2)}\}\} = s_2$$
$$\dot{a}_{37}^{(2)} = \max\{s_{-4}, \min\{s_4, a_{34}^{(2)} \oplus a_{47}^{(2)}\}\} = s_0$$
$$\dot{a}_{38}^{(2)} = \max\{s_{-4}, \min\{s_4, a_{34}^{(2)} \oplus a_{48}^{(2)}\}\} = s_1$$
$$\dot{a}_{56}^{(2)} = \max\{s_{-4}, \min\{s_4, a_{54}^{(2)} \oplus a_{46}^{(2)}\}\} = s_1$$
$$\dot{a}_{57}^{(2)} = \max\{s_{-4}, \min\{s_4, a_{54}^{(2)} \oplus a_{47}^{(2)}\}\} = s_{-1}$$
$$\dot{a}_{58}^{(2)} = \max\{s_{-4}, \min\{s_4, a_{54}^{(2)} \oplus a_{48}^{(2)}\}\} = s_0$$
$$\dot{a}_{67}^{(2)} = \max\{s_{-4}, \min\{s_4, a_{64}^{(2)} \oplus a_{47}^{(2)}\}\} = s_{-2}$$

$$\dot{a}_{68}^{(2)} = \max\{s_{-4}, \min\{s_4, a_{64}^{(2)} \oplus a_{48}^{(2)}\}\} = s_{-1}$$
$$\dot{a}_{78}^{(2)} = \max\{s_{-4}, \min\{s_4, a_{74}^{(2)} \oplus a_{48}^{(2)}\}\} = s_{1}$$
$$\dot{a}_{12}^{(3)} = \max\{s_{-4}, \min\{s_4, a_{18}^{(3)} \oplus a_{82}^{(3)}\}\} = s_{-2}$$
$$\dot{a}_{13}^{(3)} = \max\{s_{-4}, \min\{s_4, a_{18}^{(3)} \oplus a_{83}^{(3)}\}\} = s_{0}$$
$$\dot{a}_{14}^{(3)} = \max\{s_{-4}, \min\{s_4, a_{18}^{(3)} \oplus a_{84}^{(3)}\}\} = s_{-1}$$
$$\dot{a}_{15}^{(3)} = \max\{s_{-4}, \min\{s_4, a_{18}^{(3)} \oplus a_{85}^{(3)}\}\} = s_{2}$$
$$\dot{a}_{16}^{(3)} = \max\{s_{-4}, \min\{s_4, a_{18}^{(3)} \oplus a_{86}^{(3)}\}\} = s_{1}$$
$$\dot{a}_{17}^{(3)} = \max\{s_{-4}, \min\{s_4, a_{18}^{(3)} \oplus a_{87}^{(3)}\}\} = s_{0}$$
$$\dot{a}_{23}^{(3)} = \max\{s_{-4}, \min\{s_4, a_{28}^{(3)} \oplus a_{83}^{(3)}\}\} = s_{2}$$
$$\dot{a}_{24}^{(3)} = \max\{s_{-4}, \min\{s_4, a_{28}^{(3)} \oplus a_{84}^{(3)}\}\} = s_{1}$$
$$\dot{a}_{25}^{(3)} = \max\{s_{-4}, \min\{s_4, a_{28}^{(3)} \oplus a_{85}^{(3)}\}\} = s_{4}$$
$$\dot{a}_{26}^{(3)} = \max\{s_{-4}, \min\{s_4, a_{28}^{(3)} \oplus a_{86}^{(3)}\}\} = s_{3}$$
$$\dot{a}_{27}^{(3)} = \max\{s_{-4}, \min\{s_4, a_{28}^{(3)} \oplus a_{87}^{(3)}\}\} = s_{2}$$
$$\dot{a}_{34}^{(3)} = \max\{s_{-4}, \min\{s_4, a_{38}^{(3)} \oplus a_{84}^{(3)}\}\} = s_{-1}$$
$$\dot{a}_{35}^{(3)} = \max\{s_{-4}, \min\{s_4, a_{38}^{(3)} \oplus a_{85}^{(3)}\}\} = s_{2}$$
$$\dot{a}_{36}^{(3)} = \max\{s_{-4}, \min\{s_4, a_{38}^{(3)} \oplus a_{86}^{(3)}\}\} = s_{1}$$
$$\dot{a}_{37}^{(3)} = \max\{s_{-4}, \min\{s_4, a_{38}^{(3)} \oplus a_{87}^{(3)}\}\} = s_{0}$$
$$\dot{a}_{45}^{(3)} = \max\{s_{-4}, \min\{s_4, a_{48}^{(3)} \oplus a_{85}^{(3)}\}\} = s_{3}$$
$$\dot{a}_{46}^{(3)} = \max\{s_{-4}, \min\{s_4, a_{48}^{(3)} \oplus a_{86}^{(3)}\}\} = s_{2}$$
$$\dot{a}_{47}^{(3)} = \max\{s_{-4}, \min\{s_4, a_{48}^{(3)} \oplus a_{87}^{(3)}\}\} = s_{1}$$
$$\dot{a}_{56}^{(3)} = \max\{s_{-4}, \min\{s_4, a_{58}^{(3)} \oplus a_{86}^{(3)}\}\} = s_{-1}$$
$$\dot{a}_{57}^{(3)} = \max\{s_{-4}, \min\{s_4, a_{58}^{(3)} \oplus a_{87}^{(3)}\}\} = s_{-2}$$
$$\dot{a}_{67}^{(3)} = \max\{s_{-4}, \min\{s_4, a_{68}^{(3)} \oplus a_{87}^{(3)}\}\} = s_{-1}$$

Then by (3.1), we get the complete additive linguistic preference relations:

$$\dot{\boldsymbol{A}}_1 = (\dot{a}_{ij}^{(1)})_{8\times 8} = \begin{pmatrix} s_0 & s_{-2} & s_{-1} & s_{-3} & s_1 & s_0 & s_{-2} & s_1 \\ s_2 & s_0 & s_1 & s_{-1} & s_3 & s_2 & s_0 & s_3 \\ s_1 & s_{-1} & s_0 & s_{-2} & s_2 & s_1 & s_{-1} & s_2 \\ s_3 & s_1 & s_2 & s_0 & s_4 & s_3 & s_1 & s_4 \\ s_{-1} & s_{-3} & s_{-2} & s_{-4} & s_0 & s_{-1} & s_{-3} & s_0 \\ s_0 & s_{-2} & s_{-1} & s_{-3} & s_1 & s_0 & s_{-2} & s_1 \\ s_2 & s_0 & s_1 & s_{-1} & s_3 & s_2 & s_0 & s_3 \\ s_{-1} & s_{-3} & s_{-2} & s_{-4} & s_0 & s_{-1} & s_{-3} & s_0 \end{pmatrix}$$

$$\dot{A}_2 = (\dot{a}_{ij}^{(2)})_{8\times 8} = \begin{pmatrix} s_0 & s_{-1} & s_0 & s_{-2} & s_1 & s_2 & s_0 & s_1 \\ s_1 & s_0 & s_1 & s_{-1} & s_2 & s_3 & s_1 & s_2 \\ s_0 & s_{-1} & s_0 & s_{-2} & s_1 & s_2 & s_0 & s_1 \\ s_2 & s_1 & s_2 & s_0 & s_3 & s_4 & s_2 & s_3 \\ s_{-1} & s_{-2} & s_{-1} & s_{-3} & s_0 & s_1 & s_{-1} & s_0 \\ s_{-2} & s_{-3} & s_{-2} & s_{-4} & s_{-1} & s_0 & s_{-2} & s_{-1} \\ s_0 & s_{-1} & s_0 & s_{-2} & s_1 & s_2 & s_0 & s_1 \\ s_{-1} & s_{-2} & s_{-1} & s_{-3} & s_0 & s_1 & s_{-1} & s_0 \end{pmatrix}$$

$$\dot{A}_3 = (\dot{a}_{ij}^{(3)})_{8\times 8} = \begin{pmatrix} s_0 & s_{-2} & s_0 & s_{-1} & s_2 & s_1 & s_0 & s_1 \\ s_2 & s_0 & s_2 & s_1 & s_4 & s_3 & s_2 & s_3 \\ s_0 & s_{-2} & s_0 & s_{-1} & s_2 & s_1 & s_0 & s_1 \\ s_1 & s_{-1} & s_1 & s_0 & s_3 & s_2 & s_1 & s_2 \\ s_{-2} & s_{-4} & s_{-2} & s_{-3} & s_0 & s_{-1} & s_{-2} & s_{-1} \\ s_{-1} & s_{-3} & s_{-1} & s_{-2} & s_1 & s_0 & s_{-1} & s_0 \\ s_0 & s_{-2} & s_0 & s_{-1} & s_2 & s_1 & s_0 & s_1 \\ s_{-1} & s_{-3} & s_{-1} & s_{-2} & s_1 & s_0 & s_{-1} & s_0 \end{pmatrix}$$

Step 3 Utilize the LWA$_2$ operator (3.32) to aggregate all the additive linguistic preference relations $\dot{A}_k = (\dot{a}_{ij}^{(k)})_{8\times 8}$ $(k = 1, 2, 3)$ into the collective additive linguistic preference relation:

$$\dot{A} = (\dot{a}_{ij})_{8\times 8} = \begin{pmatrix} s_0 & s_{-1.7} & s_{-0.5} & s_{-2.3} & s_{1.2} & s_{0.8} & s_{-1} & s_1 \\ s_{1.7} & s_0 & s_{1.2} & s_{-0.6} & s_{2.9} & s_{2.5} & s_{0.7} & s_{2.7} \\ s_{0.5} & s_{-2} & s_0 & s_{-1.8} & s_{1.7} & s_{1.3} & s_{-0.5} & s_{1.5} \\ s_{2.3} & s_{-1.2} & s_{1.8} & s_0 & s_{3.5} & s_{3.1} & s_{1.3} & s_{3.3} \\ s_{-1.2} & s_{-2.9} & s_{-1.7} & s_{-3.5} & s_0 & s_{-0.4} & s_{-2.2} & s_{-0.2} \\ s_{-0.8} & s_{-2.5} & s_{-1.3} & s_{-3.1} & s_{0.4} & s_0 & s_{-1.8} & s_{0.2} \\ s_1 & s_{-0.7} & s_{0.5} & s_{-1.3} & s_{2.2} & s_{1.8} & s_0 & s_2 \\ s_{-1} & s_{-2.7} & s_{-1.5} & s_{-3.3} & s_{0.2} & s_{-0.2} & s_{-2} & s_0 \end{pmatrix}$$

Step 4 Utilize the LA operator (3.33) to aggregate all the elements $\dot{a}_{ij}$ $(j = 1, 2, \cdots, 8)$ in the ith line of $\dot{A}$ into the averaged preference degree $\dot{a}_i$ of the alternative x_i over all the alternatives:

$$\dot{a}_1 = s_{-0.31}, \quad \dot{a}_2 = s_{1.39}, \quad \dot{a}_3 = s_{0.19}, \quad \dot{a}_4 = s_{1.99}$$

$$\dot{a}_5 = s_{-1.51}, \quad \dot{a}_6 = s_{-1.11}, \quad \dot{a}_7 = s_{0.69}, \quad \dot{a}_8 = s_{-1.31}$$

Step 5 Use $\dot{a}_i$ $(i = 1, 2, \cdots, 8)$ to rank the alternatives x_i $(i = 1, 2, \cdots, 8)$:

$$x_4 \succ x_2 \succ x_7 \succ x_3 \succ x_1 \succ x_6 \succ x_8 \succ x_5$$

and thus, the best alternative is x_4.

Xu (2005a) investigated on general acceptable additive incomplete linguistic preference relations (each with no less than $n - 1$ judgments), and gave an approach to consrtructing complete additive linguistic preference relations:

Let $\boldsymbol{A} = (a_{ij})_{n\times n}$ be a general acceptable incomplete additive linguistic preference relation. Then we replace each missing element in $\boldsymbol{A}$ with the following:

$$\dot{a}_{ij} = \frac{1}{n_{ij}} \sum_{k\in N_{ij}} \max\{s_{-\tau},\ \min\{s_\tau,\ a_{ik} \oplus a_{kj}\}\} \tag{3.34}$$

and get a complete additive linguistic preference relation $\dot{\boldsymbol{A}} = (\dot{a}_{ij})_{n\times n}$ corresponding to $\boldsymbol{A} = (a_{ij})_{n\times n}$, where $N_{ij} = \{k|a_{ik},\ a_{kj} \in \Omega\}$, n_{ij} is the number of the elements in N_{ij}, and

$$\dot{a}_{ij} = \begin{cases} \dfrac{1}{n_{ij}} \displaystyle\sum_{k\in N_{ij}} \max\{s_{-\tau},\ \min\{s_\tau,\ a_{ik} \oplus a_{kj}\}\}, & a_{ij} \notin \Omega \\ a_{ij}, & a_{ij} \in \Omega \end{cases} \tag{3.35}$$

The additive linguistic preference relation $\dot{\boldsymbol{A}} = (\dot{a}_{ij})_{n\times n}$ derived by (3.34) contains both the direct linguistic preference information given by the decision maker and the indirect linguistic preference information derived from the known linguistic preference information.

In the following, we introduce an approach to group decision making based on general acceptable additive incomplete linguistic preference relations (Xu, 2005c):

Step 1 For a group decision making problem, the decision maker $e_k \in E$ utilizes the additive linguistic evaluation scale $S_2 = \{s_\alpha|\alpha = -\tau, \cdots, -1, 0, 1, \cdots, \tau\}$ to compare each pair of alternatives in X, and constructs a general acceptable incomplete additive linguistic preference relation $\boldsymbol{A}_k = (a_{ij}^{(k)})_{n\times n}$.

Step 2 Utilize the known elements in $\boldsymbol{A}_k$ $(k = 1, 2, \cdots, l)$ and (3.34) to determine all the missing elements in $\boldsymbol{A}_k$ $(k = 1, 2, \cdots, l)$, until the complete additive linguistic preference relations $\dot{\boldsymbol{A}}_k = (\dot{a}_{ij}^{(k)})_{n\times n}$ $(k = 1, 2, \cdots, l)$ are obtained.

Step 3 Utilize the LWA$_2$ operator and (3.32) to aggregate all the additive linguistic preference relations $\dot{\boldsymbol{A}}_k = (\dot{a}_{ij}^{(k)})_{n\times n}$ $(k = 1, 2, \cdots, l)$ into the collective additive linguistic preference relation $\dot{\boldsymbol{A}} = (\dot{a}_{ij})_{n\times n}$.

Step 4 Utilize the LA operator and (3.33) to aggregate all the elements $\dot{a}_{ij}$ $(j = 1, 2, \cdots, n)$ in the ith line of $\dot{\boldsymbol{A}}$ into the averaged preference degree $\dot{a}_i$ of the alternative x_i over all the alternatives.

Step 5 Utilize $\dot{a}_i$ $(i = 1, 2, \cdots, n)$ to rank the alternatives x_i $(i = 1, 2, \cdots, n)$, and then select the best one.

Example 3.4 Assume that, in Example 3.3, three decision makers e_k $(k = 1, 2, 3)$ utilize the additive linguistic evaluation scale S_2 to compare each pair of alternatives x_i $(i = 1, 2, \cdots, 8)$ with respect to the criterion "research", and construct, respectively, the following general additive linguistic preference relations $\boldsymbol{A}_k = (a_{ij}^{(k)})_{8\times 8}$ $(k = 1, 2, 3)$:

$$A_1 = (a_{ij}^{(1)})_{8\times 8} = \begin{pmatrix} s_0 & s_{-2} & s_{-4} & s_{-3} & x & s_1 & s_{-2} & s_2 \\ s_2 & s_0 & s_2 & s_{-1} & s_3 & s_2 & s_1 & s_2 \\ s_4 & s_{-2} & s_0 & s_{-3} & s_2 & x & s_{-1} & s_3 \\ s_3 & s_1 & s_3 & s_0 & x & s_3 & s_1 & x \\ x & s_{-3} & s_{-2} & x & s_0 & s_{-2} & s_{-3} & s_0 \\ s_{-1} & s_{-2} & x & s_{-3} & s_2 & s_0 & s_{-3} & s_2 \\ s_2 & s_{-1} & s_1 & s_{-1} & s_3 & s_3 & s_0 & x \\ s_{-2} & s_{-2} & s_{-3} & x & s_0 & s_{-2} & x & s_0 \end{pmatrix}$$

$$A_2 = (a_{ij}^{(2)})_{8\times 8} = \begin{pmatrix} s_0 & s_{-3} & x & s_{-1} & s_2 & x & s_1 & s_3 \\ s_3 & s_0 & s_2 & s_{-3} & s_3 & s_2 & x & s_2 \\ x & s_{-2} & s_0 & s_{-2} & s_1 & x & s_2 & s_2 \\ s_1 & s_3 & s_2 & s_0 & s_3 & s_4 & s_2 & s_3 \\ s_{-2} & s_{-3} & s_{-1} & s_{-3} & s_0 & s_3 & s_{-1} & x \\ x & s_{-2} & x & s_{-4} & s_{-3} & s_0 & s_{-4} & s_{-2} \\ s_{-1} & x & s_{-2} & s_{-2} & s_1 & s_4 & s_0 & s_1 \\ s_{-3} & s_{-2} & s_{-2} & s_{-3} & x & s_2 & s_{-1} & s_0 \end{pmatrix}$$

$$A_3 = (a_{ij}^{(3)})_{8\times 8} = \begin{pmatrix} s_0 & x & s_0 & s_{-1} & s_2 & s_1 & s_3 & s_4 \\ x & s_0 & s_2 & s_3 & s_4 & s_3 & s_1 & s_3 \\ s_0 & s_{-2} & s_0 & x & s_2 & s_1 & s_2 & x \\ s_1 & s_{-3} & x & s_0 & s_3 & s_1 & s_2 & s_3 \\ s_{-2} & s_{-4} & s_{-2} & s_{-3} & s_0 & x & s_{-3} & s_{-2} \\ s_{-1} & s_{-3} & s_{-1} & s_{-1} & x & s_0 & s_{-3} & x \\ s_{-3} & s_{-1} & s_{-2} & s_{-2} & s_3 & s_3 & s_0 & s_3 \\ s_{-4} & s_{-3} & x & s_{-3} & s_2 & x & s_{-3} & s_0 \end{pmatrix}$$

We first employ the elements in $\boldsymbol{A}_k$ $(k = 1, 2, 3)$ and (3.33) to determine all the missing elements in $\boldsymbol{A}_k$ $(k = 1, 2, 3)$:

$$\begin{aligned}\dot{a}_{15}^{(1)} =& \frac{1}{4}(\max\{s_{-4}, \min\{s_4, a_{12}^{(1)} \oplus a_{25}^{(1)}\}\} \oplus \max\{s_{-4}, \min\{s_4, a_{16}^{(1)} \oplus a_{65}^{(1)}\}\} \\ & \oplus \max\{s_{-4}, \min\{s_4, a_{17}^{(1)} \oplus a_{75}^{(1)}\}\} \oplus \max\{s_{-4}, \min\{s_4, a_{18}^{(1)} \oplus a_{85}^{(1)}\}\}) \\ =& \frac{1}{4}(\max\{s_{-4}, \min\{s_4, s_{-2} \oplus s_3\}\} \oplus \max\{s_{-4}, \min\{s_4, s_1 \oplus s_2\}\} \\ & \oplus \max\{s_{-4}, \min\{s_4, s_{-2} \oplus s_3\}\} \oplus \max\{s_{-4}, \min\{s_4, s_2 \oplus s_0\}\}) \\ =& \frac{1}{4}(s_1 \oplus s_3 \oplus s_1 \oplus s_2) = s_{1.75}\end{aligned}$$

Similarly, we have

$$\begin{aligned}\dot{a}_{36}^{(1)} =& \frac{1}{6}(\max\{s_{-4}, \min\{s_4, a_{31}^{(1)} \oplus a_{16}^{(1)}\}\} \oplus \max\{s_{-4}, \min\{s_4, a_{32}^{(1)} \oplus a_{26}^{(1)}\}\} \\ & \oplus \max\{s_{-4}, \min\{s_4, a_{34}^{(1)} \oplus a_{46}^{(1)}\}\} \oplus \max\{s_{-4}, \min\{s_4, a_{35}^{(1)} \oplus a_{56}^{(1)}\}\} \\ & \oplus \max\{s_{-4}, \min\{s_4, a_{37}^{(1)} \oplus a_{76}^{(1)}\}\} \oplus \max\{s_{-4}, \min\{s_4, a_{38}^{(1)} \oplus a_{86}^{(1)}\}\}) \\ =& s_{0.67}\end{aligned}$$

$$\begin{aligned}\dot{a}_{45}^{(1)} =&\frac{1}{4}(\max\{s_{-4},\min\{s_4,a_{42}^{(1)}\oplus a_{25}^{(1)}\}\}\oplus\max\{s_{-4},\min\{s_4,a_{43}^{(1)}\oplus a_{35}^{(1)}\}\}\\&\oplus\max\{s_{-4},\min\{s_4,a_{46}^{(1)}\oplus a_{65}^{(1)}\}\}\oplus\max\{s_{-4},\min\{s_4,a_{47}^{(1)}\oplus a_{75}^{(1)}\}\})\\=&s_4\\\dot{a}_{48}^{(1)} =&\frac{1}{4}(\max\{s_{-4},\min\{s_4,a_{41}^{(1)}\oplus a_{18}^{(1)}\}\}\oplus\max\{s_{-4},\min\{s_4,a_{42}^{(1)}\oplus a_{28}^{(1)}\}\}\\&\oplus\max\{s_{-4},\min\{s_4,a_{43}^{(1)}\oplus a_{38}^{(1)}\}\}\oplus\max\{s_{-4},\min\{s_4,a_{46}^{(1)}\oplus a_{68}^{(1)}\}\})\\=&s_{3.75}\\\dot{a}_{78}^{(1)} =&\frac{1}{5}(\max\{s_{-4},\min\{s_4,a_{71}^{(1)}\oplus a_{18}^{(1)}\}\}\oplus\max\{s_{-4},\min\{s_4,a_{72}^{(1)}\oplus a_{28}^{(1)}\}\}\\&\oplus\max\{s_{-4},\min\{s_4,a_{73}^{(1)}\oplus a_{38}^{(1)}\}\}\oplus\max\{s_{-4},\min\{s_4,a_{75}^{(1)}\oplus a_{58}^{(1)}\}\}\\&\oplus\max\{s_{-4},\min\{s_4,a_{76}^{(1)}\oplus a_{68}^{(1)}\}\})\\=&s_{3.20}\\\dot{a}_{13}^{(2)} =&\frac{1}{5}(\max\{s_{-4},\min\{s_4,a_{12}^{(2)}\oplus a_{23}^{(2)}\}\}\oplus\max\{s_{-4},\min\{s_4,a_{14}^{(2)}\oplus a_{43}^{(2)}\}\}\\&\oplus\max\{s_{-4},\min\{s_4,a_{15}^{(2)}\oplus a_{53}^{(2)}\}\}\oplus\max\{s_{-4},\min\{s_4,a_{17}^{(2)}\oplus a_{73}^{(2)}\}\}\\&\oplus\max\{s_{-4},\min\{s_4,a_{18}^{(2)}\oplus a_{83}^{(2)}\}\})\\=&s_{0.2}\\\dot{a}_{16}^{(2)} =&\frac{1}{5}(\max\{s_{-4},\min\{s_4,a_{12}^{(2)}\oplus a_{26}^{(2)}\}\}\oplus\max\{s_{-4},\min\{s_4,a_{14}^{(2)}\oplus a_{46}^{(2)}\}\}\\&\oplus\max\{s_{-4},\min\{s_4,a_{15}^{(2)}\oplus a_{56}^{(2)}\}\}\oplus\max\{s_{-4},\min\{s_4,a_{17}^{(2)}\oplus a_{76}^{(2)}\}\}\\&\oplus\max\{s_{-4},\min\{s_4,a_{18}^{(2)}\oplus a_{86}^{(2)}\}\})\\=&s_{2.8}\\\dot{a}_{27}^{(2)} =&\frac{1}{6}(\max\{s_{-4},\min\{s_4,a_{21}^{(2)}\oplus a_{17}^{(2)}\}\}\oplus\max\{s_{-4},\min\{s_4,a_{23}^{(2)}\oplus a_{37}^{(2)}\}\}\\&\oplus\max\{s_{-4},\min\{s_4,a_{24}^{(2)}\oplus a_{47}^{(2)}\}\}\oplus\max\{s_{-4},\min\{s_4,a_{25}^{(2)}\oplus a_{57}^{(2)}\}\}\\&\oplus\max\{s_{-4},\min\{s_4,a_{26}^{(2)}\oplus a_{67}^{(2)}\}\}\oplus\max\{s_{-4},\min\{s_4,a_{28}^{(2)}\oplus a_{87}^{(2)}\}\})\\=&s_{1.33}\\\dot{a}_{36}^{(2)} =&\frac{1}{5}(\max\{s_{-4},\min\{s_4,a_{32}^{(2)}\oplus a_{26}^{(2)}\}\}\oplus\max\{s_{-4},\min\{s_4,a_{34}^{(2)}\oplus a_{46}^{(2)}\}\}\\&\oplus\max\{s_{-4},\min\{s_4,a_{35}^{(2)}\oplus a_{56}^{(2)}\}\}\oplus\max\{s_{-4},\min\{s_4,a_{37}^{(2)}\oplus a_{76}^{(2)}\}\}\\&\oplus\max\{s_{-4},\min\{s_4,a_{38}^{(2)}\oplus a_{86}^{(2)}\}\})\\=&s_{2.8}\\\dot{a}_{58}^{(2)} =&\frac{1}{6}(\max\{s_{-4},\min\{s_4,a_{51}^{(2)}\oplus a_{18}^{(2)}\}\}\oplus\max\{s_{-4},\min\{s_4,a_{52}^{(2)}\oplus a_{28}^{(2)}\}\}\\&\oplus\max\{s_{-4},\min\{s_4,a_{53}^{(2)}\oplus a_{38}^{(2)}\}\}\oplus\max\{s_{-4},\min\{s_4,a_{54}^{(2)}\oplus a_{48}^{(2)}\}\}\\&\oplus\max\{s_{-4},\min\{s_4,a_{56}^{(2)}\oplus a_{68}^{(2)}\}\}\oplus\max\{s_{-4},\min\{s_4,a_{57}^{(2)}\oplus a_{78}^{(2)}\}\})\end{aligned}$$

$$=s_{0.33}$$

$$\begin{aligned}\dot{a}_{12}^{(3)} =&\frac{1}{6}(\max\{s_{-4},\min\{s_4,a_{13}^{(3)}\oplus a_{32}^{(3)}\}\}\oplus\max\{s_{-4},\min\{s_4,a_{14}^{(3)}\oplus a_{42}^{(3)}\}\}\\&\oplus\max\{s_{-4},\min\{s_4,a_{15}^{(3)}\oplus a_{52}^{(3)}\}\}\oplus\max\{s_{-4},\min\{s_4,a_{16}^{(3)}\oplus a_{62}^{(3)}\}\}\\&\oplus\max\{s_{-4},\min\{s_4,a_{17}^{(3)}\oplus a_{72}^{(3)}\}\}\oplus\max\{s_{-4},\min\{s_4,a_{18}^{(3)}\oplus a_{82}^{(3)}\}\})\\=&s_{-1.17}\end{aligned}$$

$$\begin{aligned}\dot{a}_{34}^{(3)} =&\frac{1}{5}(\max\{s_{-4},\min\{s_4,a_{31}^{(3)}\oplus a_{14}^{(3)}\}\}\oplus\max\{s_{-4},\min\{s_4,a_{32}^{(3)}\oplus a_{24}^{(3)}\}\}\\&\oplus\max\{s_{-4},\min\{s_4,a_{35}^{(3)}\oplus a_{54}^{(3)}\}\}\oplus\max\{s_{-4},\min\{s_4,a_{36}^{(3)}\oplus a_{64}^{(3)}\}\}\\&\oplus\max\{s_{-4},\min\{s_4,a_{37}^{(3)}\oplus a_{74}^{(3)}\}\})\\=&s_{-0.2}\end{aligned}$$

$$\begin{aligned}\dot{a}_{38}^{(3)} =&\frac{1}{4}(\max\{s_{-4},\min\{s_4,a_{31}^{(3)}\oplus a_{18}^{(3)}\}\}\oplus\max\{s_{-4},\min\{s_4,a_{32}^{(3)}\oplus a_{28}^{(3)}\}\}\\&\oplus\max\{s_{-4},\min\{s_4,a_{35}^{(3)}\oplus a_{58}^{(3)}\}\}\oplus\max\{s_{-4},\min\{s_4,a_{37}^{(3)}\oplus a_{78}^{(3)}\}\})\\=&s_{2.25}\end{aligned}$$

$$\begin{aligned}\dot{a}_{56}^{(3)} =&\frac{1}{5}(\max\{s_{-4},\min\{s_4,a_{51}^{(3)}\oplus a_{16}^{(3)}\}\}\oplus\max\{s_{-4},\min\{s_4,a_{52}^{(3)}\oplus a_{26}^{(3)}\}\}\\&\oplus\max\{s_{-4},\min\{s_4,a_{53}^{(3)}\oplus a_{36}^{(3)}\}\}\oplus\max\{s_{-4},\min\{s_4,a_{54}^{(3)}\oplus a_{46}^{(3)}\}\}\\&\oplus\max\{s_{-4},\min\{s_4,a_{57}^{(3)}\oplus a_{76}^{(3)}\}\})\\=&s_{-1}\end{aligned}$$

$$\begin{aligned}\dot{a}_{68}^{(3)} =&\frac{1}{4}(\max\{s_{-4},\min\{s_4,a_{61}^{(3)}\oplus a_{18}^{(3)}\}\}\oplus\max\{s_{-4},\min\{s_4,a_{62}^{(3)}\oplus a_{28}^{(3)}\}\}\\&\oplus\max\{s_{-4},\min\{s_4,a_{64}^{(3)}\oplus a_{48}^{(3)}\}\}\oplus\max\{s_{-4},\min\{s_4,a_{67}^{(3)}\oplus a_{78}^{(3)}\}\})\\=&s_{1.25}\end{aligned}$$

Then by (3.1), we get the corresponding complete additive linguistic preference relations $\dot{\boldsymbol{A}}_k = (\dot{a}_{ij}^{(k)})_{8\times 8}$ $(k = 1, 2, 3)$:

$$\dot{\boldsymbol{A}}_1 = (\dot{a}_{ij}^{(1)})_{8\times 8} = \begin{pmatrix} s_0 & s_{-2} & s_{-4} & s_{-3} & s_{1.75} & s_1 & s_{-2} & s_2 \\ s_2 & s_0 & s_2 & s_{-1} & s_3 & s_2 & s_1 & s_2 \\ s_4 & s_{-2} & s_0 & s_{-3} & s_2 & s_{0.67} & s_{-1} & s_3 \\ s_3 & s_1 & s_3 & s_0 & s_4 & s_3 & s_1 & s_{3.75} \\ s_{-1.75} & s_{-3} & s_{-2} & s_{-4} & s_0 & s_{-2} & s_{-3} & s_0 \\ s_{-1} & s_{-2} & s_{-0.67} & s_{-3} & s_2 & s_0 & s_{-3} & s_2 \\ s_2 & s_{-1} & s_1 & s_{-1} & s_3 & s_3 & s_0 & s_{3.2} \\ s_{-2} & s_{-2} & s_{-3} & s_{-3.75} & s_0 & s_{-2} & s_{-3.2} & s_0 \end{pmatrix}$$

$$\dot{\boldsymbol{A}}_2 = (\dot{a}_{ij}^{(2)})_{8\times 8} = \begin{pmatrix} s_0 & s_{-3} & s_{0.2} & s_{-1} & s_2 & s_{2.8} & s_1 & s_3 \\ s_3 & s_0 & s_2 & s_{-3} & s_3 & s_2 & s_{1.33} & s_2 \\ s_{-0.2} & s_{-2} & s_0 & s_{-2} & s_1 & s_{2.8} & s_2 & s_2 \\ s_1 & s_3 & s_2 & s_0 & s_3 & s_4 & s_2 & s_3 \\ s_{-2} & s_{-3} & s_{-1} & s_{-3} & s_0 & s_3 & s_{-1} & s_{0.33} \\ s_{-2.8} & s_{-2} & s_{-2.8} & s_{-4} & s_{-3} & s_0 & s_{-4} & s_{-2} \\ s_{-1} & s_{-1.33} & s_{-2} & s_{-2} & s_1 & s_4 & s_0 & s_1 \\ s_{-3} & s_{-2} & s_{-2} & s_{-3} & s_{-0.33} & s_2 & s_{-1} & s_0 \end{pmatrix}$$

$$\dot{\boldsymbol{A}}_3 = (\dot{a}_{ij}^{(3)})_{8\times 8} = \begin{pmatrix} s_0 & s_{-1.17} & s_0 & s_{-1} & s_2 & s_1 & s_3 & s_4 \\ s_{1.17} & s_0 & s_2 & s_3 & s_4 & s_3 & s_1 & s_3 \\ s_0 & s_{-2} & s_0 & s_{-0.2} & s_2 & s_1 & s_2 & s_{2.25} \\ s_1 & s_{-3} & s_{0.2} & s_0 & s_3 & s_1 & s_2 & s_3 \\ s_{-2} & s_{-4} & s_{-2} & s_{-3} & s_0 & s_{-1} & s_{-3} & s_{-2} \\ s_{-1} & s_{-3} & s_{-1} & s_{-1} & s_1 & s_0 & s_{-3} & s_{1.25} \\ s_{-3} & s_{-1} & s_{-2} & s_{-2} & s_3 & s_3 & s_0 & s_3 \\ s_{-4} & s_{-3} & s_{-2.25} & s_{-3} & s_2 & s_{-1.25} & s_{-3} & s_0 \end{pmatrix}$$

After that, we use the LWA$_2$ operator (3.32) to aggregate all the additive linguistic preference relations $\dot{\boldsymbol{A}}_k = (\dot{a}_{ij}^{(k)})_{8\times 8}$ $(k = 1, 2, 3)$ into the collective linguistic preference relation $\dot{\boldsymbol{A}} = (\dot{a}_{ij})_{8\times 8}$:

$$\dot{\boldsymbol{A}} = (\dot{a}_{ij})_{8\times 8} = \begin{pmatrix} s_0 & s_{-2.13} & s_{-1.94} & s_{-2} & s_{1.88} & s_{1.54} & s_{-0.1} & s_{2.7} \\ s_{2.13} & s_0 & s_2 & s_{-0.8} & s_{3.2} & s_{2.2} & s_{1.1} & s_{2.2} \\ s_{1.94} & s_{-2} & s_0 & s_{-2.14} & s_{1.7} & s_{1.38} & s_{0.5} & s_{2.55} \\ s_2 & s_{0.8} & s_{2.14} & s_0 & s_{1.5} & s_{2.9} & s_{1.5} & s_{3.38} \\ s_{-1.88} & s_{-3.2} & s_{-1.7} & s_{-1.5} & s_0 & s_{-0.3} & s_{-2.4} & s_{-0.3} \\ s_{-1.54} & s_{-2.2} & s_{-1.38} & s_{-2.9} & s_{0.3} & s_0 & s_{-3.3} & s_{0.65} \\ s_{0.1} & s_{-1.1} & s_{-0.5} & s_{-1.5} & s_{2.4} & s_{3.3} & s_0 & s_{2.5} \\ s_{-2.7} & s_{-2.2} & s_{-2.55} & s_{-3.38} & s_{0.3} & s_{-0.65} & s_{-2.5} & s_0 \end{pmatrix}$$

and then utilize the LA operator (3.33) to aggregate all the elements $\dot{a}_{ij}$ $(j = 1, 2, \cdots, 8)$ in the ith line of $\bar{\boldsymbol{A}}$ into the averaged preference degree $\bar{a}_i$ of the alternative x_i over all the alternatives:

$$\dot{a}_1 = s_{-0.01}, \quad \dot{a}_2 = s_{1.50}, \quad \dot{a}_3 = s_{0.49}, \quad \dot{a}_4 = s_{1.78}$$

$$\dot{a}_5 = s_{-1.32}, \quad \dot{a}_6 = s_{-1.30}, \quad \dot{a}_7 = s_{0.65}, \quad \dot{a}_8 = s_{-1.71}$$

Based on which rank the alternatives x_i $(i = 1, 2, \cdots, 8)$:

$$x_4 \succ x_2 \succ x_7 \succ x_3 \succ x_1 \succ x_6 \succ x_5 \succ x_8$$

and thus, the best alternative is x_4.

3.3 Dynamic Additive Linguistic Preference Relations

In many real-world situations, such as personnel dynamic examination, military system efficiency dynamic evaluation, etc., the linguistic preference relations provided by the decision makers in a decision making problem are usually collected from different periods, i.e., the linguistic arguments are the functions of time variable. To solve this issue, Xu (2007b) defined the concept of dynamic additive linguistic preference relation, and based on the DLWA operator (2.96), he presented an approach to multi-period decision making with dynamic additive linguistic preference relations:

Definition 3.17 (Xu, 2007b) A dynamic additive linguistic preference relation $\boldsymbol{A}(t)$ on X is given as $\boldsymbol{A}(t) = (a_{ij}(t))_{n\times n} \subset X \times X$, which satisfies

$$s_{-\tau} \leqslant a_{ij}(t) \leqslant s_{\tau}, \quad a_{ij}(t) \oplus a_{ji}(t) = s_0, \quad a_{ii}(t) = s_0, \quad i,j = 1,2,\cdots,n \tag{3.36}$$

where $a_{ij}(t) \in \bar{S}_2$ is an additive linguistic label at the time t, and can be interpreted as the degree that the alternative x_i over x_j. Especially, $a_{ij}(t) = s_0$ denotes indifference between the alternatives x_i and x_j; $a_{ij}(t) > s_0$ means that the alternative x_i is preferred to the alternative x_j, the larger $a_{ij}(t)$, the greater the degree that the alternative x_i is preferred to the alternative x_j; $a_{ij}(t) < s_0$ means that the alternative x_j is preferred to the alternative x_i, the smaller $a_{ij}(t)$, the greater the degree that the alternative x_j is preferred to the alternative x_i.

Definition 3.18 (Xu, 2007b) Let $\boldsymbol{A}(t) = (a_{ij}(t))_{n\times n}$ be a dynamic additive linguistic preference relation. If

$$a_{ij}(t) = a_{ik}(t) \oplus a_{kj}(t), \quad i,j,k = 1,2,\cdots,n \tag{3.37}$$

then $\boldsymbol{A}(t)$ is called a dynamic consistent additive linguistic preference relation.

For the dynamic additive linguistic preference relation $\boldsymbol{A}(t) = (a_{ij}(t))_{n\times n}$, if $t = t_1, t_2, \cdots, t_p$, then $\boldsymbol{A}(t_k) = (a_{ij}(t_k))_{n\times n}$ $(k = 1,2,\cdots,p)$ denote the additive linguistic preference relations collected from p different periods.

Similar to Theorem 3.5, the following conclusion can be obtained:

Theorem 3.17 (Xu, 2007d) Let $\boldsymbol{A}(t_k) = (a_{ij}(t_k))_{n\times n}$ $(k = 1,2,\cdots,p)$ be p additive linguistic preference relations collected from p different periods, and $\boldsymbol{w}(t) = (w(t_1), w(t_2), \cdots, w(t_p))^{\mathrm{T}}$ be the weight vector of the periods t_k $(k = 1,2,\cdots,p)$. Then the weighted combination $\boldsymbol{A}(t) = (a_{ij}(t))_{n\times n}$ of $\boldsymbol{A}(t_k) = (a_{ij}(t_k))_{n\times n}$ $(k = 1,2,\cdots,p)$ derived by the DLWA operator:

$$a_{ij}(t) = w(t_1)a_{ij}(t_1) \oplus w(t_2)a_{ij}(t_2) \oplus \cdots \oplus w(t_p)a_{ij}(t_p), \quad i,j = 1,2,\cdots,n \tag{3.38}$$

is also an additive linguistic preference relation, where $\boldsymbol{w}(t)$ can be determined by the BUM function based method (2.98). Especially, if all $\boldsymbol{A}(t_k) = (a_{ij}(t_k))_{n\times n}$ $(k = 1,2,\cdots,p)$ are the consistent additive linguistic preference relations, then the complex linguistic preference relation $\boldsymbol{A}(t) = (a_{ij}(t))_{n\times n}$ derived by the DLWA operator (3.38) is also a consistent additive linguistic preference relation.

Based on the above knowledge and the DLWA operator (3.38), in the following, we introduce a multi-period decision making approach (Xu, 2007b):

For a multi-period decision making problem, let $X = \{x_1, x_2, \cdots, x_n\}$ be a finite set of alternatives, and t_k $(k = 1, 2, \cdots, p)$ be p different periods, whose weight vector is $\boldsymbol{w}(t) = (w(t_1), w(t_2), \cdots, w(t_p))^{\mathrm{T}}$, where $\boldsymbol{w}(t_k) \geqslant 0\,(k = 1, 2, \cdots, p)$, and $\sum_{k=1}^{p} w(t_k) = 1$. Let $\boldsymbol{A}(t_k) = (a_{ij}(t_k))_{n\times n}$ $(k = 1, 2, \cdots, p)$ be the additive linguistic preference relations collected from p different periods, where $a_{ij}(t_k)$ denotes the preference degree that the alternative x_i over x_j at the period t_k.

We first use the DLWA operator (3.38) to aggregate the additive linguistic preference relations $\boldsymbol{A}(t_k)$ $(k = 1, 2, \cdots, p)$ into the complex additive linguistic preference relation $\boldsymbol{A}(t) = (a_{ij}(t))_{n\times n}$.

Then we utilize the DLA operator:

$$a_i(t) = \mathrm{DLA}\,(a_{i1}(t), a_{i2}(t), \cdots, a_{in}(t)) = \frac{1}{n}[a_{i1}(t) \oplus a_{i2}(t) \oplus \cdots \oplus a_{in}(t)] \quad (3.39)$$

to aggregate all the elements $a_{ij}(t)$ $(j = 1, 2, \cdots, n)$ in the i th line of $\boldsymbol{A}(t)$, and get the averaged preference degree of the alternative x_i over all the alternatives $a_i(t)$ $(i = 1, 2, \cdots, n)$.

Finally, we employ $a_i(t)$ $(i = 1, 2, \cdots, n)$ to rank the alternatives x_i $(i = 1, 2, \cdots, n)$, and then select the best one. The larger $a_i(t)$, the better the alternative x_i.

Example 3.5 (Xu, 2007b) In Example 3.2, assume that the invest company uses the additive linguistic evaluation scale:

$$S_2 = \{s_{-4} = \text{extremely poor},\ s_{-3} = \text{very poor},\ s_{-2} = \text{poor},\ s_{-1} = \text{slightly poor},$$

$$s_0 = \text{fair},\ s_1 = \text{slightly good},\ s_2 = \text{good},\ s_3 = \text{very good},\ s_4 = \text{extremely good}\}$$

to compare each pair of the alternatives: ① x_1 is a car industry; ② x_2 is a food industry; ③ x_3 is a computer industry; ④ x_4 is an arms industry; and ⑤ x_5 is a TV company, with respect to their performance in 2004~2006, and constructs the additive linguistic preference relations $\boldsymbol{A}(t_k)$ $(k = 1, 2, 3)$:

$$\boldsymbol{A}(t_1) = \begin{pmatrix} s_0 & s_1 & s_{-2} & s_2 & s_3 \\ s_{-1} & s_0 & s_{-1} & s_2 & s_1 \\ s_2 & s_1 & s_0 & s_1 & s_{-2} \\ s_{-2} & s_{-2} & s_{-1} & s_0 & s_2 \\ s_{-3} & s_{-3} & s_2 & s_{-2} & s_0 \end{pmatrix}$$

$$\boldsymbol{A}(t_2) = \begin{pmatrix} s_0 & s_2 & s_{-1} & s_1 & s_3 \\ s_{-2} & s_0 & s_{-1} & s_3 & s_2 \\ s_1 & s_1 & s_0 & s_2 & s_{-2} \\ s_{-1} & s_{-3} & s_{-2} & s_0 & s_1 \\ s_{-3} & s_{-2} & s_2 & s_{-1} & s_0 \end{pmatrix}$$

$$
\boldsymbol{A}(t_3) = \begin{pmatrix} s_0 & s_1 & s_{-3} & s_3 & s_2 \\ s_{-1} & s_0 & s_{-2} & s_1 & s_0 \\ s_3 & s_2 & s_0 & s_1 & s_{-3} \\ s_{-3} & s_{-1} & s_{-1} & s_0 & s_2 \\ s_{-2} & s_0 & s_3 & s_{-2} & s_0 \end{pmatrix}
$$

where t_1 denotes "2004", t_2 denotes "2005", t_3 denotes "2006", and $\boldsymbol{w}(t) = (0.23, 0.32, 0.45)^{\mathrm{T}}$ is the weight vector of t_k $(k = 1, 2, 3)$) (derived by (2.100) taking $\alpha = 1$).

We first employ the DLWA operator (3.38) to aggregate the additive linguistic preference relations $\boldsymbol{A}(t_k) = (a_{ij}(t_k))_{5\times 5}$ $(k = 1, 2, 3)$ into the complex additive linguistic preference relation $\boldsymbol{A}(t) = (a_{ij}(t))_{5\times 5}$:

$$
\boldsymbol{A}(t) = \begin{pmatrix} s_0 & s_{1.32} & s_{-2.13} & s_{2.13} & s_{2.55} \\ s_{-1.32} & s_0 & s_{-1.45} & s_{1.87} & s_{0.87} \\ s_{2.13} & s_{1.45} & s_0 & s_{1.32} & s_{-2.45} \\ s_{-2.13} & s_{-1.87} & s_{-1.32} & s_0 & s_{1.68} \\ s_{-2.55} & s_{-0.87} & s_{2.45} & s_{-1.68} & s_0 \end{pmatrix}
$$

Then we use the DLA operator (3.39) to aggregate all the elements $a_{ij}(t)$ $(j = 1, 2, 3, 4, 5)$ in the ith line of $\boldsymbol{A}(t)$ into the averaged preference degree $a_i(t)$ $(i = 1, 2, 3, 4, 5)$ of the alternative x_i over all the alternatives:

$$
a_1(t) = s_{0.77}, \quad a_2(t) = s_{-0.01}, \quad a_3(t) = s_{0.49}, \quad a_4(t) = s_{-0.73}, \quad a_5(t) = s_{-0.53}
$$

Finally, we utilize $a_i(t)$ $(i = 1, 2, 3, 4, 5)$ to rank the alternatives x_i $(i = 1, 2, 3, 4, 5)$:

$$
x_1 \succ x_3 \succ x_2 \succ x_5 \succ x_4
$$

Thus, the best alternative is x_1.

3.4 Multiplicative Linguistic Preference Relations

In the previous sections, we have introduced various linguistic preference relations based on the additive linguistic evaluation scale. From this section, we shall discuss situations where the linguistic preference relations are constructed based on the multiplicative linguistic evaluation scale.

For a decision making problem, the decision maker uses the multiplicative linguistic evaluation scale $S_4 = \{s_\alpha | \alpha = 1/\tau, \cdots, 1/2, 1, 2, \cdots, \tau\}$ to compare each pair of the alternatives in X, and constructs a linguistic preference relation $\boldsymbol{B} = (b_{ij})_{n\times n}$. If

$$
s_{1/\tau} \leqslant b_{ij} \leqslant s_\tau, \quad b_{ij} \otimes b_{ji} = s_1, \quad b_{ii} = s_1, \quad i, j = 1, 2, \cdots, n \tag{3.40}
$$

then $\boldsymbol{B}$ is called a multiplicative linguistic preference relation (Xu, 2004a), where b_{ij} denotes the preference degree of the alternative x_i over x_j. Especially, $b_{ij} = s_1$ denotes indifference between the alternatives x_i and x_j; $b_{ij} > s_1$ denotes the alternative x_i is preferred to the alternative x_j; $b_{ij} = s_\tau$ indicates that the alternative x_i is absolutely preferred to x_j; $b_{ij} < s_1$ means that the alternative x_j is preferred to the alternative x_i, $b_{ij} = s_{1/\tau}$ indicates that the alternative x_j is absolutely preferred to x_i.

Similar to the additive linguistic preference relation, the multiplicative linguistic preference relation has the following properties:

Definition 3.19 Let $\boldsymbol{B} = (b_{ij})_{n\times n}$ be a multiplicative linguistic preference relation. If we remove the ith line and i th column from $\boldsymbol{B}$, then the preference relation composed by the remainder $(n-1)$ lines and $(n-1)$ columns of $\boldsymbol{B}$ is also a multiplicative linguistic preference relation.

Theorem 3.18 Let $\boldsymbol{B} = (b_{ij})_{n\times n}$ be a multiplicative linguistic preference relation. Then

$$\bigotimes_{i,j=1}^{n} b_{ij} = s_1 \tag{3.41}$$

Proof By (3.40), we have

$$\bigotimes_{i,j=1}^{n} b_{ij} = \left[\bigotimes_{i<j}(b_{ij}\otimes b_{ji})\right]\otimes\left(\bigotimes_{i=1}^{n} b_{ii}\right) = \left(\bigotimes_{i<j} s_1\right)\otimes\left(\bigotimes_{i=1}^{n} s_1\right) = s_1$$

which completes the proof.

Definition 3.20 Let $\boldsymbol{B} = (b_{ij})_{n\times n}$ be a multiplicative linguistic preference relation. If $b_{ij}^{\mathrm{T}} = b_{ji}\,(i,j = 1,2,\cdots,n)$, then $\boldsymbol{B}^{\mathrm{T}} = (b_{ij}^{\mathrm{T}})_{n\times n}$ is called the transpose of $\boldsymbol{B}$.

Theorem 3.19 Let $\boldsymbol{B} = (b_{ij})_{n\times n}$ be a multiplicative linguistic preference relation. Then the transpose $\boldsymbol{B}^{\mathrm{T}} = (\boldsymbol{b}_{ij}^{\mathrm{T}})_{n\times n}$ is also a multiplicative linguistic preference relation.

Proof Since

$$s_{1/\tau} \leqslant b_{ij}^{\mathrm{T}} = b_{ji} \leqslant s_\tau, \quad b_{ij}^{\mathrm{T}}\otimes b_{ji}^{\mathrm{T}} = b_{ji}\otimes b_{ij} = s_1, \quad b_{ii}^{\mathrm{T}} = b_{ii} = s_1$$

then $\boldsymbol{B}^{\mathrm{T}} = (b_{ij}^{\mathrm{T}})_{n\times n}$ is a multiplicative linguistic preference relation, which completes the proof.

Theorem 3.20 Let $\boldsymbol{B} = (b_{ij})_{n\times n}$ be a multiplicative linguistic preference relation, and $\boldsymbol{B}^{\mathrm{T}}$ be the transpose of $\boldsymbol{B}$. Then $\boldsymbol{B}^{\mathrm{T}} = \mathrm{rec}(\boldsymbol{B})$, where $\boldsymbol{b}_{ij}^{\mathrm{T}} = \mathrm{rec}(b_{ij})$, "rec" is a reciprocal operator (see Section 1.2).

Definition 3.21 (Xu, 2004a) Let $\boldsymbol{B} = (b_{ij})_{n\times n}$ be a multiplicative linguistic preference relation. If

$$b_{ij} = b_{ik}\otimes b_{kj}, \quad i,j,k = 1,2,\cdots,n \tag{3.42}$$

then $\boldsymbol{B}$ is called a consistent multiplicative linguistic preference relation.

(3.42) is a type of multiplicative transitivity. It represents the idea as follows: Interpreting the value b_{ik} as the intensity of preference of the alternative x_i over x_k, and b_{kj} as the intensity of preference of the alternative x_k over x_k, it is reasonable to suppose that the intensity of preference of x_i over x_j should be equal to the sum of the intensities of preferences when using an intermediate alternative x_k. Especially, if the decision maker provides his/her linguistic preferences over these three alternatives in an indifference situation, that is, $x_i \sim x_k \sim x_j$, then it follows that $b_{ik} = b_{kj} = b_{ij} = s_1$, which coincides with (3.42).

Moreover, the multiplicative linguistic preference relation $\boldsymbol{B} = (b_{ij})_{n\times n}$ has the following properties:

(1) If $b_{ik} \geqslant s_1$, $b_{kj} \geqslant s_1 \Rightarrow b_{ij} \geqslant s_1$ for all $i, j, k = 1, 2, \cdots, n$, then we say $\boldsymbol{B}$ satisfies the weak transitivity property;

(2) If $b_{ij} \geqslant \min\{b_{ik}, b_{kj}\}$ for all $i, j, k = 1, 2, \cdots, n$, then we say $\boldsymbol{B}$ satisfies the max-min transitivity property;

(3) If $b_{ij} \geqslant \max\{b_{ik}, b_{kj}\}$ for all $i, j, k = 1, 2, \cdots, n$, then we say $\boldsymbol{B}$ satisfies the max-max transitivity property;

(4) If $b_{ik} \geqslant s_1$, $b_{kj} \geqslant s_1 \Rightarrow b_{ij} \geqslant \min\{b_{ik}, b_{kj}\}$ for all $i, j, k = 1, 2, \cdots, n$, then $\boldsymbol{B}$ satisfies the restricted max-min transitivity property;

(5) If $b_{ik} \geqslant s_1$, $b_{kj} \geqslant s_1 \Rightarrow b_{ij} \geqslant \max\{b_{ik}, b_{kj}\}$ for all $i, j, k = 1, 2, \cdots, n$, then we say $\boldsymbol{B}$ satisfies the restricted max-max transitivity property.

Theorem 3.21 (Xu, 2004a) Let $\boldsymbol{B}_k = (b_{ij}^{(k)})_{n\times n}$ $(k = 1, 2, \cdots, l)$ be the multiplicative linguistic preference relation given by l decision makers e_k $(k = 1, 2, \cdots, l)$, where

$$s_{1/\tau} \leqslant b_{ij}^{(k)} \leqslant s_\tau, \quad b_{ij}^{(k)} \otimes b_{ji}^{(k)} = s_1, \quad b_{ii}^{(k)} = s_1, \quad k = 1, 2, \cdots, l, \ i, j = 1, 2, \cdots, n,$$

and let $\boldsymbol{v} = (v_1, v_2, \cdots, v_l)^{\mathrm{T}}$ be the weight vector of the decision makers e_k $(k = 1, 2, \cdots, l)$, $v_k \geqslant 0\,(k = 1, 2, \cdots, l)$ and $\sum_{k=1}^{l} v_k = 1$. Then the complex linguistic preference relation $\boldsymbol{B} = (b_{ij})_{n\times n}$ of $\boldsymbol{B}_k = (b_{ij}^{(k)})_{n\times n}$ $(k = 1, 2, \cdots, l)$ is also a multiplicative linguistic preference relation, where

$$b_{ij} = \mathrm{LWG}(b_{ij}^{(1)}, b_{ij}^{(2)}, \cdots, b_{ij}^{(l)}) = (b_{ij}^{(1)})^{v_1} \otimes (b_{ij}^{(2)})^{v_2} \otimes \cdots \otimes (b_{ij}^{(l)})^{v_l}$$

$$s_{1/\tau} \leqslant b_{ij} \leqslant s_\tau, \quad b_{ij} \oplus b_{ji} = S_1, \quad b_{ii} = s_1, \quad i, j = 1, 2, \cdots, n \tag{3.43}$$

If all $\boldsymbol{B}_k = (b_{ij}^{(k)})_{n\times n}$ $(k = 1, 2, \cdots, l)$ are the consistent multiplicative linguistic preference relations, then the collective linguistic preference relation $\boldsymbol{B} = (b_{ij})_{n\times n}$ is also a consistent multiplicative linguistic preference relation.

Proof Since $\boldsymbol{B}_k = (b_{ij}^{(k)})_{n\times n}$ $(k = 1, 2, \cdots, l)$ are the multiplicative linguistic preference relations, then

$$\begin{aligned}
b_{ij} &= (b_{ij}^{(1)})^{\upsilon_1} \otimes (b_{ij}^{(2)})^{\upsilon_2} \otimes \cdots \otimes (b_{ij}^{(l)})^{\upsilon_l} \\
&\geqslant (s_{1/\tau})^{\upsilon_1} \otimes (s_{1/\tau})^{\upsilon_2} \otimes \cdots \otimes (s_{1/\tau})^{\upsilon_l} \\
&= (s_{1/\tau})^{(\upsilon_1+\upsilon_2+\cdots+\upsilon_l)} \\
&= s_{1/\tau}, \quad i, j = 1, 2, \cdots, n \\
b_{ij} &= (b_{ij}^{(1)})^{\upsilon_1} \otimes (b_{ij}^{(2)})^{\upsilon_2} \otimes \cdots \otimes (b_{ij}^{(l)})^{\upsilon_l} \\
&\leqslant (s_{\tau})^{\upsilon_1} \otimes (s_{\tau})^{\upsilon_2} \otimes \cdots \otimes (s_{\tau})^{\upsilon_l} \\
&= (s_{\tau})^{(\upsilon_1+\upsilon_2+\cdots+\upsilon_l)} \\
&= s_{\tau}, \quad i, j = 1, 2, \cdots, n \\
b_{ij} \otimes b_{ji} &= [(b_{ij}^{(1)})^{\upsilon_1} \otimes (b_{ij}^{(2)})^{\upsilon_2} \otimes \cdots \otimes (b_{ij}^{(1)})^{\upsilon_l}] \otimes [(b_{ji}^{(1)})^{\upsilon_1} \otimes (b_{ji}^{(2)})^{\upsilon_2} \otimes \cdots \otimes (b_{ji}^{(l)})^{\upsilon_l}] \\
&= (b_{ij}^{(1)} \otimes b_{ji}^{(1)})^{\upsilon_1} \otimes (b_{ij}^{(2)} \otimes b_{ji}^{(2)})^{\upsilon_2} \otimes \cdots \otimes (b_{ij}^{(l)} \otimes b_{ji}^{(l)})^{\upsilon_l} \\
&= (s_1)^{\upsilon_1} \otimes (s_1)^{\upsilon_2} \otimes \cdots \otimes (s_1)^{\upsilon_l} \\
&= (s_1)^{(\upsilon_1+\upsilon_2+\cdots+\upsilon_l)} \\
&= s_1, \quad i, j = 1, 2, \cdots, n \\
b_{ii} &= (b_{ii}^{(1)})^{\upsilon_1} \otimes (b_{ii}^{(2)})^{\upsilon_2} \otimes \cdots \otimes (b_{ii}^{(l)})^{\upsilon_l} \\
&= (s_1)^{\upsilon_1} \otimes (s_1)^{\upsilon_2} \otimes \cdots \otimes (s_1)^{\upsilon_l} \\
&= (s_1)^{(\upsilon_1+\upsilon_2+\cdots+\upsilon_l)} \\
&= s_1, \quad i = 1, 2, \cdots, n
\end{aligned}$$

Consequently, $\boldsymbol{B} = (b_{ij})_{n\times n}$ is a multiplicative linguistic preference relation.

If all $\boldsymbol{B}_k = (b_{ij}^{(k)})_{n\times n}$ $(k = 1, 2, \cdots, l)$ are the consistent multiplicative linguistic preference relations, i.e., $b_{ij}^{(k)} = b_{is}^{(k)} \otimes b_{sj}^{(k)}$ $(i, j, s = 1, 2, \cdots, n,\ k = 1, 2, \cdots, l)$, then we have

$$\begin{aligned}
b_{ij} &= (b_{ij}^{(1)})^{\upsilon_1} \otimes (b_{ij}^{(2)})^{\upsilon_2} \otimes \cdots \otimes (b_{ij}^{(l)})^{\upsilon_l} \\
&= (b_{is}^{(1)} \otimes b_{sj}^{(1)})^{\upsilon_1} \otimes (b_{is}^{(2)} \otimes b_{sj}^{(2)})^{\upsilon_2} \otimes \cdots \otimes (b_{is}^{(n)} \otimes b_{sj}^{(n)})^{\upsilon_l} \\
&= [(b_{is}^{(1)})^{\upsilon_1} \otimes (b_{is}^{(2)})^{\upsilon_2} \otimes \cdots \otimes (b_{is}^{(n)})^{\upsilon_l}] \otimes [(b_{sj}^{(1)})^{\upsilon_1} \otimes (b_{sj}^{(2)})^{\upsilon_2} \otimes \cdots \otimes (b_{sj}^{(n)})^{\upsilon_l}] \\
&= b_{is} \otimes b_{sj}, \quad i, j, s = 1, 2, \cdots, n
\end{aligned}$$

Thus, the collective linguistic preference relation $\boldsymbol{B} = (b_{ij})_{n\times n}$ is a consistent multiplicative linguistic preference relation. This completes the proof.

Similar to Herrera et al. (2003), we can prove the following theorem:

Theorem 3.22 (Xu, 2004a) Let $\boldsymbol{B}_k = (b_{ij}^{(k)})_{n\times n}$ $(k = 1, 2, \cdots, l)$ be the multiplicative linguistic preference relations given by l decision makers e_k $(k = 1, 2, \cdots, l)$, where

$$s_{1/\tau} \leqslant b_{ij}^{(k)} \leqslant s_\tau, \quad b_{ij}^{(k)} \otimes b_{ji}^{(k)} = s_1, \quad b_{ii}^{(k)} = s_1, \quad k = 1, 2, \cdots, l, \ \ i, j = 1, 2, \cdots, n$$

Then the collective linguistic preference relation $\boldsymbol{B} = (b_{ij})_{n\times n}$ of $\boldsymbol{B}_k = (b_{ij}^{(k)})_{n\times n}$ $(k = 1, 2, \cdots, l)$ is also a multiplicative linguistic preference relation, where

$$b_{ij} = \text{LOWG}(b_{ij}^{(1)}, b_{ij}^{(2)}, \cdots, b_{ij}^{(l)}) = (b_{ij}^{\sigma(1)})^{\omega_1} \otimes (b_{ij}^{\sigma(2)})^{\omega_2} \otimes \cdots \otimes (b_{ij}^{\sigma(l)})^{\omega_l}$$

$$s_{1/\tau} \leqslant b_{ij} \leqslant s_\tau, \quad b_{ij} \otimes b_{ji} = s_1, \quad b_{ii} = s_1, \quad i, j = 1, 2, \cdots, n \tag{3.44}$$

$(\sigma(1), \sigma(2), \cdots, \sigma(n))$ is any permutation of $(1, 2, \cdots, n)$, such that $b_{ij}^{\sigma(k)} \leqslant b_{ij}^{\sigma(k-1)}$ $(k = 1, 2, \cdots, l)$, and $\boldsymbol{\omega} = (\omega_1, \omega_2, \cdots, \omega_l)^{\mathrm{T}}$ can be obtained by (3.6) and (3.7). If all $\boldsymbol{B}_k = (b_{ij}^{(k)})_{n\times n}$ $(k = 1, 2, \cdots, l)$ are the consistent multiplicative linguistic preference relations, then the collective linguistic preference relation $\boldsymbol{B} = (b_{ij})_{n\times n}$ is also a consistent multiplicative linguistic preference relation.

Xu (2004a) utilized the LOWG and LG operators to develop a group decision making approach based on multiplicative linguistic preference relations, which involves the following steps:

Step 1 Given a group decision making problem with multiplicative linguistic preference relations. Each decision maker $e_k \in E$ employs the multiplicative linguistic evaluation scale $S_4 = \{s_\alpha | \alpha = 1/\tau, \cdots, 1/2, 1, 2, \cdots, \tau\}$ to compare each pair of alternatives in X, and constructs a multiplicative linguistic preference relation $\boldsymbol{B}_k = (b_{ij}^{(k)})_{n\times n}$, where

$$s_{1/\tau} \leqslant b_{ij}^{(k)} \leqslant s_\tau, \quad b_{ij}^{(k)} \otimes b_{ji}^{(k)} = s_1, \quad b_{ii}^{(k)} = s_1, \quad i, j = 1, 2, \cdots, n, \ \ k = 1, 2, \cdots, l$$

Step 2 Utilize the LOWG operator (3.44) to aggregate all the multiplicative linguistic preference relations $\boldsymbol{B}_k = (b_{ij}^{(k)})_{n\times n}$ $(k = 1, 2, \cdots, l)$ into the collective multiplicative linguistic preference relation $\boldsymbol{B} = (b_{ij})_{n\times n}$.

Step 3 Utilize the LG operator:

$$b_i = \text{LG}(b_{i1}, b_{i2}, \cdots, b_{in}) \tag{3.45}$$

to aggregate all the preferences in the ith line of the collective multiplicative linguistic preference relation $\boldsymbol{B} = (b_{ij})_{n\times n}$ into the overall preference degree b_i of the alternative x_i over all the alternatives.

Step 4 Rank and select the alternatives x_i $(i = 1, 2, \cdots, n)$ according to b_i $(i = 1, 2, \cdots, n)$.

Example 3.6 (Xu, 2004a) In Example 3.1, suppose that the three decision makers e_k $(k = 1, 2, 3)$ utilize the multiplicative linguistic evaluation scale:

$$S_4 = \{s_{1/5} = \text{extremely poor},\ \ s_{1/4} = \text{very poor},\ \ s_{1/3} = \text{poor},\ \ s_{1/2} = \text{slightly poor},$$

$$s_1 = \text{fair},\ \ s_2 = \text{slightly good},\ \ s_3 = \text{good},\ \ s_4 = \text{very good},\ \ s_5 = \text{extremely good}\}$$

to compare each pair of these five alternatives x_i $(i = 1, 2, 3, 4, 5)$ with respect to the criterion "research", and construct the multiplicative linguistic preference relations $\boldsymbol{B}_k$ $(k = 1, 2, 3)$ respectively:

$$\boldsymbol{B}_1 = \begin{pmatrix} s_1 & s_2 & s_{1/3} & s_4 & s_{1/2} \\ s_{1/2} & s_1 & s_5 & s_2 & s_{1/3} \\ s_3 & s_{1/5} & s_1 & s_3 & s_4 \\ s_{1/4} & s_{1/2} & s_{1/3} & s_1 & s_5 \\ s_2 & s_3 & s_{1/4} & s_{1/5} & s_1 \end{pmatrix}$$

$$\boldsymbol{B}_2 = \begin{pmatrix} s_1 & s_3 & s_{1/2} & s_3 & s_{1/3} \\ s_{1/3} & s_1 & s_5 & s_3 & s_{1/2} \\ s_2 & s_{1/5} & s_1 & s_4 & s_3 \\ s_{1/3} & s_{1/3} & s_{1/4} & s_1 & s_5 \\ s_3 & s_2 & s_{1/3} & s_{1/5} & s_1 \end{pmatrix}$$

$$\boldsymbol{B}_3 = \begin{pmatrix} s_1 & s_4 & s_{1/4} & s_2 & s_{1/4} \\ s_{1/4} & s_1 & s_4 & s_3 & s_{1/3} \\ s_4 & s_{1/4} & s_1 & s_4 & s_2 \\ s_{1/2} & s_{1/3} & s_{1/4} & s_1 & s_5 \\ s_4 & s_3 & s_{1/2} & s_{1/5} & s_1 \end{pmatrix}$$

The following procedure is used to select the optimal alternatives:

Step 1 Use the LOWG operator $b_{ij} = \text{LOWG}\,(b_{ij}^{(1)}, b_{ij}^{(2)}, b_{ij}^{(3)})$ $(i, j = 1, 2, 3, 4, 5)$ (whose weights can be determined by (3.6) and (3.7). Without loss of generality, we take $a = 0.25$ and $b = 0.75$, then $\boldsymbol{\omega} = (1/6, 2/3, 1/6)^{\mathrm{T}}$) to aggregate the multiplicative linguistic preference relaions $\boldsymbol{B}_k = (b_{ij}^{(k)})_{5\times5}$ $(k = 1, 2, \cdots, l)$ provided by the decision makers e_k $(k = 1, 2, 3)$ into the collective multiplicative linguistic preference relation $\boldsymbol{B} = (b_{ij})_{5\times5}$:

$$b_{11} = (s_1)^{1/6} \otimes (s_1)^{2/3} \otimes (s_1)^{1/6} = s_1$$
$$b_{12} = (s_4)^{1/6} \otimes (s_3)^{2/3} \otimes (s_2)^{1/6} = s_{2.94}$$
$$b_{13} = (s_{1/2})^{1/6} \otimes (s_{1/3})^{2/3} \otimes (s_{1/4})^{1/6} = s_{0.34}$$
$$b_{14} = (s_4)^{1/6} \otimes (s_3)^{2/3} \otimes (s_2)^{1/6} = s_{2.94}$$
$$b_{15} = (s_{1/2})^{1/6} \otimes (s_{1/3})^{2/3} \otimes (s_{1/4})^{1/6} = s_{0.34}$$
$$b_{21} = (s_{1/2})^{1/6} \otimes (s_{1/3})^{2/3} \otimes (s_{1/4})^{1/6} = s_1$$

$$b_{22} = (s_1)^{1/6} \otimes (s_1)^{2/3} \otimes (s_1)^{1/6} = s_1$$
$$b_{23} = (s_5)^{1/6} \otimes (s_5)^{2/3} \otimes (s_4)^{1/6} = s_1$$
$$b_{24} = (s_3)^{1/6} \otimes (s_3)^{2/3} \otimes (s_2)^{1/6} = s_1$$
$$b_{25} = (s_{1/2})^{1/6} \otimes (s_{1/3})^{2/3} \otimes (s_{1/3})^{1/6} = s_{0.36}$$
$$b_{31} = (s_4)^{1/6} \otimes (s_3)^{2/3} \otimes (s_2)^{1/6} = s_{2.94}$$
$$b_{32} = (s_{1/4})^{1/6} \otimes (s_{1/5})^{2/3} \otimes (s_{1/5})^{1/6} = s_{0.21}$$
$$b_{33} = (s_1)^{1/6} \otimes (s_1)^{2/3} \otimes (s_1)^{1/6} = s_1$$
$$b_{34} = (s_4)^{1/6} \otimes (s_4)^{2/3} \otimes (s_3)^{1/6} = s_{3.81}$$
$$b_{35} = (s_4)^{1/6} \otimes (s_3)^{2/3} \otimes (s_2)^{1/6} = s_{2.94}$$
$$b_{41} = (s_{1/2})^{1/6} \otimes (s_{1/3})^{2/3} \otimes (s_{1/4})^{1/6} = s_{0.34}$$
$$b_{42} = (s_{1/2})^{1/6} \otimes (s_{1/3})^{2/3} \otimes (s_{1/3})^{1/6} = s_{0.36}$$
$$b_{43} = (s_{1/3})^{1/6} \otimes (s_{1/4})^{2/3} \otimes (s_{1/4})^{1/6} = s_{0.26}$$
$$b_{44} = (s_1)^{1/6} \otimes (s_1)^{2/3} \otimes (s_1)^{1/6} = s_1$$
$$b_{45} = (s_5)^{1/6} \otimes (s_5)^{2/3} \otimes (s_5)^{1/6} = s_5$$
$$b_{51} = (s_4)^{1/6} \otimes (s_3)^{2/3} \otimes (s_2)^{1/6} = s_{2.94}$$
$$b_{52} = (s_3)^{1/6} \otimes (s_3)^{2/3} \otimes (s_2)^{1/6} = s_{2.80}$$
$$b_{53} = (s_{1/2})^{1/6} \otimes (s_{1/3})^{2/3} \otimes (s_{1/4})^{1/6} = s_{0.34}$$
$$b_{54} = (s_{1/5})^{1/6} \otimes (s_{1/5})^{2/3} \otimes (s_{1/5})^{1/6} = s_{0.20}$$
$$b_{55} = (s_1)^{1/6} \otimes (s_1)^{2/3} \otimes (s_1)^{1/6} = s_1$$

$$\boldsymbol{B} = \begin{pmatrix} s_1 & s_{2.94} & s_{0.34} & s_{2.94} & s_{0.34} \\ s_{0.34} & s_1 & s_{4.82} & s_{2.80} & s_{0.36} \\ s_{2.94} & s_{0.21} & s_1 & s_{3.81} & s_{2.94} \\ s_{0.34} & s_{0.36} & s_{0.26} & s_1 & s_5 \\ s_{2.94} & s_{2.80} & s_{0.34} & s_{0.20} & s_1 \end{pmatrix}$$

Step 2 Utilize the LG operator $b_i = \mathrm{LG}(b_{i1}, b_{i2}, b_{i3}, b_{i4}, b_{i5})$ to aggregate all the preferences in the ith line of the collective multiplicative linguistic preference relation $\boldsymbol{B} = (b_{ij})_{n\times n}$ into the overall preference degree b_i of the alternative x_i over all the alternatives:

$$b_1 = s_1, \quad b_2 = s_{1.11}, \quad b_3 = s_{1.47}, \quad b_4 = s_{0.69}, \quad b_5 = s_{0.89}$$

Then

$$b_3 > b_2 > b_1 > b_5 > b_4$$

from which we rank the alternatives x_i $(i = 1, 2, \cdots, 5)$:

$$x_3 \succ x_2 \succ x_1 \succ x_5 \succ x_4$$

Thus, the best alternative is x_3.

Zhang and Xu (2006) applied the LOWG operator with symmetric weighting vector to group decision making:

Theorem 3.23 (Zhang and Xu, 2006) Suppose that the decision maker $e_k \in E$ employs the multiplicative linguistic evaluation scale S_4 to compare each pair of the alternatives in X, and constructs the multiplicative linguistic preference relation $\boldsymbol{B}_k = (b_{ij}^{(k)})_{n\times n}$ $(k = 1, 2, \cdots, l)$, where $s_{1/\tau} \leqslant b_{ij}^{(k)} \leqslant s_\tau$, $b_{ij}^{(k)} \otimes b_{ji}^{(k)} = s_1$, $b_{ii}^{(k)} = s_1$ $(i, j = 1, 2, \cdots, n)$. Then the collective linguistic preference relation $\boldsymbol{B} = (b_{ij})_{n\times n}$ derived by the LOWG operator with the symmetric weighting vector is also a multiplicative linguistic preference relation.

Proof Let $\boldsymbol{\omega} = (\omega_1, \omega_2, \cdots, \omega_m)^{\mathrm{T}}$ be the weighting vector associated with the LOWG operator. Then we utilize the LOWG operator to aggregate the multiplicative linguistic preference relations $\boldsymbol{B}_k = (b_{ij}^{(k)})_{n\times n}$ $(k = 1, 2, \cdots, l)$, and get $\boldsymbol{B} = (b_{ij})_{n\times n}$, where

$$b_{ij} = \mathrm{LOWG}(b_{ij}^{(1)}, b_{ij}^{(2)}, \cdots, b_{ij}^{(l)}) = (b_{ij}^{\sigma(1)})^{\omega_1} \otimes (b_{ij}^{\sigma(2)})^{\omega_2} \otimes \cdots \otimes (b_{ij}^{\sigma(l)})^{\omega_l}, \quad i, j = 1, 2, \cdots, n$$

and $b_{ij}^{\sigma(k)}$ is the kth largest of $(b_{ij}^{(1)}, b_{ij}^{(2)}, \cdots, b_{ij}^{(l)})$. Since $b_{ij}^{\sigma(k)} \otimes b_{ji}^{\sigma(k)} = s_1$, then $b_{ji}^{\sigma(k)}$ is the kth smallest of $(b_{ji}^{(1)}, b_{ji}^{(2)}, \cdots, b_{ji}^{(l)})$. Considering that the weighting vector $\boldsymbol{\omega} = (\omega_1, \omega_2, \cdots, \omega_l)^{\mathrm{T}}$ is symmetric, $\omega_k = \omega_{n+1-k}$ for all k, we have

$$\begin{aligned}
b_{ij} \otimes b_{ji} =& \mathrm{LOWG}(b_{ij}^{(1)}, b_{ij}^{(2)}, \cdots, b_{ij}^{(l)}) \otimes \mathrm{LOWG}(b_{ji}^{(1)}, b_{ji}^{(2)}, \cdots, b_{ji}^{(l)}) \\
=& [(b_{ij}^{\sigma(1)})^{\omega_1} \otimes (b_{ij}^{\sigma(2)})^{\omega_2} \otimes \cdots \otimes (b_{ij}^{\sigma(l)})^{\omega_l}] \\
& \otimes [(b_{ji}^{\sigma(1)})^{\omega_l} \otimes (b_{ji}^{\sigma(2)})^{\omega_{l-1}} \otimes \cdots \otimes (b_{ji}^{\sigma(l)})^{\omega_1}] \\
=& [(b_{ij}^{\sigma(1)})^{\omega_1} \otimes (b_{ij}^{\sigma(2)})^{\omega_2} \otimes \cdots \otimes (b_{ij}^{\sigma(l)})^{\omega_l}] \\
& \otimes [(b_{ji}^{\sigma(1)})^{\omega_1} \otimes (b_{ji}^{\sigma(2)})^{\omega_2} \otimes \cdots \otimes (b_{ji}^{\sigma(l)})^{\omega_l}] \\
=& (b_{ij}^{\sigma(1)} \otimes b_{ji}^{\sigma(1)})^{\omega_1} \otimes (b_{ij}^{\sigma(2)} \otimes b_{ji}^{\sigma(2)})^{\omega_2} \otimes \cdots \otimes (b_{ij}^{\sigma(l)} \otimes b_{ji}^{\sigma(l)})^{\omega_l} \\
=& (s_1)^{\omega_1} \otimes (s_1)^{\omega_2} \otimes \cdots \otimes (s_1)^{\omega_l} \\
=& (s_1)^{(\omega_1 + \omega_2 + \cdots + \omega_l)} \\
=& s_1, \quad i, j = 1, 2, \cdots, n
\end{aligned}$$

Since $s_{1/\tau} \leqslant b_{ij}^{(k)} \leqslant s_\tau$ for all i, j and $b_{ii}^{(k)} = s_1$, then we get

$$\begin{aligned}
b_{ij} &= \mathrm{LOWG}(b_{ij}^{(1)}, b_{ij}^{(2)}, \cdots, b_{ij}^{(l)}) = (b_{ij}^{\sigma(1)})^{\omega_1} \otimes (b_{ij}^{\sigma(2)})^{\omega_2} \otimes \cdots \otimes (b_{ij}^{\sigma(l)})^{\omega_l} \\
&\geqslant (s_{1/\tau})^{\omega_1} \otimes (s_{1/\tau})^{\omega_2} \otimes \cdots \otimes (s_{1/\tau})^{\omega_l} \\
&= (s_{1/\tau})^{(\omega_1 + \omega_2 + \cdots + \omega_l)} = s_{1/\tau}, \quad i, j = 1, 2, \cdots, n
\end{aligned}$$

$$
\begin{aligned}
b_{ij} &= \mathrm{LOWG}(b_{ij}^{(1)}, b_{ij}^{(2)}, \cdots, b_{ij}^{(l)}) = (b_{ij}^{\sigma(1)})^{\omega_1} \otimes (b_{ij}^{\sigma(2)})^{\omega_2} \otimes \cdots \otimes (b_{ij}^{\sigma(l)})^{\omega_l} \\
&\leqslant (s_\tau)^{\omega_1} \otimes (s_\tau)^{\omega_2} \otimes \cdots \otimes (s_\tau)^{\omega_l} \\
&= (s_\tau)^{(\omega_1+\omega_2+\cdots+\omega_l)} = s_\tau, \quad i,j = 1,2,\cdots,n \\
b_{ii} &= \mathrm{LOWG}(b_{ii}^{(1)}, b_{ii}^{(2)}, \cdots, b_{ii}^{(l)}) = (b_{ii}^{\sigma(1)})^{\omega_1} \otimes (b_{ii}^{\sigma(2)})^{\omega_2} \otimes \cdots \otimes (b_{ii}^{\sigma(l)})^{\omega_l} \\
&= (s_1)^{\omega_1} \otimes (s_1)^{\omega_2} \otimes \cdots \otimes (s_1)^{\omega_l} \\
&= (s_1)^{(\omega_1+\omega_2+\cdots+\omega_l)} = s_1, \quad i,j = 1,2,\cdots,n
\end{aligned}
$$

Accordingly, the collective linguistic preference relation $\boldsymbol{B} = (b_{ij})_{n\times n}$ is a multiplicative linguistic preference relation. This completes the proof.

Since the weighting vector $\boldsymbol{\omega} = (\omega_1, \omega_2, \cdots, \omega_l)^{\mathrm{T}}$ obtained by (3.6) and (3.7) is symmetric, then Theorem 3.22 is a special case of Theorem 3.23.

3.5 Incomplete Multiplicative Linguistic Preference Relations

In what follows, we introduce situations where the preference information provided by the decision makers is incomplete:

Definition 3.22 (Xu, 2006d) Let $\boldsymbol{B} = (b_{ij})_{n\times n}$ be a multiplicative linguistic preference relation. If some of the elements in $\boldsymbol{B}$ are missing, then $\boldsymbol{B}$ is called an incomplete multiplicative linguistic preference relation, where the missing elelments are denoted by "x", and the known elements have the following properties:

$$
s_{1/\tau} \leqslant b_{ij} \leqslant s_\tau, \quad b_{ij} \otimes b_{ji} = s_1, \quad b_{ii} = s_1 \tag{3.46}
$$

For convenience, let Θ be the set of all the known elements in $\boldsymbol{B}$.

Definition 3.23 Let $\boldsymbol{B} = (b_{ij})_{n\times n}$ be a multiplicative linguistic preference relation. If the elements of the main diagonal are s_1, and all the other elements in $\boldsymbol{B}$ are missing, then $\boldsymbol{B} = (b_{ij})_{n\times n}$ is called a completely incomplete multiplicative linguistic preference relation.

Definition 3.24 Let $\boldsymbol{B} = (b_{ij})_{n\times n}$ be an incomplete multiplicative linguistic preference relation. If all the knwon elements satisfy

$$
b_{ij} = b_{ik} \otimes b_{kj}, \quad b_{ik}, b_{jk}, b_{ij} \in \Theta \tag{3.47}
$$

then $\boldsymbol{B}$ is called a consistent incomplete multiplicative linguistic preference relation.

Definition 3.25 (Xu, 2006d) Let $\boldsymbol{B} = (b_{ij})_{n\times n}$ be an incomplete multiplicative linguistic preference relation. If $(i,\ j) \cap (k,\ l) \neq \varnothing$, then the elements b_{ij} and b_{kl} are adjoining.

Definition 3.26 (Xu, 2006d) Let $\boldsymbol{B} = (b_{ij})_{n\times n}$ be an incomplete multiplicative linguistic preference relation, and let $b_{i_0j_0}$be an unknown element. If there exist the adjoining known elements $b_{i_0j_1}$,$b_{j_1j_2}, \cdots, b_{j_kj_0}$, then $b_{i_0j_0}$ is called available indirectly.

Definition 3.27 (Xu, 2006d) Let $\boldsymbol{B} = (b_{ij})_{n\times n}$ be an incomplete multiplicative linguistic preference relation. If any unknown elements in $\boldsymbol{B}$ can be indirectly obtained by the known elements, then $\boldsymbol{B}$ is called acceptable; otherwise, $\boldsymbol{B}$ is called unacceptable.

Definition 3.28 Let $\Phi(\boldsymbol{B})$ be the digraph of the incomplete multiplicative linguistic preference relation $\boldsymbol{B} = (b_{ij})_{n\times n}$. If every two nodes are mutually reachable, then $\Phi(B)$ is called strong connected.

Theorem 3.24 The incomplete multiplicative linguistic preference relation $\boldsymbol{B} = (b_{ij})_{n\times n}$ is acceptable if and only if the digraph $\Phi(\boldsymbol{B})$ is strong connected.

Theorem 3.25 Let $\boldsymbol{B} = (b_{ij})_{n\times n}$ be an incomplete multiplicative linguistic preference relation. If we remove the ith line and i th column from B, then the preference relation composed by the remainder $(n-1)$ lines and $(n-1)$ columns of $\boldsymbol{B}$ is also a multiplicative linguistic preference relation or incomplete multiplicative linguistic preference relation.

Theorem 3.26 Let $\boldsymbol{B} = (b_{ij})_{n\times n}$ be an incomplete multiplicative linguistic preference Relation. If $\boldsymbol{B}$ is acceptable, then there exists at least one known element (except diagonal elements) in each line or each column of $\boldsymbol{B}$, i.e., there exist at least $n-1$ judgments provided by the decision maker.

The incomplete multiplicative linguistic preference relation $\boldsymbol{B} = (b_{ij})_{n\times n}$ has the following properties:

(1) If $b_{ik} \geqslant s_1$, $b_{kj} \geqslant s_1 \Rightarrow b_{ij} \geqslant s_1$ for all $b_{ik}, b_{kj}, b_{ij} \in \Theta$, then we say $\boldsymbol{B}$ satisfies the weak transitivity property;

(2) If $b_{ij} \geqslant \min\{b_{ik}, b_{kj}\}$ for all $b_{ik}, b_{kj}, b_{ij} \in \Theta$, then we say $\boldsymbol{B}$ satisfies the max-min transitivity property;

(3) If $b_{ij} \geqslant \max\{b_{ik}, b_{kj}\}$ for all $b_{ik}, b_{kj}, b_{ij} \in \Theta$, then we say $\boldsymbol{B}$ satisfies the max-max transitivity property;

(4) If $b_{ik} \geqslant s_1$, $b_{kj} \geqslant s_1 \Rightarrow b_{ij} \geqslant \min\{b_{ik}, b_{kj}\}$ for all $b_{ik}, b_{kj}, b_{ij} \in \Theta$, then we say $\boldsymbol{B}$ satisfies the restricted max-min transitivity property;

(5) If $b_{ik} \geqslant s_1$, $b_{kj} \geqslant s_1 \Rightarrow b_{ij} \geqslant \max\{b_{ik}, b_{kj}\}$ for all $b_{ik}, b_{kj}, b_{ij} \in \Theta$, then we say $\boldsymbol{B}$ satisfies the stricted max-max transitivity property.

In what follows, based on the acceptable incomplete multiplicative linguistic preference relation with the least (i.e., $n-1$) judgments, we introduce a method for constructing a complete multiplicative linguistic preference relation:

Step 1 For a decision making problem, let $X = \{x_1, x_2, \cdots, x_n\}$ be a finite set of alternatives. The decision maker compares each pair of alternatives in X, and employs the multiplicative linguistic evaluation scale S_4 to construct an acceptable incomplete multiplicative linguistic preference relation $\boldsymbol{B} = (b_{ij})_{n\times n}$ with only $n-1$ juegments $b_{ij_1}, b_{ij_2}, \cdots, b_{ij_{n-1}}$.

Step 2 Utilize the known elements in $\boldsymbol{B}$ and the following formula to determine all the unknown elements:

$$\dot{b}_{ij} = \max\{s_{1/\tau},\ \min\{s_\tau,\ b_{ik} \otimes b_{kj}\}\},\quad b_{ik}, b_{kj} \in \Theta \tag{3.48}$$

until get a complete multiplicative linguistic preference relation $\dot{\boldsymbol{B}} = (\dot{b}_{ij})_{n\times n}$, where

$$\dot{b}_{ij} = \begin{cases} \max\{s_{1/\tau},\ \min\{s_\tau,\ b_{ik} \otimes b_{kj}\}\}, & b_{ij} \notin \Theta, b_{ik}, b_{kj} \in \Theta \\ b_{ij}, & b_{ij} \in \Theta \end{cases} \tag{3.49}$$

Now we consider group decision making problems based on incomplete multiplicative linguistic preference relations, and introduce an approach to group decision making based on the acceptable incomplete linguistic preference relations with the least judgments. Xu (2006d) developed an approach to group decision making based on the acceptable incomplete multiplicative linguistic preference relations with the least judgments, which involves the following steps:

Step 1 For a group decision making problem, let X, E and v be defined as in Section 3.2, the decision maker $e_k \in E$ utilizes the multiplicative linguistic evaluation scale $S_4 = \{s_\alpha | \alpha = 1/\tau, \cdots, 1/2, 1, 2, \cdots, \tau\}$ to compare each pair of alternatives in X, and constructs an acceptable incomplete multiplicative linguistic preference relation $\boldsymbol{B}_k = (b_{ij}^{(k)})_{n\times n}$ with only $n-1$ judgments $b_{ij_1}^{(k)}, b_{ij_2}^{(k)}, \cdots, b_{ij_{n-1}}^{(k)}$.

Step 2 Use the known elements in $\boldsymbol{B}_k$ $(k = 1, 2, \cdots, l)$ and the formula (3.48) to determine the unknown elements $\boldsymbol{B}_k$ $(k = 1, 2, \cdots, l)$, until get the complete multiplicative linguistic preference relations $\dot{\boldsymbol{B}}_k = (\dot{b}_{ij}^{(k)})_{n\times n}$ $(k = 1, 2, \cdots, l)$ respectively.

Step 3 Employ the LWG operator:

$$\dot{b}_{ij} = (\dot{b}_{ij}^{(1)})^{v_1} \otimes (\dot{b}_{ij}^{(2)})^{v_2} \otimes \cdots \otimes (\dot{b}_{ij}^{(l)})^{v_l},\quad i, j = 1, 2, \cdots, n \tag{3.50}$$

to aggregate the multiplicative linguistic preference relations $\dot{\boldsymbol{B}}_k = (\dot{b}_{ij}^{(k)})_{n\times n}$ $(k = 1, 2, \cdots, l)$ into the collective multiplicative linguistic preference relation $\dot{\boldsymbol{B}} = (\dot{b}_{ij})_{n\times n}$.

Step 4 Employ the LG operator:

$$\dot{b}_i = (\dot{b}_{i1} \otimes \dot{b}_{i2} \otimes \cdots \otimes \dot{b}_{in})^{\frac{1}{n}},\quad i = 1, 2, \cdots, n \tag{3.51}$$

to aggregate all the preferences $\dot{b}_{ij}$ $(j = 1, 2, \cdots, n)$ in the ith line of $\dot{\boldsymbol{B}}$ into the averaged preference $\dot{b}_i$ of the alternative i over all the alternatives.

Step 5 Rank and select the alternatives x_i $(i = 1, 2, \cdots, n)$ according to $\dot{b}_i$ $(i = 1, 2, \cdots, n)$.

Example 3.7 In Example 3.3, suppose that three decision makers e_k $(k = 1, 2, 3)$ employ the multiplicative linguistic evaluation scale:

$$S_4 = \{s_{1/5} = \text{extremely poor},\quad s_{1/4} = \text{very poor},\quad s_{1/3} = \text{poor},$$

$$s_{1/2} = \text{slightly poor}, \quad s_1 = \text{fair}, \quad s_2 = \text{slightly good},$$

$$s_3 = \text{good}, \quad s_4 = \text{very good}, \quad s_5 = \text{extremely good}\}$$

to compare each pair of the alternatives x_i $(i = 1, 2, \cdots, 8)$ with respect to the criterion "research", and provide the following judgments:

(1) e_1: $b_{12}^{(1)} = s_{1/3}$, $b_{13}^{(1)} = s_{1/2}$, $b_{14}^{(1)} = s_{1/4}$, $b_{15}^{(1)} = s_2$, $b_{16}^{(1)} = s_1$, $b_{17}^{(1)} = s_{1/3}$, $b_{18}^{(1)} = s_2$;

(2) e_2: $b_{41}^{(2)} = s_3$, $b_{42}^{(2)} = s_2$, $b_{43}^{(2)} = s_3$, $b_{45}^{(2)} = s_4$, $b_{46}^{(2)} = s_5$, $b_{47}^{(2)} = s_3$, $b_{48}^{(2)} = s_4$;

(3) e_3: $b_{12}^{(3)} = s_{1/3}$, $b_{23}^{(3)} = s_3$, $b_{34}^{(3)} = s_{1/2}$, $a_{45}^{(3)} = s_4$, $b_{56}^{(3)} = s_{1/2}$, $b_{67}^{(3)} = s_{1/2}$, $b_{78}^{(3)} = s_2$.

According to Definition 3.24 and the above preferences provided by the decision makers e_k $(k = 1, 2, 3)$, we construct the acceptable incomplete multiplicative linguistic preference relations $\boldsymbol{B}_k$ $(k = 1, 2, 3)$:

$$\boldsymbol{B}_1 = (b_{ij}^{(1)})_{8\times 8} = \begin{pmatrix} s_1 & s_{1/3} & s_{1/2} & s_{1/4} & s_2 & s_1 & s_{1/3} & s_2 \\ s_3 & s_1 & x & x & x & x & x & x \\ s_2 & x & s_1 & x & x & x & x & x \\ s_4 & x & x & s_1 & x & x & x & x \\ s_{1/2} & x & x & x & s_1 & x & x & x \\ s_1 & x & x & x & x & s_1 & x & x \\ s_3 & x & x & x & x & x & s_1 & x \\ s_{1/2} & x & x & x & x & x & x & s_1 \end{pmatrix}$$

$$\boldsymbol{B}_2 = (b_{ij}^{(2)})_{8\times 8} = \begin{pmatrix} s_1 & x & x & s_{1/3} & x & x & x & x \\ x & s_1 & x & s_{1/2} & x & x & x & x \\ x & x & s_1 & s_{1/3} & x & x & x & x \\ s_3 & s_2 & s_3 & s_1 & s_4 & s_5 & s_3 & s_4 \\ x & x & x & s_{1/4} & s_1 & x & x & x \\ x & x & x & s_{1/5} & x & s_1 & x & x \\ x & x & x & s_{1/3} & x & x & s_1 & x \\ x & x & x & s_{1/4} & x & x & x & s_1 \end{pmatrix}$$

$$\boldsymbol{B}_3 = (b_{ij}^{(3)})_{8\times 8} = \begin{pmatrix} s_1 & x & x & x & x & x & x & s_2 \\ x & s_1 & x & x & x & x & x & s_4 \\ x & x & s_1 & x & x & x & x & s_2 \\ x & x & x & s_1 & x & x & x & s_3 \\ x & x & x & x & s_1 & x & x & s_{1/2} \\ x & x & x & x & x & s_1 & x & s_1 \\ x & x & x & x & x & x & s_1 & s_2 \\ s_{1/2} & s_{1/4} & s_{1/2} & s_{1/3} & s_2 & s_1 & s_{1/2} & s_1 \end{pmatrix}$$

Step 1 Utilize the known elements in $\boldsymbol{B}_k$ $(k = 1, 2, 3)$ and the formula (3.48) to determine all the unknown elements in $\boldsymbol{B}_k$ $(k = 1, 2, 3)$:

$$\begin{aligned}\dot{b}_{23}^{(1)} &= \max\{s_{1/5}, \min\{s_5, b_{21}^{(1)} \otimes b_{13}^{(1)}\}\} = \max\{s_{1/5}, \min\{s_5, s_3 \otimes s_{1/2}\}\} \\ &= \max\{s_{1/5}, \min\{s_5, s_{3/2}\}\} = \max\{s_{1/5}, s_{3/2}\} = s_{3/2}\end{aligned}$$

Similarly, we have

$$\begin{aligned}
\dot{b}_{24}^{(1)} &= \max\{s_{1/5}, \min\{s_5, b_{21}^{(1)} \otimes b_{14}^{(1)}\}\} = s_{3/4} \\
\dot{b}_{25}^{(1)} &= \max\{s_{1/5}, \min\{s_5, b_{21}^{(1)} \otimes b_{15}^{(1)}\}\} = s_5 \\
\dot{b}_{26}^{(1)} &= \max\{s_{1/5}, \min\{s_5, b_{21}^{(1)} \otimes b_{16}^{(1)}\}\} = s_3 \\
\dot{b}_{27}^{(1)} &= \max\{s_{1/5}, \min\{s_5, b_{21}^{(1)} \otimes b_{17}^{(1)}\}\} = s_1 \\
\dot{b}_{28}^{(1)} &= \max\{s_{1/5}, \min\{s_5, b_{21}^{(1)} \otimes b_{18}^{(1)}\}\} = s_5 \\
\dot{b}_{34}^{(1)} &= \max\{s_{1/5}, \min\{s_5, b_{31}^{(1)} \otimes b_{14}^{(1)}\}\} = s_{1/2} \\
\dot{b}_{35}^{(1)} &= \max\{s_{1/5}, \min\{s_5, b_{31}^{(1)} \otimes b_{15}^{(1)}\}\} = s_4 \\
\dot{b}_{36}^{(1)} &= \max\{s_{1/5}, \min\{s_5, b_{31}^{(1)} \otimes b_{16}^{(1)}\}\} = s_2 \\
\dot{b}_{37}^{(1)} &= \max\{s_{1/5}, \min\{s_5, b_{31}^{(1)} \otimes b_{17}^{(1)}\}\} = s_{2/3} \\
\dot{b}_{38}^{(1)} &= \max\{s_{1/5}, \min\{s_5, b_{31}^{(1)} \otimes b_{18}^{(1)}\}\} = s_4 \\
\dot{b}_{45}^{(1)} &= \max\{s_{1/5}, \min\{s_5, b_{41}^{(1)} \otimes b_{15}^{(1)}\}\} = s_5 \\
\dot{b}_{46}^{(1)} &= \max\{s_{1/5}, \min\{s_5, b_{41}^{(1)} \otimes b_{16}^{(1)}\}\} = s_4 \\
\dot{b}_{47}^{(1)} &= \max\{s_{1/5}, \min\{s_5, b_{41}^{(1)} \otimes b_{17}^{(1)}\}\} = s_{4/3} \\
\dot{b}_{48}^{(1)} &= \max\{s_{1/5}, \min\{s_5, b_{41}^{(1)} \otimes b_{18}^{(1)}\}\} = s_5 \\
\dot{b}_{56}^{(1)} &= \max\{s_{1/5}, \min\{s_5, b_{51}^{(1)} \otimes b_{16}^{(1)}\}\} = s_{1/2} \\
\dot{b}_{57}^{(1)} &= \max\{s_{1/5}, \min\{s_5, b_{51}^{(1)} \otimes b_{17}^{(1)}\}\} = s_{1/5} \\
\dot{b}_{58}^{(1)} &= \max\{s_{1/5}, \min\{s_5, b_{51}^{(1)} \otimes b_{18}^{(1)}\}\} = s_1 \\
\dot{b}_{67}^{(1)} &= \max\{s_{1/5}, \min\{s_5, b_{61}^{(1)} \otimes b_{17}^{(1)}\}\} = s_{1/3} \\
\dot{b}_{68}^{(1)} &= \max\{s_{1/5}, \min\{s_5, b_{61}^{(1)} \otimes b_{18}^{(1)}\}\} = s_2 \\
\dot{b}_{78}^{(1)} &= \max\{s_{1/5}, \min\{s_5, b_{71}^{(1)} \otimes b_{18}^{(1)}\}\} = s_5 \\
\dot{b}_{12}^{(2)} &= \max\{s_{1/5}, \min\{s_5, b_{14}^{(2)} \otimes b_{42}^{(2)}\}\} = s_{2/3} \\
\dot{b}_{13}^{(2)} &= \max\{s_{1/5}, \min\{s_5, b_{14}^{(2)} \otimes b_{43}^{(2)}\}\} = s_1 \\
\dot{b}_{15}^{(2)} &= \max\{s_{1/5}, \min\{s_5, b_{14}^{(2)} \otimes b_{45}^{(2)}\}\} = s_{4/3} \\
\dot{b}_{16}^{(2)} &= \max\{s_{1/5}, \min\{s_5, b_{14}^{(2)} \otimes b_{46}^{(2)}\}\} = s_{5/3} \\
\dot{b}_{17}^{(2)} &= \max\{s_{1/5}, \min\{s_5, b_{14}^{(2)} \otimes b_{47}^{(2)}\}\} = s_1 \\
\dot{b}_{18}^{(2)} &= \max\{s_{1/5}, \min\{s_5, b_{14}^{(2)} \otimes b_{48}^{(2)}\}\} = s_{4/3} \\
\dot{b}_{23}^{(2)} &= \max\{s_{1/5}, \min\{s_5, b_{24}^{(2)} \otimes b_{43}^{(2)}\}\} = s_{3/2}
\end{aligned}$$

$$
\begin{aligned}
\dot{b}_{25}^{(2)} &= \max\{s_{1/5}, \min\{s_5, b_{24}^{(2)} \otimes b_{45}^{(2)}\}\} = s_2\\
\dot{b}_{26}^{(2)} &= \max\{s_{1/5}, \min\{s_5, b_{24}^{(2)} \otimes b_{46}^{(2)}\}\} = s_{5/2}\\
\dot{b}_{27}^{(2)} &= \max\{s_{1/5}, \min\{s_5, b_{24}^{(2)} \otimes b_{47}^{(2)}\}\} = s_{3/4}\\
\dot{b}_{28}^{(2)} &= \max\{s_{1/5}, \min\{s_5, b_{24}^{(2)} \otimes b_{48}^{(2)}\}\} = s_2\\
\dot{b}_{35}^{(2)} &= \max\{s_{1/5}, \min\{s_5, b_{34}^{(2)} \otimes b_{45}^{(2)}\}\} = s_{4/3}\\
\dot{b}_{36}^{(2)} &= \max\{s_{1/5}, \min\{s_5, b_{34}^{(2)} \otimes b_{46}^{(2)}\}\} = s_{5/3}\\
\dot{b}_{37}^{(2)} &= \max\{s_{1/5}, \min\{s_5, b_{34}^{(2)} \otimes b_{47}^{(2)}\}\} = s_1\\
\dot{b}_{38}^{(2)} &= \max\{s_{1/5}, \min\{s_5, b_{34}^{(2)} \otimes b_{48}^{(2)}\}\} = s_{4/3}\\
\dot{b}_{56}^{(2)} &= \max\{s_{1/5}, \min\{s_5, b_{54}^{(2)} \otimes b_{46}^{(2)}\}\} = s_{5/4}\\
\dot{b}_{57}^{(2)} &= \max\{s_{1/5}, \min\{s_5, b_{54}^{(2)} \otimes b_{47}^{(2)}\}\} = s_{3/4}\\
\dot{b}_{58}^{(2)} &= \max\{s_{1/5}, \min\{s_5, b_{54}^{(2)} \otimes b_{48}^{(2)}\}\} = s_1\\
\dot{b}_{67}^{(2)} &= \max\{s_{1/5}, \min\{s_5, b_{64}^{(2)} \otimes b_{47}^{(2)}\}\} = s_{3/5}\\
\dot{b}_{68}^{(2)} &= \max\{s_{1/5}, \min\{s_5, b_{64}^{(2)} \otimes b_{48}^{(2)}\}\} = s_{4/5}\\
\dot{b}_{78}^{(2)} &= \max\{s_{1/5}, \min\{s_5, b_{74}^{(2)} \otimes b_{48}^{(2)}\}\} = s_{4/3}\\
\dot{b}_{12}^{(3)} &= \max\{s_{1/5}, \min\{s_5, b_{18}^{(3)} \otimes b_{82}^{(3)}\}\} = s_{1/2}\\
\dot{b}_{13}^{(3)} &= \max\{s_{1/5}, \min\{s_5, b_{18}^{(3)} \otimes b_{83}^{(3)}\}\} = s_1\\
\dot{b}_{14}^{(3)} &= \max\{s_{1/5}, \min\{s_5, b_{18}^{(3)} \otimes b_{84}^{(3)}\}\} = s_{3/2}\\
\dot{b}_{15}^{(3)} &= \max\{s_{1/5}, \min\{s_5, b_{18}^{(3)} \otimes b_{85}^{(3)}\}\} = s_4\\
\dot{b}_{16}^{(3)} &= \max\{s_{1/5}, \min\{s_5, b_{18}^{(3)} \otimes b_{86}^{(3)}\}\} = s_2\\
\dot{b}_{17}^{(3)} &= \max\{s_{1/5}, \min\{s_5, b_{18}^{(3)} \otimes b_{87}^{(3)}\}\} = s_1\\
\dot{b}_{23}^{(3)} &= \max\{s_{1/5}, \min\{s_5, b_{28}^{(3)} \otimes b_{83}^{(3)}\}\} = s_2\\
\dot{b}_{24}^{(3)} &= \max\{s_{1/5}, \min\{s_5, b_{28}^{(3)} \otimes b_{84}^{(3)}\}\} = s_{4/3}\\
\dot{b}_{25}^{(3)} &= \max\{s_{1/5}, \min\{s_5, b_{28}^{(3)} \otimes b_{85}^{(3)}\}\} = s_5\\
\dot{b}_{26}^{(3)} &= \max\{s_{1/5}, \min\{s_5, b_{28}^{(3)} \otimes b_{86}^{(3)}\}\} = s_4\\
\dot{b}_{27}^{(3)} &= \max\{s_{1/5}, \min\{s_5, b_{28}^{(3)} \otimes b_{87}^{(3)}\}\} = s_2\\
\dot{b}_{34}^{(3)} &= \max\{s_{1/5}, \min\{s_5, b_{38}^{(3)} \otimes b_{84}^{(3)}\}\} = s_{2/3}\\
\dot{b}_{35}^{(3)} &= \max\{s_{1/5}, \min\{s_5, b_{38}^{(3)} \otimes b_{85}^{(3)}\}\} = s_4\\
\dot{b}_{36}^{(3)} &= \max\{s_{1/5}, \min\{s_5, b_{38}^{(3)} \otimes b_{86}^{(3)}\}\} = s_2\\
\dot{b}_{37}^{(3)} &= \max\{s_{1/5}, \min\{s_5, b_{38}^{(3)} \otimes b_{87}^{(3)}\}\} = s_1\\
\dot{b}_{45}^{(3)} &= \max\{s_{1/5}, \min\{s_5, b_{48}^{(3)} \otimes b_{85}^{(3)}\}\} = s_5\\
\dot{b}_{46}^{(3)} &= \max\{s_{1/5}, \min\{s_5, b_{48}^{(3)} \otimes b_{86}^{(3)}\}\} = s_3\\
\dot{b}_{47}^{(3)} &= \max\{s_{1/5}, \min\{s_5, b_{48}^{(3)} \otimes b_{87}^{(3)}\}\} = s_{3/2}
\end{aligned}
$$

$$\dot{b}_{56}^{(3)} = \max\{s_{1/5}, \min\{s_5, b_{58}^{(3)} \otimes b_{86}^{(3)}\}\} = s_{1/2}$$
$$\dot{b}_{57}^{(3)} = \max\{s_{1/5}, \min\{s_5, b_{58}^{(3)} \otimes b_{87}^{(3)}\}\} = s_{1/4}$$
$$\dot{b}_{67}^{(3)} = \max\{s_{1/5}, \min\{s_5, b_{68}^{(3)} \otimes b_{87}^{(3)}\}\} = s_{1/2}$$

Then by (3.47), we get the incomplete multiplicative linguistic preference relations:

$$\dot{\boldsymbol{B}}_1 = (\dot{b}_{ij}^{(1)})_{8\times 8} = \begin{pmatrix} s_1 & s_{1/3} & s_{1/2} & s_{1/4} & s_2 & s_1 & s_{1/3} & s_2 \\ s_3 & s_1 & s_{3/2} & s_{3/4} & s_5 & s_3 & s_1 & s_5 \\ s_2 & s_{2/3} & s_1 & s_{1/2} & s_4 & s_2 & s_{2/3} & s_4 \\ s_4 & s_{4/3} & s_2 & s_1 & s_5 & s_4 & s_{4/3} & s_5 \\ s_{1/2} & s_{1/5} & s_{1/4} & s_{1/5} & s_1 & s_{1/2} & s_{1/5} & s_1 \\ s_1 & s_{1/3} & s_{1/2} & s_{1/4} & s_2 & s_1 & s_{1/3} & s_2 \\ s_3 & s_1 & s_{3/2} & s_{3/4} & s_5 & s_3 & s_1 & s_5 \\ s_{1/2} & s_{1/5} & s_{1/4} & s_{1/5} & s_1 & s_{1/2} & s_{1/5} & s_1 \end{pmatrix}$$

$$\dot{\boldsymbol{B}}_2 = (\dot{b}_{ij}^{(2)})_{8\times 8} = \begin{pmatrix} s_1 & s_{2/3} & s_1 & s_{1/3} & s_{4/3} & s_{5/3} & s_1 & s_{4/3} \\ s_{3/2} & s_1 & s_{3/2} & s_{1/2} & s_2 & s_{5/2} & s_{3/4} & s_2 \\ s_1 & s_{2/3} & s_1 & s_{1/3} & s_{4/3} & s_{5/3} & s_1 & s_{4/3} \\ s_3 & s_2 & s_3 & s_1 & s_4 & s_5 & s_3 & s_4 \\ s_{3/4} & s_{1/2} & s_{3/4} & s_{1/4} & s_1 & s_{5/4} & s_{3/4} & s_1 \\ s_{3/5} & s_{2/5} & s_{3/5} & s_{1/5} & s_{4/5} & s_1 & s_{3/5} & s_{4/5} \\ s_1 & s_{4/3} & s_1 & s_{1/3} & s_{4/3} & s_{5/3} & s_1 & s_{4/3} \\ s_{3/4} & s_{1/2} & s_{3/4} & s_{1/4} & s_1 & s_{5/4} & s_{3/4} & s_1 \end{pmatrix}$$

$$\dot{\boldsymbol{B}}_3 = (\dot{b}_{ij}^{(3)})_{8\times 8} = \begin{pmatrix} s_1 & s_{1/2} & s_1 & s_{3/2} & s_4 & s_2 & s_1 & s_2 \\ s_2 & s_1 & s_2 & s_{4/3} & s_5 & s_4 & s_2 & s_4 \\ s_1 & s_{1/2} & s_1 & s_{2/3} & s_4 & s_2 & s_1 & s_2 \\ s_{2/3} & s_{3/4} & s_{3/2} & s_1 & s_5 & s_3 & s_{3/2} & s_3 \\ s_{1/4} & s_{1/5} & s_{1/4} & s_{1/5} & s_1 & s_{1/2} & s_{1/4} & s_{1/2} \\ s_{1/2} & s_{1/4} & s_{1/2} & s_{1/3} & s_2 & s_1 & s_{1/2} & s_1 \\ s_1 & s_{1/2} & s_1 & s_{2/3} & s_4 & s_2 & s_1 & s_2 \\ s_{1/2} & s_{1/4} & s_{1/2} & s_{1/3} & s_2 & s_1 & s_{1/2} & s_1 \end{pmatrix}$$

Step 2 Employ the LWG operator (3.50) to aggregate all the multiplicative linguistic preference relations $\dot{\boldsymbol{B}}_k = (\dot{b}_{ij}^{(k)})_{8\times 8}$ $(k = 1, 2, 3)$ into the collective multiplicative linguistic preference relation:

$$\dot{\boldsymbol{B}} = (\dot{b}_{ij})_{8\times 8} = \begin{pmatrix} s_1 & s_{0.45} & s_{0.71} & s_{0.39} & s_{2.03} & s_{1.34} & s_{0.58} & s_{1.77} \\ s_{2.22} & s_1 & s_{1.59} & s_{0.75} & s_{3.80} & s_{3.01} & s_{1.05} & s_{3.63} \\ s_{1.41} & s_{0.63} & s_1 & s_{0.47} & s_{2.88} & s_{1.89} & s_{0.82} & s_{2.50} \\ s_{2.56} & s_{1.33} & s_{2.13} & s_1 & s_{4.68} & s_{4.04} & s_{1.74} & s_{4.22} \\ s_{0.49} & s_{0.26} & s_{0.35} & s_{0.21} & s_1 & s_{0.66} & s_{0.31} & s_{0.87} \\ s_{0.75} & s_{0.33} & s_{0.53} & s_{0.25} & s_{1.52} & s_1 & s_{0.43} & s_{1.32} \\ s_{1.72} & s_{0.95} & s_{1.22} & s_{0.57} & s_{3.23} & s_{2.33} & s_1 & s_{2.80} \\ s_{0.56} & s_{0.28} & s_{0.40} & s_{0.24} & s_{1.15} & s_{0.76} & s_{0.36} & s_1 \end{pmatrix}$$

Step 3 Utilize the LG operator (3.51) to aggregate all the preferences $\dot{b}_{ij}$ $(j = 1, 2, \cdots, 8)$ in the ith line in $\dot{\boldsymbol{B}}$ into the averaged preference degree $\dot{b}_i$ of the alternative x_i over all the alternatives:

$$\dot{b}_1 = s_{0.88}, \quad \dot{b}_2 = s_{1.81}, \quad \dot{b}_3 = s_{1.21}, \quad \dot{b}_4 = s_{2.37}$$

$$\dot{b}_5 = s_{0.45}, \quad \dot{b}_6 = s_{0.64}, \quad \dot{b}_7 = s_{1.49}, \quad \dot{b}_8 = s_{0.51}$$

Step 4 Rank x_i $(i = 1, 2, \cdots, 8)$ according to $\dot{b}_i$ $(i = 1, 2, \cdots, 8)$:

$$x_4 \succ x_2 \succ x_7 \succ x_3 \succ x_1 \succ x_6 \succ x_8 \succ x_5$$

Thus, the optimal anternative is x_4.

In what follows, we develop an approach to consrtructing the complete multiplicative linguistic preference relation based on a general acceptable additive incomplete linguistic preference relation (with no less than $n-1$ judgments):

Let $\boldsymbol{B} = (b_{ij})_{n\times n}$ be a general acceptable incomplete multiplicative linguistic preference relation. We replace each unknown element in $\boldsymbol{B}$ with

$$\dot{b}_{ij} = \left(\prod_{k\in M_{ij}} \max\{s_{1/\tau},\ \min\{s_\tau,\ b_{ik}\otimes b_{kj}\}\}\right)^{\frac{1}{m_{ij}}} \tag{3.52}$$

until get a complete multiplicative linguistic preference relation $\dot{\boldsymbol{B}} = (\dot{b}_{ij})_{n\times n}$ corresponding to $\boldsymbol{B} = (b_{ij})_{n\times n}$, where $M_{ij} = \{k | b_{ik} \in \Theta, b_{kj} \in \Theta\}$, m_{ij} is the number of the elements in M_{ij}, and

$$\dot{b}_{ij} = \begin{cases} \left(\prod_{k\in M_{ij}} \max\{s_{1/\tau},\ \min\{s_\tau,\ b_{ik}\otimes b_{kj}\}\}\right)^{\frac{1}{m_{ij}}}, & b_{ij} \notin \Theta, b_{ik}, b_{kj} \in \Theta \\ b_{ij}, & b_{ij} \in \Theta \end{cases} \tag{3.53}$$

The multiplicative linguistic preference relation $\dot{\boldsymbol{B}} = (\dot{b}_{ij})_{n\times n}$ derived by (3.53) contains both the direct linguistic preference information given by the decision maker and the indirect linguistic preference information derived from the known linguistic preference information.

In the following, we introduce an approach to group decision making based on general acceptable multiplicative incomplete linguistic preference relations:

Step 1 For a group decision making problem, the decision maker $e_k \in E$ utilizes the multiplicative linguistic evaluation scale $S_4 = \{s_\alpha | \alpha = 1/\tau, \cdots, 1/2, 1, 2, \cdots, \tau\}$ to compare each pair of the alternatives in X, and constructs a general acceptable incomplete multiplicative linguistic preference relation $\boldsymbol{B}_k = (b_{ij}^{(k)})_{n\times n}$.

then

$$v_k = \frac{e^{d(\boldsymbol{A}_k,\boldsymbol{A})}}{\sum\limits_{k=1}^{l_1} e^{d(\boldsymbol{A}_k,\boldsymbol{A})} + \sum\limits_{k=l_1+1}^{l} e^{d(\boldsymbol{E}(\tilde{\boldsymbol{A}}_k),\boldsymbol{A})}}, \quad k = 1, 2, \cdots, l_1 \tag{3.79}$$

$$v_k = \frac{e^{d(\boldsymbol{E}(\tilde{\boldsymbol{A}}_k),\boldsymbol{A})}}{\sum\limits_{k=1}^{l_1} e^{d(\boldsymbol{A}_k,\boldsymbol{A})} + \sum\limits_{k=l_1+1}^{l} e^{d(\boldsymbol{E}(\tilde{\boldsymbol{A}}_k),\boldsymbol{A})}}, \quad k = l_1 + 1, \cdots, l \tag{3.80}$$

Case 3 If both the majority and minority opinions should be emphasized simultaneously, then we can use the following formulas to determine the decision makers' weights:

$$\begin{aligned} v_k =& (1-\alpha)\frac{f_1\left(d(\boldsymbol{A}_k,\boldsymbol{A})\right)}{\sum\limits_{k=1}^{l_1} f_1\left(d(\boldsymbol{A}_k,\boldsymbol{A})\right) + \sum\limits_{k=l_1+1}^{l} f_1\left(d(\boldsymbol{E}(\tilde{\boldsymbol{A}}_k),\boldsymbol{A})\right)} \\ &+ \alpha\frac{f_2\left(d(\boldsymbol{A}_k,\boldsymbol{A})\right)}{\sum\limits_{k=1}^{l_1} f_2\left(d(\boldsymbol{A}_k,\boldsymbol{A})\right) + \sum\limits_{k=l_1+1}^{l} f_2\left(d(\boldsymbol{E}(\tilde{\boldsymbol{A}}_k),\boldsymbol{A})\right)}, \quad k = 1, 2, \cdots, l_1 \end{aligned} \tag{3.81}$$

$$\begin{aligned} v_k =& (1-\alpha)\frac{f_1\left(d(\boldsymbol{E}(\tilde{\boldsymbol{A}}_k),\boldsymbol{A})\right)}{\sum\limits_{k=1}^{l_1} f_1\left(d(\boldsymbol{A}_k,\boldsymbol{A})\right) + \sum\limits_{k=l_1+1}^{l} f_1\left(d(\boldsymbol{E}(\tilde{\boldsymbol{A}}_k),\boldsymbol{A})\right)} \\ &+ \alpha\frac{f_2\left(d(\boldsymbol{E}(\tilde{\boldsymbol{A}}_k),\boldsymbol{A})\right)}{\sum\limits_{k=1}^{l_1} f_2\left(d(\boldsymbol{A}_k,\boldsymbol{A})\right) + \sum\limits_{k=l_1+1}^{l} f_2\left(d(\boldsymbol{E}(\tilde{\boldsymbol{A}}_k),\boldsymbol{A})\right)}, \quad k = l_1 + 1, \cdots, l \end{aligned} \tag{3.82}$$

where $\alpha \in [0, 1]$ (which can be given by the decision makers directly), $y_1 = f_1(d(\boldsymbol{A}_k, \boldsymbol{A}))$ (or $y_1 = f_1(d(\boldsymbol{E}(\tilde{\boldsymbol{A}}_k), \boldsymbol{A}))$) and $y_2 = f_2(d(\boldsymbol{A}_k, \boldsymbol{A}))$ (or $y_2 = f_2(d(\boldsymbol{E}(\tilde{\boldsymbol{A}}_k), \boldsymbol{A}))$) are, respectively, the decreasing and increasing functions of the deviation $d(\boldsymbol{A}_k, \boldsymbol{A})$ (or $d(\boldsymbol{E}(\tilde{\boldsymbol{A}}_k), \boldsymbol{A})$). For example, if $f_1\left(d(\boldsymbol{A}_k, \boldsymbol{A})\right)$ (or $f_1\left(d(\boldsymbol{E}(\tilde{\boldsymbol{A}}_k), \boldsymbol{A})\right)$) and $f_2\left(d(\boldsymbol{A}_k, \boldsymbol{A})\right)$ ($f_2\left(d(\boldsymbol{E}(\tilde{\boldsymbol{A}}_k), \boldsymbol{A})\right)$) are, respectively, taken from (3.75) and (3.78), then (3.81) and (3.82) can be rewritten as

$$v_k = (1-\alpha)\frac{e^{-d(\boldsymbol{A}_k,\boldsymbol{A})}}{\sum\limits_{k=1}^{l_1} e^{-d(\boldsymbol{A}_k,\boldsymbol{A})} + \sum\limits_{k=l_1+1}^{l} e^{-d(\boldsymbol{E}(\tilde{\boldsymbol{A}}_k),\boldsymbol{A})}} + \alpha\frac{e^{d(\boldsymbol{A}_k,\boldsymbol{A})}}{\sum\limits_{k=1}^{l_1} e^{d(\boldsymbol{A}_k,\boldsymbol{A})} + \sum\limits_{k=l_1+1}^{l} e^{d(\boldsymbol{E}(\tilde{\boldsymbol{A}}_k),\boldsymbol{A})}}$$

$$k = 1, 2, \cdots, l_1 \tag{3.83}$$

$$v_k = (1-\alpha)\frac{\mathrm{e}^{-d(\boldsymbol{E}(\tilde{\boldsymbol{A}}_k),\boldsymbol{A})}}{\sum_{k=1}^{l_1}\mathrm{e}^{-d(\boldsymbol{A}_k,\boldsymbol{A})}+\sum_{k=l_1+1}^{l}\mathrm{e}^{-d(\boldsymbol{E}(\tilde{\boldsymbol{A}}_k),\boldsymbol{A})}} + \alpha\frac{\mathrm{e}^{d(\boldsymbol{E}(\tilde{\boldsymbol{A}}_k),\boldsymbol{A})}}{\sum_{k=1}^{l_1}\mathrm{e}^{d(\boldsymbol{A}_k,\boldsymbol{A})}+\sum_{k=l_1+1}^{l}\mathrm{e}^{d(\boldsymbol{E}(\tilde{\boldsymbol{A}}_k),\boldsymbol{A})}}$$

$$k = l_1+1, \cdots, l \tag{3.84}$$

In particular, if $\alpha = 0$, then (3.83) and (3.84) reduce to (3.76) and (3.77) respectively; if $\alpha = 1$, then (3.83) and (3.84) reduce to (3.79) and (3.80) respectively. Thus, both Cases 1 and 2 are the special cases of Case 3.

Moreover, if

$$d(\boldsymbol{A}_k, \boldsymbol{A}) \leqslant \tau_0 \quad (\text{or } d(\boldsymbol{E}(\tilde{\boldsymbol{A}}_k), \boldsymbol{A}) \leqslant \tau_0) \tag{3.85}$$

then the additive linguistic preference relation $\boldsymbol{A}_k$ (or the expected additive linguistic preference relation $\boldsymbol{E}(\tilde{\boldsymbol{A}}_k)$) and the collective additive linguistic preference relation $\boldsymbol{A}$ is called of acceptable deviation degree, where τ_0 is the threshold of acceptable deviation degree, which is predefined by all the decision makers (who participate in determining the decision makers' original weights) in practical applications. Especially, if $d(\boldsymbol{A}_k, \boldsymbol{A}) = 0$ (or $d(\boldsymbol{E}(\tilde{\boldsymbol{A}}_k), \boldsymbol{A}) = 0$), then the additive linguistic preference relation $\boldsymbol{A}_k$ (or the expected additive linguistic preference relation $\boldsymbol{E}(\tilde{\boldsymbol{A}}_k)$) is equal to the collective additive linguistic preference relation. However, if $d(\boldsymbol{A}_k, \boldsymbol{A}) > \tau_0$ (or $d(\boldsymbol{E}(\tilde{\boldsymbol{A}}_k), \boldsymbol{A}) > \tau_0$), then we shall return $\boldsymbol{A}_k$ (or the uncertain additive linguistic prefrence relation $\tilde{\boldsymbol{A}}_k$ together with the expected additive linguistic preference relation $\boldsymbol{E}(\tilde{\boldsymbol{A}}_k)$) and the collective additive linguistic preference relation $\boldsymbol{A}$ to the decision maker for revaluation. We repeat this process (in which the decision makers' original weights are replaced by the new weights derived from (3.83) and (3.84) for obtaining the new collective additive linguistic preference relation $\boldsymbol{A}$ by using the LWA_2 operator (3.69) until the additive linguistic preference relation $\boldsymbol{A}_k$ (or the expected additive linguistic preference relation $\boldsymbol{E}(\tilde{\boldsymbol{A}}_k)$) and the collective additive linguistic preference relation are of acceptable deviation degree or the process will stop as the repetition times reach the maximum number predefined by the decision makers, and then go to the next step.

Step 4 Use the LA operator:

$$a_i = \text{LA}(a_{i1}, a_{i2}, \cdots, a_{in}), \quad i = 1, 2, \cdots, n \tag{3.86}$$

to aggregate all the preferences in the ith line of the collective additive linguistic preference relation $\boldsymbol{A}$ into the averaged preference a_i of the alternative x_i over all the alternatives.

Step 5 Rank and select x_i $(i = 1, 2, \cdots, n)$ according to a_i $(i = 1, 2, \cdots, n)$.

Example 3.13 (Xu, 2008) In this example, a group decision making problem involves the prioritization of a set of six information technology improvement projects (adapted from Ngwenyama and Bryson (1999)) is used to illustrate the developed procedures. The information management steering committee of Midwest American Manufacturing Corp. (MAMC), which comprises: ① d1——the chief executive officer; ② d2——the chief information officer; and ③ d3——the chief operating officer, must prioritize for development and implementation a set of six information technology improvement projects x_i $(i = 1, 2, \cdots, 6)$, which have been proposed by area managers. The committee is concerned that the projects are prioritized from highest to lowest potential contribution to the firm's strategic goal of gaining competitive advantage in the industry. In assessing the potential contribution of each project, one main factor considered is productivity. The productivity factor assesses the potential of a proposed project to increase the effectiveness and efficiency of the firm's manufacturing and service operations. The following is the list of proposed information systems projects: ① x_1——quality management information; ② x_2——inventory control; ③ x_3——customer order tracking; ④ x_4——materials purchasing management; ⑤ x_5——fleet management; and ⑥ x_6——design change management.

Suppose that the committee members e_k $(k = 1, 2, 3, 4)$ compare each pair of these projects x_i $(i = 1, 2, \cdots, 6)$ with respect to the factor "productivity" by using the additive linguistic labels in the set:

$$S_2 = \{s_{-4} = \text{extremely poor}, \quad s_{-3} = \text{very poor}, \quad s_{-2} = \text{poor},$$
$$s_{-1} = \text{slightly poor}, \quad s_0 = \text{fair}, \quad s_1 = \text{slightly good},$$
$$s_2 = \text{good}, \quad s_3 = \text{very good}, \quad s_4 = \text{extremely good}\}$$

and construct the traditional additive linguistic preference relations $\boldsymbol{A}_1$, $\boldsymbol{A}_2$ and the uncertain additive linguistic preference relations as $\tilde{\boldsymbol{A}}_3$ and $\tilde{\boldsymbol{A}}_4$ respectively:

$$\boldsymbol{A}_1 = \begin{pmatrix} s_0 & s_1 & s_{-2} & s_3 & s_1 & s_{-1} \\ s_{-1} & s_0 & s_{-1} & s_2 & s_0 & s_1 \\ s_2 & s_1 & s_0 & s_1 & s_{-2} & s_2 \\ s_{-3} & s_{-2} & s_{-1} & s_0 & s_2 & s_3 \\ s_{-1} & s_0 & s_2 & s_{-2} & s_0 & s_{-2} \\ s_1 & s_{-1} & s_{-2} & s_{-3} & s_2 & s_0 \end{pmatrix}$$

$$\boldsymbol{A}_2 = \begin{pmatrix} s_0 & s_2 & s_{-3} & s_1 & s_2 & s_{-2} \\ s_{-2} & s_0 & s_{-2} & s_3 & s_1 & s_2 \\ s_3 & s_2 & s_0 & s_2 & s_{-1} & s_0 \\ s_{-1} & s_{-3} & s_{-2} & s_0 & s_3 & s_1 \\ s_{-2} & s_{-1} & s_1 & s_{-3} & s_0 & s_1 \\ s_2 & s_{-2} & s_0 & s_{-1} & s_{-1} & s_0 \end{pmatrix}$$

$$\tilde{\boldsymbol{A}}_3=\begin{pmatrix}[s_0,s_0] & [s_1,s_2] & [s_{-2},s_{-1}] & [s_2,s_4] & [s_{-1},s_1] & [s_{-1},s_0]\\ [s_{-2},s_{-1}] & [s_0,s_0] & [s_{-1},s_0] & [s_1,s_3] & [s_{-1},s_1] & [s_1,s_2]\\ [s_1,s_2] & [s_0,s_1] & [s_0,s_0] & [s_{-1},s_2] & [s_{-2},s_{-1}] & [s_1,s_2]\\ [s_{-4},s_{-2}] & [s_{-3},s_{-1}] & [s_{-2},s_1] & [s_0,s_0] & [s_1,s_2] & [s_2,s_3]\\ [s_{-1},s_1] & [s_{-1},s_1] & [s_1,s_2] & [s_{-2},s_{-1}] & [s_0,s_0] & [s_{-2},s_{-1}]\\ [s_0,s_1] & [s_{-2},s_{-1}] & [s_{-2},s_{-1}] & [s_{-3},s_{-2}] & [s_1,s_2] & [s_0,s_0]\end{pmatrix}$$

$$\tilde{\boldsymbol{A}}_4=\begin{pmatrix}[s_0,s_0] & [s_2,s_3] & [s_{-2},s_0] & [s_1,s_2] & [s_0,s_2] & [s_{-2},s_{-1}]\\ [s_{-3},s_{-2}] & [s_0,s_0] & [s_{-2},s_{-1}] & [s_2,s_4] & [s_{-2},s_0] & [s_0,s_1]\\ [s_0,s_2] & [s_1,s_2] & [s_0,s_0] & [s_0,s_1] & [s_{-1},s_1] & [s_2,s_3]\\ [s_{-2},s_{-1}] & [s_{-4},s_{-2}] & [s_{-1},s_0] & [s_0,s_0] & [s_2,s_3] & [s_1,s_2]\\ [s_{-2},s_0] & [s_0,s_2] & [s_{-1},s_1] & [s_{-3},s_{-2}] & [s_0,s_0] & [s_0,s_1]\\ [s_1,s_2] & [s_{-1},s_0] & [s_{-3},s_{-2}] & [s_{-2},s_{-1}] & [s_{-1},s_0] & [s_0,s_0]\end{pmatrix}$$

To prioritize the projects x_i $(i=1,2,\cdots,6)$, the following steps are involved:

Step 1 Suppose that the committee members e_k $(k=3,4)$ give the indices of their risk-bearing attitudes $\theta_3=0.3$ and $\theta_4=0.6$ respectively. Then we utilize (3.66) to transform the uncertain additive linguistic preference relations $\tilde{\boldsymbol{A}}_k=(\tilde{a}_{ij}^{(k)})_{6\times6}$ $(k=3,4)$ into the expected additive linguistic preference relations $\boldsymbol{E}(\tilde{\boldsymbol{A}}_k)=E(\tilde{a}_{ij}^{(k)})_{6\times6}$ $(k=3,4)$:

$$\boldsymbol{E}(\tilde{\boldsymbol{A}}_3)=\begin{pmatrix}s_0 & s_{1.3} & s_{-1.7} & s_{2.6} & s_{-0.4} & s_{-0.7}\\ s_{-1.3} & s_0 & s_{-0.7} & s_{1.6} & s_{-0.4} & s_{1.3}\\ s_{1.7} & s_{0.7} & s_0 & s_{-0.1} & s_{-1.7} & s_{1.3}\\ s_{-2.6} & s_{-1.6} & s_{-0.1} & s_0 & s_{1.3} & s_{2.3}\\ s_{0.4} & s_{0.4} & s_{1.7} & s_{-1.3} & s_0 & s_{-1.7}\\ s_{0.7} & s_{-1.3} & s_{-1.3} & s_{-2.3} & s_{1.7} & s_0\end{pmatrix}$$

$$\boldsymbol{E}(\tilde{\boldsymbol{A}}_4)=\begin{pmatrix}s_0 & s_{2.6} & s_{-0.8} & s_{1.6} & s_{1.2} & s_{-1.4}\\ s_{-2.6} & s_0 & s_{-1.4} & s_{3.2} & s_{-0.8} & s_{0.6}\\ s_{0.8} & s_{1.4} & s_0 & s_{0.6} & s_{0.2} & s_{2.6}\\ s_{-1.6} & s_{-3.2} & s_{-0.6} & s_0 & s_{2.6} & s_{1.6}\\ s_{-1.2} & s_{0.8} & s_{-0.2} & s_{-2.6} & s_0 & s_{0.6}\\ s_{1.4} & s_{-0.6} & s_{-2.6} & s_{-1.6} & s_{-0.6} & s_0\end{pmatrix}$$

Step 2 Let the original weight vector of the committee members e_k $(k=1,2,3,4)$ be $\boldsymbol{v}=(0.35,0.20,0.15,0.30)^{\mathrm{T}}$, and assume that these four members predefine the threshold of acceptable deviation degree as $\tau_0=0.07$. Then, we utilize the LWA_2 operator (3.69) to aggregate the expected additive linguistic preference relations $\boldsymbol{E}(\tilde{\boldsymbol{A}}_k)$ $(k=3,4)$ and the additive linguistic preference relations $\boldsymbol{A}_k$ $(k=1,2)$ into the collective additive linguistic preference relation $\boldsymbol{A}=(a_{ij})_{6\times6}$:

$$A = \begin{pmatrix} s_0 & s_{1.73} & s_{-1.80} & s_{2.32} & s_{1.05} & s_{-1.28} \\ s_{-1.73} & s_0 & s_{-1.28} & s_{2.5} & s_{-0.10} & s_{0.93} \\ s_{1.80} & s_{1.28} & s_0 & s_{0.92} & s_{-1.10} & s_{2.08} \\ s_{-2.32} & s_{-2.5} & s_{-0.92} & s_0 & s_{2.28} & s_{2.08} \\ s_{-1.05} & s_{0.10} & s_{1.10} & s_{-2.28} & s_0 & s_{-0.58} \\ s_{1.28} & s_{-0.93} & s_{-2.08} & s_{-2.08} & s_{0.58} & s_0 \end{pmatrix}$$

Step 3 Utilize (3.17), (3.70) and (3.71) to calculate the deviation degree between the collective additive linguistic prefrence relation $\boldsymbol{A}$ and each of the individual additive linguistic preference relations $\boldsymbol{A}_k$ $(k = 1, 2)$, and the deviation degree between the collective additive linguistic prefrence relation $\boldsymbol{A}$ and each of the expected additive linguistic preference relations $\boldsymbol{E}(\tilde{\boldsymbol{A}}_k)$ $(k = 3, 4)$:

$$d(\boldsymbol{A}_1, \boldsymbol{A}) = 0.049, \quad d(\boldsymbol{A}_2, \boldsymbol{A}) = 0.095$$

$$d(\boldsymbol{E}(\tilde{\boldsymbol{A}}_3), \boldsymbol{A}) = 0.065, \quad d(\boldsymbol{E}(\tilde{\boldsymbol{A}}_4), \boldsymbol{A}) = 0.064$$

Since $d(\boldsymbol{A}_2, \boldsymbol{A}) > 0.07$, we need to recalculate the weights of the committee members e_k $(k = 1, 2, 3, 4)$ by using (3.83) and (3.84) (without loss of generality, let $\alpha = 0.3$),

$$\boldsymbol{v} = (0.2659,\ 0.2321,\ 0.2508,\ 0.2512)^{\mathrm{T}}$$

and return the additive linguistic preferenc relation $\boldsymbol{A}_2$ together with the collective additive linguistic preference relation $\boldsymbol{A}$ to the committee member e_2 for revaluation. Suppose that the committee member e_2 provides the revaluated additive linguistic preference relation as follows:

$$\boldsymbol{A}_2' = \begin{pmatrix} s_0 & s_2 & s_{-2} & s_2 & s_1 & s_{-1} \\ s_{-2} & s_0 & s_{-1} & s_3 & s_0 & s_1 \\ s_2 & s_1 & s_0 & s_1 & s_{-1} & s_2 \\ s_{-2} & s_{-3} & s_{-1} & s_0 & s_3 & s_2 \\ s_{-1} & s_0 & s_1 & s_{-3} & s_0 & s_{-1} \\ s_1 & s_{-1} & s_{-2} & s_{-2} & s_1 & s_0 \end{pmatrix}$$

Then we utilize the recalculated weight vector $\boldsymbol{v}$ and the LWA$_2$ operator (3.69) to aggregate the expected additive linguistic preference relations $\boldsymbol{E}(\tilde{\boldsymbol{A}}_k)$ $(k = 3, 4)$ and the additive linguistic preference relations $\boldsymbol{A}_1$ and $\boldsymbol{A}_2'$ into the collective additive linguistic preference relation $\boldsymbol{A}' = (a_{ij}')_{6\times 6}$:

$$\boldsymbol{A}' = \begin{pmatrix} s_0 & s_{1.71} & s_{-1.62} & s_{2.32} & s_{0.70} & s_{-1.03} \\ s_{-1.71} & s_0 & s_{-1.03} & s_{2.45} & s_{-0.30} & s_{0.97} \\ s_{1.62} & s_{1.03} & s_0 & s_{0.62} & s_{-1.14} & s_{1.98} \\ s_{-2.32} & s_{-2.45} & s_{-0.62} & s_0 & s_{2.21} & s_{2.24} \\ s_{-0.70} & s_{0.30} & s_{1.14} & s_{-2.21} & s_0 & s_{-1.04} \\ s_{1.03} & s_{-0.97} & s_{-1.98} & s_{-2.24} & s_{1.04} & s_0 \end{pmatrix}$$

By (3.17), (3.70) and (3.71), we get

$$d(\boldsymbol{A}_1, \boldsymbol{A}') = 0.042, \quad d(\boldsymbol{A}_2', \boldsymbol{A}') = 0.027$$
$$d(\boldsymbol{E}(\tilde{\boldsymbol{A}}_3), \boldsymbol{A}') = 0.051, \quad d(\boldsymbol{E}(\tilde{\boldsymbol{A}}_4), \boldsymbol{A}') = 0.069$$

Since

$$d(\boldsymbol{A}_1, \boldsymbol{A}') < 0.6, \quad d(\boldsymbol{A}_2', \boldsymbol{A}') < 0.6, \quad d(\boldsymbol{E}(\tilde{\boldsymbol{A}}_j), \boldsymbol{A}') < 0.6, \quad j = 3, 4$$

i.e., all the deviation degrees are acceptable, then go to Step 4.

Step 4 Use the LA operator (3.86) to aggregate all the preferences in the ith line of the collective additive linguistic preference relation $\boldsymbol{A}'$ into the averaged preference a_i' of the alternative x_i over all the alternatives:

$$a_1' = s_{0.35}, \quad a_2' = s_{0.06}, \quad a_3' = s_{0.69}, \quad a_4' = s_{-0.16}, \quad a_5' = s_{-0.42}, \quad a_6' = s_{-0.52}$$

Step 5 Rank $x_i\ (i = 1, 2, \cdots, 6)$ according to $a_i'\ (i = 1, 2, \cdots, 6)$:

$$x_3 \succ x_1 \succ x_2 \succ x_4 \succ x_5 \succ x_6$$

and thus, the project x_3 has the highest potential contribution to the firm's strategic goal of gaining competitive advantage in the industry.

3.7.2 Uncertain Multiplicative Linguistic Preference Relations

Definition 3.33 (Xu, 2010) An uncertain multiplicative linguistic preference relation $\tilde{\boldsymbol{B}} = (\tilde{b}_{ij})_{n\times n}$ is given by the decision maker by using the multiplicative linguistic evaluation scale S_4(or S_5, see Section 1.2) to compare each pair of the alternatives in X, where $\tilde{b}_{ij}$ is an uncertain linguistic variable, indicating the degree range that the alternative x_i is preferred to x_j, and satisfies

$$\tilde{b}_{ij} = [b_{ij}^{\mathrm{L}}, b_{ij}^{\mathrm{U}}], \quad s_{1/\tau} \leqslant b_{ij}^{\mathrm{L}} \leqslant b_{ij}^{\mathrm{U}} \leqslant s_\tau, \quad b_{ij}^{\mathrm{L}} \otimes \mathrm{b}_{ji}^{\mathrm{U}} = s_1, \quad b_{ji}^{\mathrm{L}} \otimes b_{ij}^{\mathrm{U}} = s_1$$
$$i, j = 1, 2, \cdots, n \tag{3.87}$$

Similar to Theorem 3.21, we can prove the following conclusion:

Theorem 3.29 (Xu, 2010) Let $\tilde{\boldsymbol{B}}_k = (\tilde{b}_{ij}^{(k)})_{n\times n}\ (k = 1, 2, \cdots, l)$ be the uncertain multiplicative linguistic preference relations provided by the decision makers $e_k\ (k = 1, 2, \cdots, l)$, where

$$\tilde{b}_{ij}^{(k)} = [b_{ij}^{\mathrm{L}(k)}, b_{ij}^{\mathrm{U}(k)}], \quad s_{1/\tau} \leqslant b_{ij}^{\mathrm{L}(k)} \leqslant b_{ij}^{\mathrm{U}(k)} \leqslant s_\tau, \quad b_{ij}^{\mathrm{L}(k)} \otimes b_{ji}^{\mathrm{U}(k)} = s_1, \quad b_{ji}^{\mathrm{L}(k)} \otimes b_{ij}^{\mathrm{U}(k)} = s_1,$$
$$k = 1, 2, \cdots, l, \quad i, j = 1, 2, \cdots, n$$

Then the collective linguistic preference relation $\tilde{\boldsymbol{B}} = (\tilde{b}_{ij})_{n\times n}$ derived by the ULWG operator:

$$\tilde{b}_{ij} = \mathrm{ULWG}\,(\tilde{b}_{ij}^{(1)}, \tilde{b}_{ij}^{(2)}, \cdots, \tilde{b}_{ij}^{(l)}) = (\tilde{b}_{ij}^{(1)})^{v_1} \otimes (\tilde{b}_{ij}^{(2)})^{v_2} \otimes \cdots \otimes (\tilde{b}_{ij}^{(l)})^{v_l}, \quad i, j = 1, 2, \cdots, n \tag{3.88}$$

is also an uncertain multiplicative linguistic preference relation, where $\boldsymbol{v} = (v_1, v_2, \cdots, v_l)^{\mathrm{T}}$ is the weight vector of the decision makers e_k $(k = 1, 2, \cdots, l)$, $v_i \geqslant 0$ $(i = 1, 2, \cdots, l)$ and $\sum_{i=1}^{l} v_i = 1$.

Definition 3.34 (Xu, 2010) Let $\tilde{s}_1 = [s_{\alpha_1}, s_{\beta_1}]$ and $\tilde{s}_2 = [s_{\alpha_2}, s_{\beta_2}]$ be two uncertain linguistic variables. Then

$$\rho(\tilde{s}_1, \tilde{s}_2) = \frac{\alpha_1\beta_1 + \alpha_2\beta_2}{\max\{\alpha_1^2 + \alpha_2^2,\ \beta_1^2 + \beta_2^2\}} \tag{3.89}$$

is called the similarity degree between $\tilde{s}_1$ and $\tilde{s}_2$.

Example 3.14 Suppose that $\tilde{s}_1 = [s_{1/2}, s_{3/4}]$ and $\tilde{s}_2 = [s_{1/2}, s_1]$. Then

$$\rho(\tilde{s}_1, \tilde{s}_2) = \frac{\frac{1}{2} \times \frac{1}{2} + \frac{1}{2} \times 1}{\max\left\{\left(\frac{1}{2}\right)^2 + \left(\frac{3}{4}\right)^2,\ \left(\frac{1}{2}\right)^2 + 1^2\right\}} = \frac{\frac{3}{4}}{\max\left\{\frac{13}{16}, \frac{5}{4}\right\}} = \frac{3}{5}$$

Definition 3.35 (Xu, 2010) Let $\tilde{\boldsymbol{B}}_k = (\tilde{b}_{ij}^{(k)})_{n\times n}$ $(k = 1, 2, \cdots, l)$ be the uncertain multiplicative linguistic preference relations provided by the decision makers $e_k (k = 1, 2, \cdots, l)$, and $\tilde{\boldsymbol{B}} = (\tilde{b}_{ij})_{n\times n}$ is the collective uncertain multiplicative linguistic preference relation obtained by (3.69). Then

$$\rho(\tilde{\boldsymbol{B}}_k, \tilde{\boldsymbol{B}}) = \frac{1}{n^2}\sum_{i=1}^{n}\sum_{j=1}^{n}\rho(\tilde{b}_{ij}^{(k)}, \tilde{b}_{ij}) \tag{3.90}$$

is called the similarity degree between $\tilde{\boldsymbol{B}}_k$ and $\tilde{\boldsymbol{B}}$, where $\rho(\tilde{b}_{ij}^{(k)}, \tilde{b}_{ij})$ is the similarity degree between $\tilde{b}_{ij}^{(k)}$ and $\tilde{b}_{ij}$, which can be determined by (3.89).

Furthermore, if

$$\rho(\tilde{\boldsymbol{B}}_k, \tilde{\boldsymbol{B}}) > \beta_0 \tag{3.91}$$

then $\tilde{\boldsymbol{B}}_k$ and $\tilde{\boldsymbol{B}}$ are called of acceptable similarity degree, where β_0 is the threshold of acceptable similarity degree. In practical applications, β_0 can be predefined by the decision makers. In particular, if $\rho(\tilde{\boldsymbol{B}}_k, \tilde{\boldsymbol{B}}) = 1$, then $\tilde{\boldsymbol{B}}_k$ is completely similar to $\tilde{\boldsymbol{B}}$ (Xu, 2010).

$\rho(\tilde{\boldsymbol{B}}_k, \tilde{\boldsymbol{B}})$ has the following properties:

(1) $0 \leqslant \rho(\tilde{\boldsymbol{B}}_k, \tilde{\boldsymbol{B}}) \leqslant 1$;

(2) $\rho(\tilde{\boldsymbol{B}}_k, \tilde{\boldsymbol{B}}) = 1$ if and only if $\tilde{\boldsymbol{B}}_k = \tilde{\boldsymbol{B}}$;

(3) $\rho(\tilde{\boldsymbol{B}}_k, \tilde{\boldsymbol{B}}) = \rho(\tilde{\boldsymbol{B}}, \tilde{\boldsymbol{B}}_k)$.

In the process of group decision making, if $\rho(\tilde{\boldsymbol{B}}_k, \tilde{\boldsymbol{B}}) \leqslant \beta_0$, then we shall return the uncertain multiplicative linguistic preference relation $\tilde{\boldsymbol{B}}_k$ together with the collective uncertain multiplicative linguistic preference relation $\tilde{\boldsymbol{B}}$ to the decision maker e_k, and at the same time, inform him/her of some elements of $\boldsymbol{R}^{(k)}$ with small degrees of similarity, and ask him/her to revaluate them. We repeat this procedure until $\tilde{\boldsymbol{B}}_k$ and $\tilde{\boldsymbol{B}}$ are of acceptable similarity degree.

In general, the similarity degree $\rho(\tilde{\boldsymbol{B}}_k, \tilde{\boldsymbol{B}})$ reflects the degree of consensus between the individual uncertain multiplicative linguistic preference relation $\tilde{\boldsymbol{B}}_k$ and the collective uncertain multiplicative linguistic preference relation $\tilde{\boldsymbol{B}}$. In many actual situations, some decision makers may provide unduly high or unduly low preference arguments for their preferred or repugnant objects, which may result in the low degrees of consensus among the collective uncertain multiplicative linguistic preference relation and the individual uncertain multiplicative linguistic preference relations. In such cases, we shall assign low weights to these decision makers in the process of decision making. That is, the more the similarity degree $\rho(\tilde{\boldsymbol{B}}_k, \tilde{\boldsymbol{B}})$, the more the weight of the decision maker e_k. As a result, we introduce a formula for determining the decision makers' weights as follows (Xu, 2010):

$$v_k = \frac{\rho(\tilde{\boldsymbol{B}}_k, \tilde{\boldsymbol{B}})}{\sum\limits_{k=1}^{l} \rho(\tilde{\boldsymbol{B}}_k, \tilde{\boldsymbol{B}})}, \quad k = 1, 2, \cdots, l \tag{3.92}$$

Based on the above analysis, Xu (2010) utilized the ULWG and ULG operators to develop a practical interactive procedure for group decision making based on uncertain multiplicative linguistic preference relations, which involves the following steps:

Step 1 Consider a group decision making problem with uncertain multiplicative linguistic preference relations. Let X, E and $\boldsymbol{v}$ defined as in Section 3.2. The decision maker $e_k \in E$ compares each pair of alternatives in X by using the multiplicative linguistic evaluation scale $S_5 = \left\{ s_\alpha \middle| \alpha = \frac{1}{\tau}, \frac{2}{\tau}, \cdots, \frac{\tau-1}{\tau}, 1, \frac{\tau}{\tau-1}, \cdots, \frac{\tau}{2}, \tau \right\}$, and constructs an uncertain multiplicative linguistic preference relation $\tilde{\boldsymbol{B}}_k = (\tilde{b}_{ij}^{(k)})_{n\times n}$. Suppose that the weights of the decision makers e_k $(k = 1, 2, \cdots, l)$ are completely unknown. In such a case, all decision makers are assigned originally the same weight, i.e., $\boldsymbol{\lambda} = (1/l, 1/l, \cdots, 1/l)^{\mathrm{T}}$.

Step 2 Utilize the ULWG operator (3.88) to aggregate all individual uncertain multiplicative preference relations $\tilde{\boldsymbol{B}}_k = (\tilde{b}_{ij}^{(k)})_{n\times n}$ $(k = 1, 2, \cdots, l)$ into the collective uncertain multiplicative preference relation $\tilde{\boldsymbol{B}} = (\tilde{b}_{ij})_{n\times n}$.

Step 3 Utilize (3.90) to calculate the similarity degree $\rho(\tilde{\boldsymbol{B}}_k, \tilde{\boldsymbol{B}})$ between the individual uncertain multiplicative preference relation $\tilde{\boldsymbol{B}}_k$ and the collective uncertain multiplicative linguistic preference relation $\tilde{\boldsymbol{B}}$. Assume that the decision makers predefine the dead line of acceptable similarity degree as β_0.

Step 4 If $\rho(\tilde{\boldsymbol{B}}_k, \tilde{\boldsymbol{B}}) > \beta_0$, then go to Step 5; otherwise, we recalculate the weights of the decision makers e_k $(k = 1, 2, \cdots, l)$ by using (3.92), and return $\tilde{\boldsymbol{B}}_k$ together with $\tilde{\boldsymbol{B}}$ to the decision maker e_k, and at the same time, inform him/her of some elements of $\tilde{\boldsymbol{B}}_k$ with small degrees of similarity, which are needed to be revaluated. For convenience, we also denote $\tilde{\boldsymbol{B}}_k$ as the revaluated uncertain multiplicative linguistic preference relation. Then go to Step 2.

Step 5 Utilize the ULG operator:

$$\tilde{a}_i = \text{ULG}(\tilde{a}_{i1}, \tilde{a}_{i2}, \cdots, \tilde{a}_{in}) = (\tilde{a}_{i1} \otimes \tilde{a}_{i2} \otimes \cdots \otimes \tilde{a}_{in})^{1/n} \tag{3.93}$$

to aggregate all the preference information $\tilde{a}_{ij}$ $(i = 1, 2, \cdots, n)$ in the ith line of the collective uncertain multiplicative linguistic preference relation $\tilde{\boldsymbol{B}}$ into the overall preference degree $\tilde{a}_i$ of the alternative x_i over all the alternatives.

Step 6 Utilize (2.69) and (2.82) to rank the averaged preferences $\tilde{a}_i$ $(i = 1, 2, \cdots,$
$n)$, by which we rank and select the alternatives x_i $(i = 1, 2, \cdots, n)$.

Example 3.15 (Xu, 2010) The problem discussed here is concerned with a manufacturing company, searching the best global supplier for one of its most critical parts used in assembling process (adapted from Chan and Kumar (2007)).

The company considers "overall cost of the product" as one of the important criteria in assessing six potential global suppliers x_j $(j = 1, 2, \cdots, 6)$. An expert group is formed which consists of four experts e_k $(k = 1, 2, 3, 4)$ from each strategic decision area. The experts compare these six alternatives with respect to the criterion "overall cost of the product" by using the multiplicative linguistic labels in the set:

$$S_5 = \{s_{1/5} = \text{extremely poor}, \quad s_{2/5} = \text{very poor}, \quad s_{3/5} = \text{poor},$$
$$s_{4/5} = \text{slightly poor}, \quad s_1 = \text{fair}, \quad s_{5/4} = \text{slightly good},$$
$$s_{5/3} = \text{good}, \quad s_{5/2} = \text{very good}, \quad s_5 = \text{extremely good}\}$$

and construct, respectively, the uncertain multiplicative linguistic preference relations $\tilde{\boldsymbol{B}}_k$ $(k = 1, 2, 3, 4)$:

$$\tilde{\boldsymbol{B}}_1 = \begin{pmatrix} [s_1, s_1] & [s_{2/5}, s_{3/5}] & [s_1, s_{5/4}] & [s_{5/2}, s_5] & [s_{5/3}, s_{5/2}] & [s_{2/5}, s_{4/5}] \\ [s_{5/3}, s_{5/2}] & [s_1, s_1] & [s_{3/5}, s_1] & [s_1, s_{5/3}] & [s_{5/3}, s_{5/2}] & [s_{3/5}, s_{4/5}] \\ [s_{4/5}, s_1] & [s_1, s_{5/3}] & [s_1, s_1] & [s_{5/4}, s_{5/3}] & [s_{2/5}, s_{4/5}] & [s_{5/3}, s_{5/2}] \\ [s_{1/5}, s_{2/5}] & [s_{3/5}, s_1] & [s_{3/5}, s_{4/5}] & [s_1, s_1] & [s_{5/4}, s_{5/3}] & [s_{5/2}, s_5] \\ [s_{2/5}, s_{3/5}] & [s_{2/5}, s_{3/5}] & [s_{5/4}, s_{5/2}] & [s_{3/5}, s_{4/5}] & [s_1, s_1] & [s_{2/5}, s_1] \\ [s_{5/4}, s_{5/2}] & [s_{5/4}, s_{5/3}] & [s_{2/5}, s_{3/5}] & [s_{1/5}, s_{2/5}] & [s_1, s_{5/2}] & [s_1, s_1] \end{pmatrix}$$

$$\tilde{\boldsymbol{B}}_2=\begin{pmatrix} [s_1,s_1] & [s_{5/3},s_{5/2}] & [s_{2/5},s_{3/5}] & [s_1,s_{5/4}] & [s_{5/4},s_{5/3}] & [s_{4/5},s_1] \\ [s_{2/5},s_{3/5}] & [s_1,s_1] & [s_1,s_{5/3}] & [s_{5/4},s_{5/3}] & [s_1,s_{5/4}] & [s_{4/5},s_{5/4}] \\ [s_{5/3},s_{5/2}] & [s_{3/5},s_1] & [s_1,s_1] & [s_1,s_{5/3}] & [s_{3/5},s_1] & [s_{5/4},s_{5/3}] \\ [s_{4/5},s_1] & [s_{3/5},s_{4/5}] & [s_{3/5},s_1] & [s_1,s_1] & [s_{5/3},s_{5/2}] & [s_{4/5},s_1] \\ [s_{3/5},s_{4/5}] & [s_{4/5},s_1] & [s_1,s_{5/3}] & [s_{2/5},s_{3/5}] & [s_1,s_1] & [s_{5/2},s_5] \\ [s_1,s_{5/4}] & [s_{4/5},s_{5/4}] & [s_{3/5},s_{4/5}] & [s_1,s_{5/4}] & [s_{1/5},s_{2/5}] & [s_1,s_1] \end{pmatrix}$$

$$\tilde{\boldsymbol{B}}_3=\begin{pmatrix} [s_1,s_1] & [s_{3/5},s_{4/5}] & [s_1,s_{5/4}] & [s_{5/3},s_{5/2}] & [s_{5/4},s_{5/2}] & [s_{3/5},s_{4/5}] \\ [s_{5/4},s_{5/3}] & [s_1,s_1] & [s_{3/5},s_{4/5}] & [s_{5/4},s_{5/3}] & [s_{5/3},s_{5/2}] & [s_{3/5},s_1] \\ [s_{4/5},s_{3/5}] & [s_{5/4},s_{5/3}] & [s_1,s_1] & [s_{5/4},s_{5/3}] & [s_{3/5},s_{4/5}] & [s_{5/3},s_{5/2}] \\ [s_{2/5},s_{3/5}] & [s_{3/5},s_{4/5}] & [s_{3/5},s_{4/5}] & [s_1,s_1] & [s_1,s_{5/4}] & [s_{5/2},s_5] \\ [s_{2/5},s_{4/5}] & [s_{2/5},s_{3/5}] & [s_{5/4},s_{5/3}] & [s_{4/5},s_1] & [s_1,s_1] & [s_{3/5},s_{4/5}] \\ [s_{5/4},s_{5/3}] & [s_1,s_{5/3}] & [s_{2/5},s_{3/5}] & [s_{1/5},s_{2/5}] & [s_{5/4},s_{5/3}] & [s_1,s_1] \end{pmatrix}$$

$$\tilde{\boldsymbol{B}}_4=\begin{pmatrix} [s_1,s_1] & [s_{5/4},s_{5/3}] & [s_{2/5},s_{4/5}] & [s_1,s_{5/3}] & [s_{5/4},s_{5/3}] & [s_{4/5},s_1] \\ [s_{3/5},s_{4/5}] & [s_1,s_1] & [s_1,s_{5/3}] & [s_{5/4},s_{5/2}] & [s_1,s_{5/4}] & [s_{4/5},s_1] \\ [s_{5/4},s_{3/5}] & [s_{3/5},s_1] & [s_1,s_1] & [s_1,s_{5/3}] & [s_{4/5},s_1] & [s_{5/4},s_{5/2}] \\ [s_{3/5},s_1] & [s_{2/5},s_{4/5}] & [s_{3/5},s_1] & [s_1,s_1] & [s_{5/4},s_{5/2}] & [s_{4/5},s_{5/4}] \\ [s_{3/5},s_{4/5}] & [s_{4/5},s_1] & [s_1,s_{5/4}] & [s_{2/5},s_{4/5}] & [s_1,s_1] & [s_{5/3},s_{5/2}] \\ [s_1,s_{5/4}] & [s_1,s_{5/4}] & [s_{2/5},s_{4/5}] & [s_{4/5},s_{5/4}] & [s_{2/5},s_{3/5}] & [s_1,s_1] \end{pmatrix}$$

In what follows, we apply the developed procedure to the selection of best global supplier for the manufacturing firm from the potential global suppliers x_j $(j = 1,2,\cdots,6)$:

Step 1 Set the original weight vector of the experts e_k $(k = 1,2,3,4)$, $\boldsymbol{\lambda} = (1/4,1/4,1/4,1/4)^{\mathrm{T}}$, and assume that the experts predefine the threshold of acceptable similarity degree as $\alpha_0 = 0.8$. Then we utilize the ULWG (3.88) to aggregate all individual uncertain multiplicative preference relations $\tilde{\boldsymbol{B}}^{(k)} = (\tilde{b}_{ij}^{(k)})_{6\times 6}$ $(k = 1,2,3,4)$ into the collective uncertain multiplicative preference relation $\tilde{\boldsymbol{B}} = (\tilde{b}_{ij})_{6\times 6}$:

$$\tilde{\boldsymbol{B}}=\begin{pmatrix} [s_1,s_1] & [s_{0.84},s_{1.19}] & [s_{0.63},s_{0.93}] & [s_{1.14},s_{2.26}] & [s_{1.34},s_{2.04}] & [s_{0.63},s_{0.89}] \\ [s_{0.84},s_{1.19}] & [s_1,s_1] & [s_{0.77},s_{1.22}] & [s_{1.18},s_{1.72}] & [s_{1.29},s_{1.77}] & [s_{0.60},s_1] \\ [s_{1.08},s_{1.59}] & [s_{0.82},s_{1.30}] & [s_1,s_1] & [s_1,s_{5/3}] & [s_{0.58},s_{0.89}] & [s_{1.44},s_{2.26}] \\ [s_{0.44},s_{0.88}] & [s_{0.58},s_{0.85}] & [s_{0.60},s_{0.89}] & [s_1,s_1] & [s_{1.26},s_{1.85}] & [s_{1.39},s_{2.36}] \\ [s_{0.49},s_{0.75}] & [s_{0.56},s_{0.78}] & [s_{1.12},s_{1.72}] & [s_{0.54},s_{0.79}] & [s_1,s_1] & [s_1,s_{1.78}] \\ [s_{1.12},s_{1.59}] & [s_1,s_{1.67}] & [s_{0.44},s_{0.69}] & [s_{0.49},s_{0.72}] & [s_{0.58},s_1] & [s_1,s_1] \end{pmatrix}$$

Step 2 Utilize (3.89) to calculate the similarity degree $\rho(\tilde{b}_{ij}^{(k)},\tilde{b}_{ij})$ $(i,j = 1,2,\cdots,6, k = 1,2,3,4)$, as listed in Tables 3.1∼3.4.

Table 3.1 Similarity degrees $\rho(\tilde{b}_{ij}^{(1)}, \tilde{b}_{ij})$ $(i, j = 1, 2, \cdots, 6)$

	x_1	x_2	x_3	x_4	x_5	x_6
x_1	1	0.4948	0.6994	0.4528	0.8125	0.8108
x_2	0.4846	1	0.8083	0.9301	0.7283	0.8529
x_3	0.6641	0.7905	1	0.9639	0.8361	0.8917
x_4	0.4545	0.8809	0.9306	1	0.9298	0.4888
x_5	0.8045	0.7505	0.7296	0.9560	1	0.5230
x_6	0.6880	0.9293	0.8806	0.5092	0.4248	1

Table 3.2 Similarity degrees $\rho(\tilde{b}_{ij}^{(2)}, \tilde{b}_{ij})$ $(i, j = 1, 2, \cdots, 6)$

	x_1	x_2	x_3	x_4	x_5	x_6
x_1	1	0.4846	0.6418	0.6189	0.8519	0.8500
x_2	0.4948	1	0.7420	0.9979	0.7301	0.7853
x_3	0.6397	0.7587	1	0.9655	0.9103	0.7752
x_4	0.7512	0.9707	0.9191	1	0.7449	0.4628
x_5	0.8940	0.7488	0.9463	0.7533	1	0.3648
x_6	0.8214	0.7621	0.8160	0.5423	0.3862	1

Table 3.3 Similarity degrees $\rho(\tilde{b}_{ij}^{(3)}, \tilde{b}_{ij})$ $(i, j = 1, 2, \cdots, 6)$

	x_1	x_2	x_3	x_4	x_5	x_6
x_1	1	0.6861	0.6994	0.8363	0.8671	0.9167
x_2	0.6989	1	0.6910	0.9979	0.7283	1.000
x_3	0.6641	0.7354	1	0.9639	0.9389	0.8917
x_4	0.7273	0.9707	0.9306	1	0.7131	0.4888
x_5	0.9913	0.7505	0.6396	0.7451	1	0.4856
x_6	0.8881	0.9985	0.8806	0.5092	0.5511	1

Table 3.4 Similarity degrees $\rho(\tilde{b}_{ij}^{(4)}, \tilde{b}_{ij})$ $(i, j = 1, 2, \cdots, 6)$

	x_1	x_2	x_3	x_4	x_5	x_6
x_1	1	0.6989	0.7892	0.7658	0.8519	0.8500
x_2	0.6861	1	0.7420	0.7392	0.7301	0.9024
x_3	0.6816	0.6486	1	0.9655	0.8256	0.9535
x_4	0.8412	0.8612	0.9191	1	0.7935	0.5415
x_5	0.8940	0.7488	0.7762	0.9258	1	0.6775
x_6	0.8214	0.8149	0.9100	0.5865	0.6228	1

Then we utilize (3.90) to aggregate the similarity degrees $\rho(\tilde{b}_{ij}^{(k)}, \tilde{b}_{ij})$ $(i, j = 1, 2, \cdots, 6, k = 1, 2, 3, 4)$ listed in Tables 3.1~3.4, and get the similarity degrees $\rho(\tilde{\boldsymbol{B}}_k, \tilde{\boldsymbol{B}})$ $(k = 1, 2, 3, 4)$:

$$\rho(\tilde{\boldsymbol{B}}_1, \tilde{\boldsymbol{B}}) = 0.7724, \quad \rho(\tilde{\boldsymbol{B}}_2, \tilde{\boldsymbol{B}}) = 0.7814$$

$$\rho(\tilde{\boldsymbol{B}}_3, \tilde{\boldsymbol{B}}) = 0.8218, \quad \rho(\tilde{\boldsymbol{B}}_4, \tilde{\boldsymbol{B}}) = 0.8201$$

Step 3 Since $\rho(\tilde{\boldsymbol{B}}_k, \tilde{\boldsymbol{B}}) < \beta_0$ $(k = 1, 2)$, we utilize (3.92) to recalculate the weight vector of the decision makers e_k $(k = 1, 2, 3, 4)$:

$$\boldsymbol{v} = (0.2436,\ 0.2439,\ 0.2565,\ 0.2560)^{\mathrm{T}}$$

and return $\tilde{\boldsymbol{B}}_1$ and $\tilde{\boldsymbol{B}}_2$ together with $\tilde{\boldsymbol{B}}$ to the experts e_k $(k = 1, 2)$ respectively, and suggest them to revaluate the elements in $\tilde{\boldsymbol{B}}_1$ and $\tilde{\boldsymbol{B}}_2$, especially those with the similarity degrees less than α_0, including $\tilde{b}_{12}^{(1)}$, $\tilde{b}_{14}^{(1)}$, $\tilde{b}_{41}^{(1)}$, $\tilde{b}_{46}^{(1)}$, $\tilde{b}_{56}^{(1)}$, $\tilde{b}_{64}^{(1)}$, $\tilde{b}_{65}^{(1)}$, $\tilde{b}_{12}^{(2)}$, $\tilde{b}_{21}^{(2)}$, $\tilde{b}_{46}^{(2)}$, $\tilde{b}_{56}^{(2)}$ and $\tilde{b}_{65}^{(2)}$, etc.

Assume that the experts e_k $(k = 1, 2)$ provide the revaluated uncertain multiplicative linguistic preference relations as below (for convenience, we also denote $\tilde{\boldsymbol{B}}_1$ and $\tilde{\boldsymbol{B}}_2$ as the two revaluated uncertain multiplicative linguistic preference relations respectively):

$$\tilde{\boldsymbol{B}}_1 = \begin{pmatrix} [s_1, s_1] & [s_1, s_{5/4}] & [s_1, s_{5/4}] & [s_{5/4}, s_{5/2}] & [s_{5/3}, s_{5/2}] & [s_{2/5}, s_{4/5}] \\ [s_{4/5}, s_1] & [s_1, s_1] & [s_{3/5}, s_1] & [s_1, s_{5/3}] & [s_{5/4}, s_{5/3}] & [s_{3/5}, s_{4/5}] \\ [s_{2/5}, s_{4/5}] & [s_1, s_{5/3}] & [s_1, s_1] & [s_{5/4}, s_{5/3}] & [s_{2/5}, s_{4/5}] & [s_{5/3}, s_{5/2}] \\ [s_{2/5}, s_{3/5}] & [s_{3/5}, s_1] & [s_{3/5}, s_{4/5}] & [s_1, s_1] & [s_{5/4}, s_{5/3}] & [s_{5/3}, s_{5/2}] \\ [s_{2/5}, s_{3/5}] & [s_{3/5}, s_{4/5}] & [s_{5/4}, s_{5/2}] & [s_{3/5}, s_{4/5}] & [s_1, s_1] & [s_1, s_{5/3}] \\ [s_{5/4}, s_{5/2}] & [s_{5/4}, s_{5/3}] & [s_{2/5}, s_{3/5}] & [s_{2/5}, s_{3/5}] & [s_{3/5}, s_1] & [s_1, s_1] \end{pmatrix}$$

$$\tilde{\boldsymbol{B}}_2 = \begin{pmatrix} [s_1, s_1] & [s_1, s_{5/4}] & [s_{4/5}, s_1] & [s_1, s_{5/2}] & [s_{5/4}, s_{5/3}] & [s_{4/5}, s_1] \\ [s_{4/5}, s_1] & [s_1, s_1] & [s_{4/5}, s_{5/4}] & [s_{5/4}, s_{5/3}] & [s_{5/4}, s_{5/3}] & [s_{3/5}, s_1] \\ [s_1, s_{5/4}] & [s_{4/5}, s_{5/4}] & [s_1, s_1] & [s_1, s_{5/3}] & [s_{3/5}, s_1] & [s_{5/3}, s_{5/2}] \\ [s_{2/5}, s_1] & [s_{3/5}, s_{4/5}] & [s_{3/5}, s_1] & [s_1, s_1] & [s_{5/4}, s_{5/3}] & [s_{5/3}, s_{5/2}] \\ [s_{3/5}, s_{4/5}] & [s_{3/5}, s_{4/5}] & [s_1, s_{5/3}] & [s_{3/5}, s_{4/5}] & [s_1, s_1] & [s_1, s_{5/3}] \\ [s_1, s_{5/4}] & [s_1, s_{5/3}] & [s_{2/5}, s_{3/5}] & [s_{2/5}, s_{3/5}] & [s_{3/5}, s_1] & [s_1, s_1] \end{pmatrix}$$

Step 4 Utilize (3.88) to aggregate $\tilde{\boldsymbol{B}}_3$, $\tilde{\boldsymbol{B}}_4$ and the revaluated $\tilde{\boldsymbol{B}}_1$ and $\tilde{\boldsymbol{B}}_2$ into the collective uncertain multiplicative preference relation $\tilde{\boldsymbol{B}} = (\tilde{b}_{ij})_{6\times 6}$:

$$\tilde{\boldsymbol{B}} = \begin{pmatrix} [s_1, s_1] & [s_{0.96}, s_{1.24}] & [s_{0.75}, s_{1.06}] & [s_{1.20}, s_{2.25}] & [s_{1.34}, s_{2.04}] & [s_{0.63}, s_{0.89}] \\ [s_{0.81}, s_{1.04}] & [s_1, s_1] & [s_{0.73}, s_{1.14}] & [s_{1.18}, s_{1.85}] & [s_{1.27}, s_{1.72}] & [s_{0.65}, s_{0.95}] \\ [s_{0.94}, s_{1.33}] & [s_{0.88}, s_{1.37}] & [s_1, s_1] & [s_{1.12}, s_{1.67}] & [s_{0.59}, s_{0.89}] & [s_{1.55}, s_{2.50}] \\ [s_{0.44}, s_{0.83}] & [s_{0.54}, s_{0.85}] & [s_{0.60}, s_{0.89}] & [s_1, s_1] & [s_{1.18}, s_{1.72}] & [s_{1.53}, s_{2.50}] \\ [s_{0.49}, s_{0.75}] & [s_{0.58}, s_{0.79}] & [s_{1.12}, s_{1.69}] & [s_{0.58}, s_{0.85}] & [s_1, s_1] & [s_1, s_{1.53}] \\ [s_{1.12}, s_{1.59}] & [s_{1.05}, s_{1.54}] & [s_{0.40}, s_{0.65}] & [s_{0.40}, s_{0.65}] & [s_{0.65}, s_1] & [s_1, s_1] \end{pmatrix}$$

Then by (3.90) and (3.91), we get the similarity degrees $\rho(\tilde{\boldsymbol{B}}_k, \tilde{\boldsymbol{B}})$ $(k = 1, 2, 3, 4)$:

$$\rho(\tilde{\boldsymbol{B}}_1, \tilde{\boldsymbol{B}}) = 0.9128, \quad \rho(\tilde{\boldsymbol{B}}_2, \tilde{\boldsymbol{B}}) = 0.9520, \quad \rho(\tilde{\boldsymbol{B}}_3, \tilde{\boldsymbol{B}}) = 0.8499, \quad \rho(\tilde{\boldsymbol{B}}_4, \tilde{\boldsymbol{B}}) = 0.8118$$

Since $\rho(\tilde{\boldsymbol{B}}_k, \tilde{\boldsymbol{B}}) > \beta_0$ $(k = 1, 2, 3, 4)$, then $\tilde{\boldsymbol{B}}_k$ $(k = 1, 2, 3, 4)$ and $\tilde{\boldsymbol{B}}$ are of acceptable similarity degrees.

Step 5 Utilize the ULG operator (3.93) to aggregate all $\tilde{a}_{ij}$ $(i = 1, 2, \cdots, 6)$ in the ith line of $\tilde{\boldsymbol{B}}$ to derive the overall uncertain preference degree $\tilde{a}_i$ of the alternative x_i over all the alternatives:

$$\tilde{a}_1 = [s_{0.96}, s_{1.32}], \quad \tilde{a}_2 = [s_{0.91}, s_{1.24}], \quad \tilde{a}_3 = [s_{0.97}, s_{1.38}]$$

$$\tilde{a}_4 = [s_{0.80}, s_{1.18}], \quad \tilde{a}_5 = [s_{0.75}, s_{1.04}], \quad \tilde{a}_6 = [s_{0.70}, s_1]$$

Step 6 Use (2.69) to compare each pair of the overall uncertain preferences $\tilde{a}_i$ $(i = 1, 2, \cdots, 6)$, and construct a possibility degree matrix:

$$\boldsymbol{P} = \begin{pmatrix} 0.5000 & 0.5942 & 0.4545 & 0.7027 & 0.8769 & 0.9394 \\ 0.4058 & 0.5000 & 0.3649 & 0.6197 & 0.7903 & 0.8571 \\ 0.5455 & 0.6351 & 0.5000 & 0.7342 & 0.9000 & 0.9577 \\ 0.2973 & 0.3803 & 0.2658 & 0.5000 & 0.6418 & 0.7059 \\ 0.1231 & 0.2097 & 0.1000 & 0.3582 & 0.5000 & 0.5763 \\ 0.0606 & 0.1429 & 0.0423 & 0.2941 & 0.4237 & 0.5000 \end{pmatrix}$$

Then we employ (2.82) to derive the priority vector of the possibility degree matrix $\boldsymbol{P}$:

$$\boldsymbol{\zeta} = (0.2023, 0.1846, 0.2091, 0.1597, 0.1297, 0.1155)^{\mathrm{T}}$$

Step 7 Rank the overall uncertain preferences $\tilde{a}_i$ $(i = 1, 2, \cdots, 6)$ according to the priority vector $\boldsymbol{\zeta}$:

$$\tilde{a}_3 > \tilde{a}_1 > \tilde{a}_2 > \tilde{a}_4 > \tilde{a}_5 > \tilde{a}_6$$

Thus

$$x_3 \succ x_1 \succ x_2 \succ x_4 \succ x_5 \succ x_6$$

Consequently, the best global supplier is x_3.

Based on the expected value principle, Xu (2008) defined the expected multiplicative linguistic preference relation of an uncertain multiplicative linguistic preference relation:

Definition 3.36 (Xu, 2008) Let $\tilde{\boldsymbol{B}} = (\tilde{b}_{ij})_{n\times n}$ be an uncertain multiplicative linguistic preference relation. Then the corresponding expected multiplicative linguistic preference relation $\boldsymbol{E}(\tilde{\boldsymbol{B}}) = (E(\tilde{b}_{ij}))_{n\times n}$, where

$$E(\tilde{b}_{ij}) = (b_{ij}^{\mathrm{L}})^{1-\theta} \otimes (b_{ij}^{\mathrm{U}})^{\theta}, \quad E(\tilde{b}_{ji}) = (E(\tilde{b}_{ij}))^{-1}, \quad i < j \tag{3.94}$$

θ is an index that reflects the decision maker's risk-bearing attitude.

According to the operational laws of the multiplicative linguistic evaluation scale and Definition 3.36, we can see

$$\begin{aligned} E(\tilde{b}_{ij}) \otimes E(\tilde{b}_{ji}) &= E(\tilde{b}_{ij}) \otimes [E(\tilde{b}_{ij})]^{-1} \\ &= (b_{ij}^{\mathrm{L}})^{1-\theta} \otimes (b_{ij}^{\mathrm{U}})^{\theta} \otimes \left[(b_{ij}^{\mathrm{L}})^{1-\theta} \otimes (b_{ij}^{\mathrm{U}})^{\theta}\right]^{-1} \\ &= (b_{ij}^{\mathrm{L}})^{1-\theta} \otimes (b_{ij}^{\mathrm{U}})^{\theta} \otimes (b_{ij}^{\mathrm{L}})^{-(1-\theta)} \otimes (b_{ij}^{\mathrm{U}})^{-\theta} \\ &= (b_{ij}^{\mathrm{L}})^{1-\theta-(1-\theta)} \otimes (b_{ij}^{\mathrm{U}})^{\theta-\theta} \\ &= s_1 \otimes s_1 \\ &= s_1, \quad i,j = 1,2,\cdots,n \end{aligned} \tag{3.95}$$

$$\begin{aligned} E(\tilde{b}_{ii}) &= (b_{ii}^{\mathrm{L}})^{1-\theta} \otimes (b_{ii}^{\mathrm{U}})^{\theta} \\ &= (s_1)^{1-\theta} \otimes (s_1)^{\theta} \\ &= (s_1)^{1-\theta+\theta} \\ &= s_1, \quad i = 1,2,\cdots,n \end{aligned} \tag{3.96}$$

To measure the deviation degree between the multiplicative linguistic labels, Xu (2008) gave a deviation measure of the multiplicative linguistic labels as follows:

Definition 3.37 (Xu, 2008) Let $s_\alpha, s_\beta \in \bar{S}_4$. Then we call

$$d'(s_\alpha, s_\beta) = \max\left\{\frac{\alpha}{\beta}, \frac{\beta}{\alpha}\right\} \tag{3.97}$$

the deviation degree between s_α and s_β.

Clearly, by (3.97), we have the following theorem:

Theorem 3.30 (Xu, 2008) Let $s_\alpha, s_\beta, s_\gamma \in \bar{S}_4$. Then

(1) $d'(s_\alpha, s_\beta) \geqslant 1$;

(2) $d'(s_\alpha, s_\beta) = 1$ if and only if $s_\alpha = s_\beta$;

(3) $d'(s_\alpha, s_\beta) = d'(s_\beta, s_\alpha)$.

Considering that in the process of group decision making, the decision makers may use different preference structures to express their preferences over alternatives, and based on Definitions 3.36 and 3.37, Xu (2008) further developed an approach to group decision making with distinct types of multiplicative linguistic preference relations:

Step 1 For a group decision making problem, let X, E and $\boldsymbol{v}$ as defined in Section 3.2. Each decision maker employs the multiplicative linguistic evaluation scale $S_4 = \{s_\alpha | \alpha = 1/\tau, \cdots, 1/2, 1, 2, \cdots, \tau\}$ to compare each pair of the alternatives in X, assume that the decision makers e_k $(k = 1, \cdots, l_1)$ construct the multiplicative linguistic preference relations $\boldsymbol{B}_k = (b_{ij}^{(k)})_{n\times n}$ $(k = 1, \cdots, l_1)$, and the decision makers e_k $(k = l_1 + 1, \cdots, l)$ construct the uncertain multiplicative linguistic preference relations $\tilde{\boldsymbol{B}}_k = (\tilde{b}_{ij}^{(k)})_{n\times n}$ $(k = l_1 + 1, \cdots, l)$.

Step 2 Use (3.94) to transform the uncertain multiplicative linguistic preference relations $\tilde{\boldsymbol{B}}_k = (\tilde{b}_{ij}^{(k)})_{n\times n}$ $(k = l_1+1, \cdots, l)$ into the corresponding expected multiplicative linguistic preference relations $\boldsymbol{E}(\tilde{\boldsymbol{B}}_k) = (E(\tilde{b}_{ij}^{(k)}))_{n\times n}$ $(k = l_1+1, \cdots, l)$.

Step 3 Utilize the LWG operator:

$$\begin{aligned} b_{ij} =& \mathrm{LWG}(b_{ij}^{(1)}, \cdots, b_{ij}^{(l_1)}, E(\tilde{b}_{ij}^{(l_1+1)}), \cdots, E(\tilde{b}_{ij}^{(l)})) \\ =& (b_{ij}^{(1)})^{v_1} \otimes \cdots \otimes (b_{ij}^{(l_1)})^{v_{l_1}} \otimes (E(\tilde{b}_{ij}^{(l_1+1)}))^{v_{l_1+1}} \\ & \otimes \cdots \otimes (E(\tilde{b}_{ij}^{(l)}))^{v_l}, \quad i,j = 1,2,\cdots,n \end{aligned} \tag{3.98}$$

to aggregate all the multiplicative linguistic preference relations $\boldsymbol{B}_k = (b_{ij}^{(k)})_{n\times n}$ $(k = 1, 2, \cdots, l_1)$ and the expected multiplicative linguistic preference relations $\boldsymbol{E}(\tilde{\boldsymbol{B}}_k) = (E(\tilde{b}_{ij}^{(k)}))_{n\times n}$ $(k = l_1+1, \cdots, l)$ into the collective multiplicative linguistic preference relation $\boldsymbol{B} = (b_{ij})_{n\times n}$.

To measure the deviation degree between the collective multiplicative linguistic preference relation $\boldsymbol{B} = (b_{ij})_{n\times n}$ and each of the multiplicative linguistic preference relation $\boldsymbol{B}_k = (b_{ij}^{(k)})_{n\times n}$ $(k = 1, 2, \cdots, l_1)$, and the deviation degree between the collective multiplicative linguistic preference relation $\boldsymbol{B} = (b_{ij})_{n\times n}$ and each of the expected multiplicative linguistic preference relation $\boldsymbol{E}(\tilde{\boldsymbol{B}}_k) = (E(\tilde{b}_{ij}^{(k)}))_{n\times n}$ $(k = l_1+1, \cdots, l)$, Xu (2008) defined

$$d'(\boldsymbol{B}_k, \boldsymbol{B}) = \left[\prod_{i,j=1}^{n} d'(b_{ij}^{(k)}, b_{ij})\right]^{1/n^2}, \quad k = 1, 2, \cdots, l_1 \tag{3.99}$$

as the deviation degree between the collective multiplicative linguistic preference relation $\boldsymbol{B}$ and each of the multiplicative linguistic preference relations $\boldsymbol{B}_k = (b_{ij}^{(k)})_{n\times n}$ $(k = 1, 2, \cdots, l_1)$, and defined

$$d'(\boldsymbol{E}(\tilde{\boldsymbol{B}}_k), \boldsymbol{B}) = \left(\prod_{i,j=1}^{n} d'(E(\tilde{b}_{ij}^{(k)}), b_{ij})\right)^{1/n^2}, \quad k = l_1+1, \cdots, l \tag{3.100}$$

as the deviation degree between the collective multiplicative linguistic preference relation $\boldsymbol{B}$ and each of the expected multiplicative linguistic preference relations $\boldsymbol{E}(\tilde{\boldsymbol{B}}_k) = (E(\tilde{b}_{ij}^{(k)}))_{n\times n}$ $(k = l_1+1, \cdots, l)$, where $d'(b_{ij}^{(k)}, b_{ij})$ $(k = 1, 2, \cdots, l_1,\ i, j = 1, 2, \cdots, n)$ and $d'(E(\tilde{b}_{ij}^{(k)}), b_{ij})$ $(k = l_1+1, \cdots, l,\ i, j = 1, 2, \cdots, n)$ can be calculated by (3.97).

$d'(\boldsymbol{B}_k, \boldsymbol{B})$ has the following properties (Xu, 2008):

(1) $d'(\boldsymbol{B}_k, \boldsymbol{B}) \geqslant 1$;

(2) $d'(\boldsymbol{B}_k, \boldsymbol{B}) = 1$ if and only if $\boldsymbol{B}_k = \boldsymbol{B}$;

(3) $d'(\boldsymbol{B}_k, \boldsymbol{B}) = d'(\boldsymbol{B}, \boldsymbol{B}_k)$.

Similarly, $d'(\boldsymbol{E}(\tilde{\boldsymbol{B}}_k), \boldsymbol{B})$ also has the properties above.

In what follows, we define a deviation function $d'(\boldsymbol{B}_k, \boldsymbol{B})$(or $d'(\boldsymbol{E}(\tilde{\boldsymbol{B}}_k), \boldsymbol{B})$) as

$$y = g\left(d'(\tilde{\boldsymbol{B}}_k, \boldsymbol{B})\right) \quad \left(\text{or } y = g\left(d'(\boldsymbol{E}(\tilde{\boldsymbol{B}}_k), \boldsymbol{B})\right)\right) \tag{3.101}$$

and based on (3.101), we give a simple formula for determining the weights of the decision makers:

$$v_k = \frac{g\left(d'(\boldsymbol{B}_k, \boldsymbol{B})\right)}{\sum_{k=1}^{l_1} g\left(d'(\boldsymbol{B}_k, \boldsymbol{B})\right) + \sum_{k=l_1+1}^{l} g\left(d'(\boldsymbol{E}(\tilde{\boldsymbol{B}}_k), \boldsymbol{B})\right)}, \quad k = 1, 2, \cdots, l_1 \tag{3.102}$$

$$v_k = \frac{g\left(d'(\boldsymbol{E}(\tilde{\boldsymbol{B}}_k), \boldsymbol{B})\right)}{\sum_{k=1}^{l_1} g\left(d'(\boldsymbol{B}_k, \boldsymbol{B})\right) + \sum_{k=l_1+1}^{l} g\left(d'(\boldsymbol{E}(\tilde{\boldsymbol{B}}_k), \boldsymbol{B})\right)}, \quad k = l_1 + 1, \cdots, l \tag{3.103}$$

Now we discuss three cases (Xu, 2008):

Case 1 If the majority opinion should be emphasized, then the low weights should be assigned to the decision maker e_k with high deviation degree $d'(\boldsymbol{B}_k, \boldsymbol{B})$ (or $d'(\boldsymbol{E}(\tilde{\boldsymbol{B}}_k), \boldsymbol{B})$). In this case, $y = g\left(d'(\boldsymbol{B}_k, \boldsymbol{B})\right)$ (or $y = g(d'(\boldsymbol{E}(\tilde{\boldsymbol{B}}_k), \boldsymbol{B}))$) is a decreasing function of the deviation. For example, if

$$g\left(d'(\boldsymbol{B}_k, \boldsymbol{B})\right) = \frac{1}{d'(\boldsymbol{B}_k, \boldsymbol{B})} \quad \left(\text{or } g\left(d'(\boldsymbol{E}(\tilde{\boldsymbol{B}}_k), \boldsymbol{B})\right) = \frac{1}{d'(\boldsymbol{E}(\tilde{\boldsymbol{B}}_k), \boldsymbol{B})}\right) \tag{3.104}$$

then

$$v_k = \frac{\dfrac{1}{d'(\boldsymbol{B}_k, \boldsymbol{B})}}{\sum_{k=1}^{l_1} \dfrac{1}{d'(\boldsymbol{B}_k, \boldsymbol{B})} + \sum_{k=l_1+1}^{l} \dfrac{1}{d'(\boldsymbol{E}(\tilde{\boldsymbol{B}}_k), \boldsymbol{B})}}, \quad k = 1, 2, \cdots, l_1 \tag{3.105}$$

$$v_k = \frac{\dfrac{1}{d'(\boldsymbol{E}(\tilde{\boldsymbol{B}}_k), \boldsymbol{B})}}{\sum_{k=1}^{l_1} \dfrac{1}{d'(\boldsymbol{B}_k, \boldsymbol{B})} + \sum_{k=l_1+1}^{l} \dfrac{1}{d'(\boldsymbol{E}(\tilde{B}_k), \boldsymbol{B})}}, \quad k = l_1 + 1, \cdots, l \tag{3.106}$$

Case 2 If the minority opinion should be emphasized, then the high weights should be assigned to the decision maker e_k with high deviation degree $d'(\boldsymbol{B}_k, \boldsymbol{B})$ (or $d'(\boldsymbol{E}(\tilde{\boldsymbol{B}}_k), \boldsymbol{B})$). In this case, $y = g\left(d'(\boldsymbol{B}_k, \boldsymbol{B})\right)$ (or $y = g(d'(\boldsymbol{E}(\tilde{\boldsymbol{B}}_k), \boldsymbol{B}))$) is an increasing function of the deviation. For example, if

$$g\left(d'(\boldsymbol{B}_k, \boldsymbol{B})\right) = d'(\boldsymbol{B}_k, \boldsymbol{B}) \tag{3.107}$$

then

$$\upsilon_k = \frac{d'(\boldsymbol{B}_k, \boldsymbol{B})}{\sum_{k=1}^{l_1} d'(\boldsymbol{B}_k, \boldsymbol{B}) + \sum_{k=l_1+1}^{l} d'(\boldsymbol{E}(\tilde{\boldsymbol{B}}_k), \boldsymbol{B})}, \quad k = 1, 2, \cdots, l_1 \tag{3.108}$$

$$\upsilon_k = \frac{d'(\boldsymbol{E}(\tilde{\boldsymbol{B}}_k), \boldsymbol{B})}{\sum_{k=1}^{l_1} d'(\boldsymbol{B}_k, \boldsymbol{B}) + \sum_{k=l_1+1}^{l} d'(\boldsymbol{E}(\tilde{\boldsymbol{B}}_k), \boldsymbol{B})}, \quad k = l_1 + 1, \cdots, l \tag{3.109}$$

Case 3 If both the majority and minority opinions should be emphasized simultaneously, then we can use the following formula to determine the decision makers' weights:

$$\begin{aligned} \upsilon_k =& (1-\alpha)\frac{g_1\left(d'(\boldsymbol{B}_k, \boldsymbol{B})\right)}{\sum_{k=1}^{l_1} g_1\left(d'(\boldsymbol{B}_k, \boldsymbol{B})\right) + \sum_{k=l_1+1}^{l} g_1\left(d'(\boldsymbol{E}(\tilde{\boldsymbol{B}}_k), \boldsymbol{B})\right)} \\ &+ \alpha\frac{g_2\left(d'(\boldsymbol{B}_k, \boldsymbol{B})\right)}{\sum_{k=1}^{l_1} g_2\left(d'(\boldsymbol{B}_k, \boldsymbol{B})\right) + \sum_{k=l_1+1}^{l} g_2\left(d'(\boldsymbol{E}(\tilde{\boldsymbol{B}}_k), \boldsymbol{B})\right)}, \quad k = 1, 2, \cdots, l_1 \end{aligned} \tag{3.110}$$

$$\begin{aligned} \upsilon_k =& (1-\alpha)\frac{g_1\left(d'(\boldsymbol{E}(\tilde{\boldsymbol{B}}_k), \boldsymbol{B})\right)}{\sum_{k=1}^{l_1} g_1\left(d'(\boldsymbol{B}_k, \boldsymbol{B})\right) + \sum_{k=l_1+1}^{l} g_1\left(d'(\boldsymbol{E}(\tilde{\boldsymbol{B}}_k), \boldsymbol{B})\right)} \\ &+ \alpha\frac{g_2\left(d'(E(\tilde{\boldsymbol{B}}_k), \boldsymbol{B})\right)}{\sum_{k=1}^{l_1} g_2\left(d'(\boldsymbol{B}_k, \boldsymbol{B})\right) + \sum_{k=l_1+1}^{l} g_2\left(d'(\boldsymbol{E}(\tilde{\boldsymbol{B}}_k), \boldsymbol{B})\right)}, \quad k = l_1 + 1, \cdots, l \end{aligned} \tag{3.111}$$

where $\alpha \in [0, 1]$ (which can be given by the decision makers directly), $y_1 = g_1(d'(\tilde{\boldsymbol{B}}_k, \boldsymbol{B}))$ (or $y_1 = g_1(d'(\boldsymbol{E}(\boldsymbol{B}_k), \boldsymbol{B}))$) and $y_2 = g_2(d'(\boldsymbol{B}_k, \boldsymbol{B}))$ (or $y_2 = g_2(d'(\boldsymbol{E}(\tilde{\boldsymbol{B}}_k), \boldsymbol{B}))$) are the decreasing and increasing functions of the deviation $d'(\boldsymbol{B}_k, \boldsymbol{B})$ (or $d'(\boldsymbol{E}(\tilde{\boldsymbol{B}}_k), \boldsymbol{B})$). For example, if $g_1(d(\boldsymbol{B}_k, \boldsymbol{B}))$ (or $g_1(d'(\boldsymbol{E}(\tilde{\boldsymbol{B}}_k), \boldsymbol{B}))$) and $g_2(d'(\boldsymbol{B}_k, \boldsymbol{B}))$ ($g_2(d'(\boldsymbol{E}(\boldsymbol{B}_k), \boldsymbol{B}))$) are (3.104) and (3.107) respectively, then (3.110) and (3.111) can be rewritten as

$$\upsilon_k = (1-\alpha)\frac{\dfrac{1}{d'(\boldsymbol{B}_k, \boldsymbol{B})}}{\sum_{k=1}^{l_1} \dfrac{1}{d'(\boldsymbol{B}_k, \boldsymbol{B})} + \sum_{k=l_1+1}^{l} \dfrac{1}{d'(\boldsymbol{E}(\tilde{\boldsymbol{B}}_k), \boldsymbol{B})}}$$

$$+\alpha\frac{d'(\boldsymbol{B}_k,\boldsymbol{B})}{\sum\limits_{k=1}^{l_1}d'(\boldsymbol{B}_k,\boldsymbol{B})+\sum\limits_{k=l_1+1}^{l}d'(\boldsymbol{E}(\tilde{\boldsymbol{B}}_k),\boldsymbol{B})},\quad k=1,2,\cdots,l_1 \tag{3.112}$$

$$\upsilon_k=(1-\alpha)\frac{\dfrac{1}{d'(\boldsymbol{E}(\tilde{\boldsymbol{B}}_k),\boldsymbol{B})}}{\sum\limits_{k=1}^{l_1}\dfrac{1}{d'(\boldsymbol{B}_k,\boldsymbol{B})}+\sum\limits_{k=l_1+1}^{l}\dfrac{1}{d'(\boldsymbol{E}(\tilde{\boldsymbol{B}}_k),\boldsymbol{B})}}$$

$$+\alpha\frac{d'(\boldsymbol{E}(\tilde{\boldsymbol{B}}_k),\boldsymbol{B})}{\sum\limits_{k=1}^{l_1}d'(\boldsymbol{B}_k,\boldsymbol{B})+\sum\limits_{k=l_1+1}^{l}d'(\boldsymbol{E}(\tilde{\boldsymbol{B}}_k),\boldsymbol{B})},\quad k=l_1+1,\cdots,l \tag{3.113}$$

In particular, if $\alpha=0$, then (3.112) and (3.113) reduce to (3.105) and (3.106) respectively; if $\alpha=1$, then (3.112) and (3.113) reduce to (3.108) and (3.109) respectively. Therefore, Cases 1 and 2 are the special cases of Case 3.

Moreover, if

$$d'\left(\boldsymbol{B}_k,\boldsymbol{B}\right)\leqslant\eta_0\quad(\text{or } d'(\boldsymbol{E}(\tilde{\boldsymbol{B}}_k),\boldsymbol{B})\leqslant\eta_0) \tag{3.114}$$

then the collective multiplicative linguistic preference relation $\boldsymbol{B}$ and the multiplicative linguistic preference relation $\boldsymbol{B}_k$ (or the expected multiplicative linguistic preference relation $\boldsymbol{E}(\tilde{\boldsymbol{B}}_k)$) are called of acceptable deviation degree, where η_0 is the threshold of acceptable deviation degree, which can be predefined by all the decision makers (who participate in determining the decision makers' original weights) in practical applications. Especially, if $d'(\boldsymbol{B}_k,\boldsymbol{B})=1$, then the multiplicative linguistic preference relation $\boldsymbol{B}_k$ (or the expected multiplicative linguistic preference relation $E(\tilde{\boldsymbol{B}}_k)$) are equal. However, if $d'(\boldsymbol{B}_k,\boldsymbol{B})>\eta_0$ (or $d'(\boldsymbol{E}(\tilde{\boldsymbol{B}}_k),\boldsymbol{B})>\eta_0$) , then we shall return $\boldsymbol{B}_k$ (or the uncertain multiplicative preference relation $\tilde{\boldsymbol{B}}_k$ together with its expected multiplicative preference relation $\boldsymbol{E}(\tilde{\boldsymbol{B}}_k)$) and the collective multiplicative linguistic preference relation $\boldsymbol{B}$ to the decision maker e_k for revaluation. We repeat this process (in which the decision makers' original weights are replaced by the new weights derived from (3.112) and (3.113) for obtaining the new collective multiplicative linguistic preference relation $\boldsymbol{B}$ by using the LWG operator (3.98) until $\boldsymbol{B}_k$ (or the expected multiplicative linguistic preference relation $\boldsymbol{E}(\tilde{\boldsymbol{B}}_k)$) and the collective multiplicative linguistic preference relation $\boldsymbol{B}$ are of acceptable deviation degree or the process will stop as the repetition times reach the maximum number predefined by the decision makers, and then go to Step 4.

Step 4 Utilize the LG operator:

$$b_i=\mathrm{LG}(b_{i1},b_{i2},\cdots,b_{in}),\quad i=1,2,\cdots,n \tag{3.115}$$

to aggregate the preference b_{ij} in the ith line of $\boldsymbol{B}$, and then get the mean preference b_i of the alternative x_i over all the alternatives.

Step 5 Rank and select the alternatives x_i $(i=1,2,\cdots,n)$ according to b_i $(i=1,2,\cdots,n)$.

Example 3.16 In Example 3.13, assume that the members e_k $(k=1,2,3,4)$ of information management steering committee compare each pair of these projects x_j $(j=1,2,\cdots,6)$ with respect to the factor productivity by using the multiplicative linguistic labels in the set:

$$S_4=\{s_{1/5}=\text{extremely poor},\ s_{2/5}=\text{very poor},\ s_{3/5}=\text{poor},\ s_{4/5}=\text{slightly poor},$$
$$s_1=\text{fair},\ s_{5/4}=\text{slightly good},\ s_{5/3}=\text{good},$$
$$s_{5/2}=\text{very good},\ s_5=\text{extremely good}\}$$

and construct the multiplicative linguistic preference relations $\boldsymbol{B}_1$, $\boldsymbol{B}_2$ and the uncertain multiplicative linguistic preference relations $\tilde{\boldsymbol{B}}_3$ and $\tilde{\boldsymbol{B}}_4$ respectively as follows:

$$\boldsymbol{B}_1=\begin{pmatrix} s_1 & s_2 & s_{1/3} & s_4 & s_2 & s_{1/2}\\ s_{1/2} & s_1 & s_{1/2} & s_3 & s_1 & s_2\\ s_3 & s_2 & s_1 & s_2 & s_{1/3} & s_3\\ s_{1/4} & s_{1/3} & s_{1/2} & s_1 & s_3 & s_4\\ s_{1/2} & s_1 & s_3 & s_{1/3} & s_1 & s_{1/3}\\ s_2 & s_{1/2} & s_{1/3} & s_{1/4} & s_3 & s_1 \end{pmatrix}$$

$$\boldsymbol{B}_2=\begin{pmatrix} s_1 & s_3 & s_{1/4} & s_2 & s_3 & s_{1/3}\\ s_{1/3} & s_1 & s_{1/3} & s_4 & s_2 & s_3\\ s_4 & s_3 & s_1 & s_3 & s_{1/2} & s_1\\ s_{1/2} & s_{1/4} & s_{1/3} & s_1 & s_4 & s_2\\ s_{1/3} & s_{1/2} & s_2 & s_{1/4} & s_1 & s_2\\ s_3 & s_{1/3} & s_1 & s_{1/2} & s_{1/2} & s_1 \end{pmatrix}$$

$$\tilde{\boldsymbol{B}}_3=\begin{pmatrix} [s_1,s_1] & [s_2,s_3] & [s_{1/3},s_{1/2}] & [s_3,s_5] & [s_{1/2},s_2] & [s_{1/2},s_1]\\ [s_{1/3},s_{1/2}] & [s_1,s_1] & [s_{1/2},s_1] & [s_2,s_4] & [s_{1/2},s_2] & [s_2,s_3]\\ [s_2,s_3] & [s_1,s_2] & [s_1,s_1] & [s_{1/2},s_3] & [s_{1/3},s_{1/2}] & [s_1,s_2]\\ [s_{1/5},s_{1/3}] & [s_{1/4},s_{1/2}] & [s_{1/3},s_2] & [s_1,s_1] & [s_2,s_3] & [s_3,s_4]\\ [s_{1/2},s_2] & [s_{1/2},s_2] & [s_2,s_3] & [s_{1/3},s_{1/2}] & [s_1,s_1] & [s_{1/3},s_{1/2}]\\ [s_1,s_2] & [s_{1/3},s_{1/2}] & [s_{1/3},s_{1/2}] & [s_{1/4},s_{1/3}] & [s_2,s_3] & [s_1,s_1] \end{pmatrix}$$

$$\tilde{\boldsymbol{B}}_4=\begin{pmatrix} [s_1,s_1] & [s_3,s_4] & [s_{1/3},s_1] & [s_2,s_3] & [s_1,s_3] & [s_{1/3},s_{1/2}]\\ [s_{1/4},s_{1/3}] & [s_1,s_1] & [s_{1/3},s_{1/2}] & [s_3,s_5] & [s_{1/3},s_1] & [s_1,s_2]\\ [s_1,s_3] & [s_2,s_3] & [s_1,s_1] & [s_1,s_2] & [s_{1/2},s_2] & [s_3,s_4]\\ [s_{1/3},s_{1/2}] & [s_{1/5},s_{1/3}] & [s_{1/2},s_1] & [s_1,s_1] & [s_3,s_4] & [s_2,s_3]\\ [s_{1/3},s_1] & [s_1,s_3] & [s_{1/2},s_2] & [s_{1/4},s_{1/3}] & [s_1,s_1] & [s_1,s_2]\\ [s_2,s_3] & [s_{1/2},s_1] & [s_{1/4},s_{1/3}] & [s_{1/3},s_{1/2}] & [s_{1/2},s_1] & [s_1,s_1] \end{pmatrix}$$

Based on the known preference information, we give the following decision making steps:

Step 1 Suppose that the members e_k $(k = 3, 4)$ of information management steering committee provide the indices of their risk-bearing attitude as $\theta_3 = 0.3$ and $\theta_4 = 0.6$. Then we utilize (3.94) to transform the uncertain multiplicative preference relations $\tilde{\boldsymbol{B}}_k = (\tilde{b}_{ij}^{(k)})_{6\times 6}$ $(k = 3, 4)$ into the corresponding expected multiplicative linguistic preference relations $\boldsymbol{E}(\tilde{\boldsymbol{B}}_k) = E(\tilde{b}_{ij}^{(k)})_{6\times 6}$ $(k = 3, 4)$:

$$\boldsymbol{E}(\tilde{\boldsymbol{B}}_3) = \begin{pmatrix} s_1 & s_{2.26} & s_{0.38} & s_{3.50} & s_{0.76} & s_{0.62} \\ s_{0.44} & s_1 & s_{0.62} & s_{2.46} & s_{0.76} & s_{2.26} \\ s_{2.63} & s_{1.61} & s_1 & s_{0.86} & s_{0.38} & s_{1.23} \\ s_{0.29} & s_{0.41} & s_{1.16} & s_1 & s_{2.26} & s_{3.27} \\ s_{1.32} & s_{1.32} & s_{2.63} & s_{0.44} & s_1 & s_{0.38} \\ s_{1.61} & s_{0.44} & s_{0.81} & s_{0.31} & s_{2.63} & s_1 \end{pmatrix}$$

$$\boldsymbol{E}(\tilde{\boldsymbol{B}}_4) = \begin{pmatrix} s_1 & s_{3.27} & s_{0.46} & s_{2.26} & s_{1.39} & s_{0.38} \\ s_{0.31} & s_1 & s_{0.38} & s_{3.50} & s_{0.46} & s_{1.23} \\ s_{2.17} & s_{2.63} & s_1 & s_{1.23} & s_{0.76} & s_{3.27} \\ s_{0.44} & s_{0.29} & s_{0.81} & s_1 & s_{3.27} & s_{2.26} \\ s_{0.72} & s_{2.17} & s_{1.32} & s_{0.31} & s_1 & s_{1.23} \\ s_{2.63} & s_{0.81} & s_{0.31} & s_{0.44} & s_{0.81} & s_1 \end{pmatrix}$$

Step 2 Let the original weight vector of the members e_k $(k = 1, 2, 3, 4)$ of information management steering committee be

$$\boldsymbol{v} = (0.35,\ 0.20,\ 0.15,\ 0.30)^{\mathrm{T}}$$

and suppose that these four members predefine the threshold of acceptable deviation degree, as $\eta_0 = 1.4$. Then we utilize the LWG operator (3.98) to aggregate all the multiplicative linguistic preference relations $\boldsymbol{B}_k$ $(k = 1, 2)$ and the expected multiplicative linguistic preference relations $\boldsymbol{E}(\tilde{\boldsymbol{B}}_k)$ $(k = 3, 4)$ into the collective multiplicative linguistic preference relation $\boldsymbol{B} = (b_{ij})_{6\times 6}$:

$$\boldsymbol{B} = \begin{pmatrix} s_1 & s_{2.56} & s_{0.35} & s_{2.88} & s_{1.68} & s_{0.44} \\ s_{0.39} & s_1 & s_{0.44} & s_{3.23} & s_{0.87} & s_{1.91} \\ s_{2.86} & s_{2.27} & s_1 & s_{1.65} & s_{0.47} & s_{2.16} \\ s_{0.35} & s_{0.31} & s_{0.61} & s_1 & s_{3.13} & s_{2.85} \\ s_{0.60} & s_{1.15} & s_{2.13} & s_{0.32} & s_1 & s_{0.72} \\ s_{2.27} & s_{0.52} & s_{0.46} & s_{0.35} & s_{1.39} & s_1 \end{pmatrix}$$

Step 3 Use (3.97), (3.99) and (3.100) to calculate the deviation degree between $\boldsymbol{B}$ and each of $\boldsymbol{B}_k (k = 1, 2)$ and $\boldsymbol{E}(\tilde{\boldsymbol{B}}_k)$ $(k = 3, 4)$:

$$d'(\boldsymbol{B}_1, \boldsymbol{B}) = 1.190, \quad d'(\boldsymbol{B}_2, \boldsymbol{B}) = 1.439$$

$$d'(\boldsymbol{E}(\tilde{\boldsymbol{B}}_3), \boldsymbol{B}) = 1.318, \quad d'(\boldsymbol{E}(\tilde{\boldsymbol{B}}_4), \boldsymbol{B}) = 1.274$$

Since $d'(\boldsymbol{B}_2, \boldsymbol{B}) > 1.4$, then we need to utilize (3.112) and (3.113) (let $\alpha = 0.3$) to recalculate the weights of the members e_k $(k = 1, 2, 3, 4)$:

$$\boldsymbol{v} = (0.2594,\ 0.2407,\ 0.2482,\ 0.2517)^{\mathrm{T}}$$

and return the multiplicative linguistic preference relation $\boldsymbol{B}_2$ together with the collective multiplicative linguistic preference relation $\tilde{\boldsymbol{B}}$ to the member e_2, and suggest him or her to revaluate the elements in $\boldsymbol{B}_2$. Assume that the member e_2 provides the revaluated multiplicative linguistic preference relation as $\boldsymbol{B}_2' = (b_{ij}^{(2)})_{6\times 6}$:

$$\boldsymbol{B}_2' = \begin{pmatrix} s_1 & s_3 & s_{1/3} & s_3 & s_2 & s_{1/3} \\ s_{1/3} & s_1 & s_{1/2} & s_3 & s_2 & s_2 \\ s_3 & s_2 & s_1 & s_2 & s_{1/2} & s_1 \\ s_{1/3} & s_{1/3} & s_{1/2} & s_1 & s_3 & s_3 \\ s_{1/2} & s_{1/2} & s_2 & s_{1/3} & s_1 & s_2 \\ s_3 & s_{1/2} & s_1 & s_{1/3} & s_{1/2} & s_1 \end{pmatrix}$$

Then we can use the recalculated weight vector $\boldsymbol{v}$ and the LWG operator (3.98) to aggregate the expected multiplicative linguistic preference relations $\boldsymbol{E}(\tilde{\boldsymbol{B}}_k)$ $(k = 3, 4)$ and the multiplicative linguistic preference relations $\boldsymbol{B}_1$ and $\boldsymbol{B}_2'$ into the collective multiplicative linguistic preference relation $\boldsymbol{B}' = (b_{ij}')_{6\times 6}$:

$$\boldsymbol{B}' = \begin{pmatrix} s_1 & s_{2.57} & s_{0.37} & s_{3.13} & s_{1.44} & s_{0.45} \\ s_{0.39} & s_1 & s_{0.49} & s_{2.97} & s_{0.91} & s_{1.82} \\ s_{2.70} & s_{2.04} & s_1 & s_{1.44} & s_{0.47} & s_{1.89} \\ s_{0.32} & s_{0.34} & s_{0.69} & s_1 & s_{2.86} & s_{3.08} \\ s_{0.69} & s_{1.10} & s_{2.13} & s_{0.35} & s_1 & s_{0.74} \\ s_{2.22} & s_{0.55} & s_{0.53} & s_{0.32} & s_{1.35} & s_1 \end{pmatrix}$$

By (3.97), (3.99) and (3.100), we have

$$d'(\boldsymbol{B}_1, \boldsymbol{B}') = 1.214, \quad d'(\boldsymbol{B}_2', \boldsymbol{B}') = 1.245$$
$$d'(\boldsymbol{E}(\tilde{\boldsymbol{B}}_3), \boldsymbol{B}') = 1.259, \quad d'(\boldsymbol{E}(\tilde{\boldsymbol{B}}_4), \boldsymbol{B}') = 1.291$$

Since all the deviation degrees $d'(\boldsymbol{B}_1, \boldsymbol{B}')$, $d'(\boldsymbol{B}_2', \boldsymbol{B}')$ and $d'(\boldsymbol{E}(\tilde{\boldsymbol{B}}_k), \boldsymbol{B}')$ $(k = 3, 4)$ are acceptable, then go to the next step.

Step 4 Utilize the LG operator (3.115) to aggregate all the preferences in the ith line of the collective uncertain multiplicative linguistic preference relation $\boldsymbol{B}'$ into the overall preference degree b_i' of the alternative x_i over all the alternatives:

$$b_1' = s_{1.12}, \quad b_2' = s_{0.99}, \quad b_3' = s_{1.38}, \quad b_4' = s_{0.93}, \quad b_5' = s_{0.86}, \quad b_6' = s_{0.81}$$

Step 5 Rank the alternatives x_i $(i = 1, 2, \cdots, 6)$ according to b_i' $(i = 1, 2, \cdots, 6)$:

$$x_3 \succ x_1 \succ x_2 \succ x_4 \succ x_5 \succ x_6$$

Accordingly, the project x_3 has the highest potential contribution to the firm's strategic goal of gaining competitive advantage in the industry.

References

Cai L C, Deng X J. 1996. An approach to constructing judgment matrix with desirable consistency in AHP. *In*: Sun H C. ed. Decision Science and Its Application. Beijing: Haiyang Press: 88-91.

Chan F T S, Kumar N. 2007. Global supplier development considering risk factors using fuzzy extended AHP-based approach. Omega, 35: 417-431.

Chen Y, Fan Z P. 2005. Study on the adverse judgment problem for group decision making based on linguistic judgment matrices. Journal of Systems Engineering, 20(2): 211-215.

Chiclana F, Herrera F, Herrera-Viedma E, et al. 2003. A note on the reciprocity in the aggregation of fuzzy preference relations using OWA operators. Fuzzy Sets and Systems, 137: 71-83.

Delgado M, Herrera F, Herrera-Viedma E, et al. 1998. Combining numerical and linguistic information in group decision making. Information Sciences, 107: 177-194.

Fan Z P, Xiao S H. 2002. The consistency and ranking method for comparison matrix with linguistic assessment. Systems Engineering-Theory & Practice, 22(5): 87-91.

Herrera F, Herrera-Viedma E, Chiclana F. 2003. A study of the origin and uses of the ordered weighted geometric operator in multicriteria decision making. International Journal of Intelligent Systems, 18: 689-707.

Herrera F, Herrera-Viedma E, Verdegay J L. 1995. A sequential selection process in group decision making with a linguistic assessment approach. Information Sciences, 85: 223-239.

Herrera F, Herrera-Viedma E, Verdegay J L. 1996a. A model of consensus in group decision making under linguistic assessments. Fuzzy Sets and Systems, 78: 73-87.

Herrera F, Herrera-Viedma E, Verdegay J L. 1996b. Direct approach processes in group decision making using linguistic OWA operators. Fuzzy Sets and Systems, 79: 175-190.

Herrera F, Herrera-Viedma E, Verdegay J L. 1997. A rational consensus model in group decision making using linguistic assessments. Fuzzy Sets and Systems, 88: 31-49.

Herrera F, Herrera-Viedma E. 1997. Aggregation operators for linguistic weighted information. IEEE Transactions on Systems, Man, and Cybernetics-Part A, 27: 646-656.

Herrera F, Herrera-Viedma E. 2000. Choice functions and mechanisms for linguistic preference relations. European Journal of Operational Research, 120: 144-161.

Herrera F, Herrera-Viedma E. 2003. Linguistic preference modeling: foundation models and new trends. *In*: Proceedings of 4th International Workshop on Preferences and Decisions, Trento, Italy 47-51.

Herrera F, Martínez L, Sánchez P J. 2005. Managing non-homogeneous information in group decision making. European Journal of Operational Research, 166: 115-132.

Ngwenyama O, Bryson N. 1999. Eliciting and mapping qualitative preferences to numeric rankings in group decision making. European Journal of Operational Research, 116: 487-497.

Xu Z S, Wu D. 2004. Method based on fuzzy linguistic judgment matrix and fuzzy induced ordered weighted averaging (FIOWA) operator for decision-making problems with a limited set of alternatives. Fuzzy Systems and Mathematics, 18(1): 76-80.

Xu Z S. 2004a. EOWA and EOWG operators for aggregating linguistic labels based on linguistic preference relations. International Journal of Uncertainty, Fuzziness and Knowledge-Based Systems, 12: 791-810.

Xu Z S. 2004b. A method based on linguistic aggregation operators for group decision making with linguistic preference relations. Information Sciences, 166: 19-30.

Xu Z S. 2004c. Goal programming models for obtaining the priority vector of incomplete fuzzy preference relation. International Journal of Approximate Reasoning, 36: 261-270.

Xu Z S. 2004d. Incomplete complementary judgment matrix. Systems Engineering-Theory & Practice, 24(6): 91-97.

Xu Z S. 2005a. An approach to group decision making based on incomplete linguistic preference relations. International Journal of Information Technology and Decision Making, 4: 153-160.

Xu Z S. 2005b. Deviation measures of linguistic preference relations in group decision making. Omega, 33: 249-254.

Xu Z S. 2005c. An overview of methods for determining OWA weights. International Journal of Intelligent Systems, 20: 843-865.

Xu Z S. 2006a. A direct approach to group decision making with uncertain additive linguistic preference relations. Fuzzy Optimization and Decision Making, 5: 23-35.

Xu Z S. 2006b. An approach based on the uncertain LOWG and the induced uncertain LOWG operators to group decision making with uncertain multiplicative linguistic preference relations. Decision Support Systems, 41: 488-499.

Xu Z S. 2006c. Incomplete linguistic preference relations and their fusion. Information Fusion, 7: 331-337.

Xu Z S. 2006d. A practical procedure for group decision making under incomplete multiplicative linguistic preference relations. Group Decision and Negotiation, 15: 593-604.

Xu Z S. 2007a. A survey of preference relations. International Journal of General Systems, 36: 179-203.

Xu Z S. 2007b. Dynamic linguistic preference relations and their use in multi-period decision making. *In*: reduplicative. Proceedings of 2007 International Conference on Management Science and Engineering. Harbin, China, 345-350.

Xu Z S. 2008. Group decision making based on multiple types of linguistic preference relations. Information Sciences, 178: 452-467.

Xu Z S. 2010. Interactive group decision making procedure based on uncertain multiplicative linguistic preference relations. Journal of Systems Engineering and Electronics, 21: 408-415.

Yager R R. 1993. Families of OWA operators. Fuzzy Sets and Systems, 59: 125-145.

Zhang H M, Xu Z S. 2006. Study on the symmetry properties of weighting vectors of information aggregation operators. Systems Engineering-Theory & Practice, 26(3): 75-82.

Chapter 4

Approaches to Linguistic Multi-Attribute Decision Making

A variety types of multi-attribute decision making (MADM) problems occur in our daily life, such as investment decision making, project appraisal, maintenance and repair services, weapon system performance evaluation, plant siting, tendering and bidding, and comprehensive evaluation of economic benefits, etc. (Wang and Fu, 1993). MADM mainly consists of the following two parts:

(1) Collect decision information. The decision information generally includes the attribute weights and the attribute values. In a MADM problem, there are generally a finite set of alternatives and a collection of attributes. The attributes are the indices used to measure the given alternatives, and each attribute has its importance, which is to be determined in the process of decision making. The attribute values are usually the measure values for the alternatives with respect to each attribute, which mainly take the form of real numbers, interval numbers, triangular fuzzy numbers, intuitionistic fuzzy numbers and linguistic variables, etc.

(2) Aggregate the decision information through some proper approaches, and then rank and select the given alternatives.

MADM is an important part of modern decision science, and has been receiving great attention from researchers and practitioners over the last decades, and achieved a wealth of research results (Chen and Hwang, 1992; Fodor and Roubens, 1994; Hwang and Yoon, 1981; Keeney and Raiffa, 1976; Saaty, 1980; Xu and Wu, 2006; Xu, 2004; Yager and Kacprzyk, 1997; Zeeny, 1982). Xu(2004) gave a thorough survey on the linguistic decision making techniques (before 2004) based on various linguistic aggregation operators (such as the LWA, LOWA, LHA, ULWA, LWG, LOWG, LHG and ULHG operators, etc.). Recently, Xu(2007a; 2007b; 2006a; 2006b; 2005, Zhang and Xu (2005) further investigated the approaches to multi-attribute decision making problems where the attribute values are expressed in linguistic labels or uncertain linguistic variables. In this chapter, we shall give a detailed introduction to these research results (after 2004).

4.1 Maximizing Deviation Procedure for Linguistic Multi-Attribute Decision Making

For a group multi-attribute decision making problem with linguistic information, let $X = \{x_1, x_2, \cdots, x_n\}$ be a finite set of alternatives, $E = \{e_1, e_2, \cdots, e_l\}$ be the set of decision makers e_k $(k = 1, 2, \cdots, l)$, whose weight vector is $\boldsymbol{v} = (v_1, v_2, \cdots, v_l)^{\mathrm{T}}$, where $v_k \geqslant 0$ $(k = 1, 2, \cdots, l)$ and $\sum\limits_{k=1}^{l} v_k = 1$. Let $G = \{G_1, G_2, \cdots, G_m\}$ be the set of attrributes G_i $(i = 1, 2, \cdots, m)$, whose weight vector is $\boldsymbol{w} = (w_1, w_2, \cdots, w_m)^{\mathrm{T}}$, where $w_i \geqslant 0$ $(i = 1, 2, \cdots, m)$, $\sum\limits_{i=1}^{m} w_i^2 = 1$, and the information on attribute weights is unknwon completely. The decision makers e_k $(k = 1, 2, \cdots, l)$ provide their preferences over the alternatives x_j $(j = 1, 2, \cdots, n)$ with respect to the attributes $G_i \in G$ $(i = 1, 2, \cdots, m)$ by using the additional evaluation scale $S_2 = \{s_\alpha | \alpha = -\tau, \cdots, -1, 0, 1, \cdots, \tau\}$, and get the attribute values $r_{ij}^{(k)} \in S_2$ $(k = 1, 2, \cdots, l,\ i = 1, 2, \cdots, m,\ j = 1, 2, \cdots, n)$, which are contained in a linguistic decision matrix $\boldsymbol{R}_k = (r_{ij}^{(k)})_{m \times n}$.

To obtain the group opinion, we utilize the LWA$_2$ operator:

$$r_{ij} = \mathrm{LWA}_2\left(r_{ij}^{(1)}, r_{ij}^{(2)}, \cdots, r_{ij}^{(l)}\right) = v_1 r_{ij}^{(1)} \oplus v_2 r_{ij}^{(2)} \oplus \cdots \oplus v_l r_{ij}^{(l)}$$
$$i = 1, 2, \cdots, m, \quad j = 1, 2, \cdots, n \tag{4.1}$$

to agregate all the individual linguistic decision matrices $\boldsymbol{R}_k = (r_{ij}^{(k)})_{m \times n}$ $(k = 1, 2, \cdots, l)$ into the collective linguistic decision matrix $\boldsymbol{R} = (r_{ij})_{m \times n}$.

Based on the decision information in $\boldsymbol{R} = (r_{ij})_{m \times n}$, the overall attribute value of the alternative x_j can be expressed as

$$z_j(\boldsymbol{w}) = w_1 r_{1j} \oplus w_2 r_{2j} \oplus \cdots \oplus w_m r_{mj}, \quad j = 1, 2, \cdots, n \tag{4.2}$$

Clearly, the greater $z_j(\boldsymbol{w})$, the better the alternative x_j.

If the information on attribute weights is known completely, then we can rank the alternatives x_j $(j = 1, 2, \cdots, n)$ according to the overall attribute values $z_j(\boldsymbol{w})$ $(j = 1, 2, \cdots, n)$. However, the weight information in the considered decision making problems is unknown completely. Thus, we need to determine the attribute weights in advance.

According to the information theory, if all alternatives have similar attribute values with respect to an attribute, then a small weight should be assigned to the attribute, this is due to that such an attribute does not help in differentiating alternatives (Xu, 2004; Zeeny, 1982). Then, by (3.17), Xu (2005) gave a deviation measure between the alternative x_j and all the other alternatives with respect to the attribute G_i:

$$d_{ij}(\boldsymbol{w}) = \sum_{k \neq j} d(r_{ij}, r_{ik}) w_i, \quad i = 1, 2, \cdots, m, \ j = 1, 2, \cdots, n \tag{4.3}$$

Let

$$d_i(\boldsymbol{w}) = \sum_{j=1}^{n} \sum_{k \neq j} d(r_{ij}, r_{ik})\, w_i, \quad i = 1, 2, \cdots, m \tag{4.4}$$

denote the sum of all the deviation degrees $d_{ij}(\boldsymbol{w})$ $(j = 1, 2, \cdots, n)$, and construct the deviation function:

$$d(\boldsymbol{w}) = \sum_{i=1}^{m} d_i(\boldsymbol{w}) = \sum_{i=1}^{m} \sum_{j=1}^{n} d_{ij}(\boldsymbol{w}) = \sum_{i=1}^{m} \sum_{j=1}^{n} \sum_{k \neq j} d(r_{ij}, r_{ik}) w_i \tag{4.5}$$

Obviously, the reasonable attrribute weight vector $\boldsymbol{w} = (w_1, w_2, \cdots, w_m)^{\mathrm{T}}$ should make the deviation degree $d(\boldsymbol{w})$ as much as possible. Thus, we can establish the following single-objective optimization model (Xu. 2005):

(M-4.1)
$$\max \quad d(\boldsymbol{w}) = \sum_{i=1}^{m} \sum_{j=1}^{n} \sum_{k \neq j} d(r_{ij}, r_{ik})\, w_i$$
$$\text{s.t.} \quad \sum_{i=1}^{m} w_i^2 = 1, w_i \geqslant 0, \ \ i = 1, 2, \cdots, m$$

To solve the model (M-4.1), we can construct the following Lagrange function:

$$L(\boldsymbol{w}, \lambda) = d(\boldsymbol{w}) + \frac{1}{2} \lambda \left(\sum_{i=1}^{m} w_i^2 - 1 \right) \tag{4.6}$$

where λ is the Lagrange parameter.

Differentiating (4.6) with respect to w_i $(i = 1, 2, \cdots, m)$ and λ, and setting these partial derivatives equal to zero, i.e., $\frac{\partial L(\boldsymbol{w}, \lambda)}{\partial w_i} = 0$ $(i = 1, 2, \cdots, m)$ and $\frac{\partial L(\boldsymbol{\omega}, \lambda)}{\partial \lambda} = 0$, the following set of equations can be obtained:

$$\begin{cases} \dfrac{\partial L(\boldsymbol{w}, \lambda)}{\partial w_i} = \displaystyle\sum_{j=1}^{n} \sum_{k \neq j} d(r_{ij}, r_{ik}) + \lambda\, w_i = 0, \quad i = 1, 2, \cdots, m \\ \dfrac{\partial L(\boldsymbol{w}, \lambda)}{\partial \lambda} = \displaystyle\sum_{i=1}^{m} w_i^2 - 1 = 0 \end{cases} \tag{4.7}$$

Solving (4.7), we get the optimal solution $\boldsymbol{w} = (w_1, w_2, \cdots, w_m)^{\mathrm{T}}$, where

$$w_i = \frac{\displaystyle\sum_{j=1}^{n} \sum_{k \neq j} d(r_{ij}, r_{ik})}{\sqrt{\displaystyle\sum_{i=1}^{m} \left(\sum_{j=1}^{n} \sum_{k \neq j} d(r_{ij}, r_{ik}) \right)^2}}, \quad i = 1, 2, \cdots, m \tag{4.8}$$

It is clear that $w_i \geqslant 0$ $(i = 1, 2, \cdots, m)$. Normalizing (4.8), we have

$$w_i = \frac{\sum_{j=1}^{n} \sum_{k \neq j} d(r_{ij}, r_{ik})}{\sum_{i=1}^{n} \sum_{j=1}^{n} \sum_{k \neq j} d(r_{ij}, r_{ik})}, \quad i = 1, 2, \cdots, m \tag{4.9}$$

From (4.2) and (4.9), we derive the overall attribute values $z_j(\boldsymbol{w})$ $(j = 1, 2, \cdots, n)$, by which we rank and select the alternatives x_j $(j = 1, 2, \cdots, n)$.

In what follows, we extend the above results to uncertain linguistic environments:

Suppose that the decision makers e_k $(k = 1, 2, \cdots, l)$ provide their uncertain preferences over the alternatives x_j $(j = 1, 2, \cdots, n)$ with respect to the attributes $G_i \in G$ $(i = 1, 2, \cdots, m)$ by using the additive linguistic evaluation scale $S_2 = \{s_\alpha | \alpha = -\tau, \cdots, -1, 0, 1, \cdots, \tau\}$, and construct the uncertain linguistic decision matrix $\tilde{\boldsymbol{R}}_k = (\tilde{r}_{ij}^{(k)})_{m \times n}$, where $\tilde{r}_{ij}^{(k)} = [r_{ij}^{L(k)}, r_{ij}^{U(k)}]$, $r_{ij}^{L(k)}, r_{ij}^{U(k)} \in S_2$ $(k = 1, 2, \cdots, l$, $i = 1, 2, \cdots, m$, $j = 1, 2, \cdots, n)$.

Then we utilize the ULWA operator:

$$\tilde{r}_{ij} = \text{ULWA}\,(\tilde{r}_{ij}^{(1)}, \tilde{r}_{ij}^{(2)}, \cdots, \tilde{r}_{ij}^{(l)}) = v_1 \tilde{r}_{ij}^{(1)} \oplus v_2 \tilde{r}_{ij}^{(2)} \oplus \cdots \oplus v_l \tilde{r}_{ij}^{(l)}$$
$$i = 1, 2, \cdots, m, \quad j = 1, 2, \cdots, n \tag{4.10}$$

to aggregate the uncertain linguistic decision matrices $\tilde{\boldsymbol{R}}_k = (\tilde{r}_{ij}^{(k)})_{m \times n}$ $(k = 1, 2, \cdots, l)$ into the collective uncertain linguistic decision matrix $\tilde{\boldsymbol{R}} = (\tilde{r}_{ij})_{m \times n}$.

Based on $\tilde{\boldsymbol{R}} = (\tilde{r}_{ij})_{m \times n}$, the uncertain overall attribute value of the alternative x_j can be expressed as

$$\tilde{z}_j(\boldsymbol{w}) = w_1 \tilde{r}_{1j} \oplus w_2 \tilde{r}_{2j} \oplus \cdots \oplus w_m \tilde{r}_{mj}, \quad j = 1, 2, \cdots, n \tag{4.11}$$

Clearly, the greater $\tilde{z}_j(\boldsymbol{w})$, the better the alternative x_j.

In cases where the information on attribute weights is known completely, by (2.69), we compare each pair of $\tilde{z}_j(\boldsymbol{w})$ $(j = 1, 2, \cdots, n)$, and construct the possibility degree matrix $\boldsymbol{P} = (p_{ij})_{n \times n}$. Then we use (2.82) to derive the priority vector $\boldsymbol{\zeta} = (\zeta_1, \zeta_2, \cdots, \zeta_n)^{\mathrm{T}}$ of $\boldsymbol{P}$, by which the uncertain overall attribute values $\tilde{z}_j(\boldsymbol{w})$ $(j = 1, 2, \cdots, n)$ can be ranked according to ζ_i $(i = 1, 2, \cdots, n)$. After that, we rank and select the alternatives x_j $(j = 1, 2, \cdots, n)$ according the ranking of $\tilde{z}_j(\boldsymbol{w})$ $(j = 1, 2, \cdots, n)$.

If the information on the attribute weights is unknown completely, then similar to Definition 3.5, we first define the deviation degree between the uncertain linguistic variables:

Definition 4.1 (Xu. 2005) For any two uncertain linguistic variables $\tilde{s}_1 = [s_{\alpha_1}, s_{\beta_1}]$ and $\tilde{s}_2 = [s_{\alpha_2}, s_{\beta_2}]$ $(s_{\alpha_i}, s_{\beta_i} \in S_2$, $i = 1, 2)$, we call

$$d(\tilde{s}_1, \tilde{s}_2) = \frac{|\alpha_1 - \beta_1| + |\alpha_2 - \beta_2|}{4\tau} \tag{4.12}$$

the deviation degree between $\tilde{s}_1$ and $\tilde{s}_2$, where $S_2 = \{s_\alpha | \alpha = -\tau, \cdots, -1, 0, 1, \cdots, \tau\}$ is the additive linguistic evaluation scale.

By (4.12), we define the deviation degree between the alternative x_j and all the other alternatives with respect to the attribute G_i:

$$d_{ij}(\boldsymbol{w}) = \sum_{k \neq j} d(\tilde{r}_{ij}, \tilde{r}_{ik}) w_i, \quad i = 1, 2, \cdots, m, \; j = 1, 2, \cdots, n \tag{4.13}$$

Let

$$d_i(\boldsymbol{w}) = \sum_{j=1}^{n} \sum_{k \neq j} d(\tilde{r}_{ij}, \tilde{r}_{ik})\, w_i, \quad i = 1, 2, \cdots, m \tag{4.14}$$

be the sum of all the deviation degrees $d_{ij}(\boldsymbol{w})$ $(j = 1, 2, \cdots, n)$, and construct the deviation function:

$$d(\boldsymbol{w}) = \sum_{i=1}^{m} d_i(\boldsymbol{w}) = \sum_{i=1}^{m} \sum_{j=1}^{n} d_{ij}(\boldsymbol{w}) = \sum_{i=1}^{m} \sum_{j=1}^{n} \sum_{k \neq j} d(\tilde{r}_{ij}, \tilde{r}_{ik}) w_i \tag{4.15}$$

by which we establish the following single-objective optimization model (Xu, 2005):

$$\text{(M-4.2)} \qquad \begin{aligned} \max \quad & d(\boldsymbol{w}) = \sum_{i=1}^{m} \sum_{j=1}^{n} \sum_{k \neq j} d(\tilde{r}_{ij}, \tilde{r}_{ik})\, w_i \\ \text{s.t.} \quad & \sum_{i=1}^{m} w_i^2 = 1, w_i \geqslant 0, \quad i = 1, 2, \cdots, m \end{aligned}$$

Similar to the model (M-4.1), by the model (M-4.2), we get the optimal soluation $\boldsymbol{w} = (w_1, w_2, \cdots, w_m)^{\mathrm{T}}$, where

$$w_i = \frac{\displaystyle\sum_{j=1}^{n} \sum_{k \neq j} d(\tilde{r}_{ij}, \tilde{r}_{ik})}{\sqrt{\displaystyle\sum_{i=1}^{m} \left[\sum_{j=1}^{n} \sum_{k \neq j} d(\tilde{r}_{ij}, \tilde{r}_{ik}) \right]^2}}, \quad i = 1, 2, \cdots, m \tag{4.16}$$

Obviously, $w_i \geqslant 0$ $(i = 1, 2, \cdots, m)$. Normalizing (4.16), we have

$$w_i = \frac{\displaystyle\sum_{j=1}^{n} \sum_{k \neq j} d(\tilde{r}_{ij}, \tilde{r}_{ik})}{\displaystyle\sum_{i=1}^{n} \sum_{j=1}^{n} \sum_{k \neq j} d(\tilde{r}_{ij}, \tilde{r}_{ik})}, \quad i = 1, 2, \cdots, m \tag{4.17}$$

By (4.11) and (4.17), we get the uncertain overall attribute values $\tilde{z}_j(\boldsymbol{w})$ $(j = 1,2,\cdots,n)$ of the alternatives x_j $(j = 1,2,\cdots,n)$. Then, we utilize (2.69) to compare each pair of $\tilde{z}_j(\boldsymbol{w})$ $(j = 1,2,\cdots,n)$, and construct a possibility degree matrix $\boldsymbol{P} = (p_{ij})_{n\times n}$. After that, we derive the priority vector $\boldsymbol{\zeta} = (\zeta_1,\zeta_2,\cdots,\zeta_n)^{\mathrm{T}}$ of $\boldsymbol{P}$ by using (2.82), and utilize ζ_i $(i = 1,2,\cdots,n)$ to rank the uncertain overall attribute values $\tilde{z}_j(\boldsymbol{w})$ $(j = 1,2,\cdots,n)$, by which all the alternatives x_j $(j = 1,2,\cdots,n)$ can be ranked and selected.

Example 4.1 (Xu, 2005) A practical use of the above approach involves the evaluation of university faculty for tenure and promotion (adapted from Bryson and Mobolurin (1995)). The attributes used at some universities are ① G_1——teaching; ② G_2——research; and ③ G_3——service. Three decision makers e_k $(k = 1,2,3)$(whose weight vector is $\boldsymbol{v} = (0.2,0.5,0.3)^{\mathrm{T}}$) utilize the additive linguistic evaluation scale:

$$S_2=\{s_{-4}=\text{extremely poor},\ \ s_{-3}=\text{very poor},\ \ s_{-2}=\text{poor},\ \ s_{-1}=\text{slightly poor},$$
$$s_0=\text{fair},\ \ s_1=\text{slightly good},\ \ s_2=\text{good},\ \ s_3=\text{very good},\ \ s_4=\text{extremely good}\}$$

to evaluate five faculty candidates (alternatives) x_j $(j = 1,2,\cdots,5)$, and construct the following decision matrices $\boldsymbol{R}_k$ $(k = 1,2,3)$ (Tables 4.1~4.3):

Table 4.1 Linguistic decision matrix R_1 (Xu, 2005)

	x_1	x_2	x_3	x_4	x_5
G_1	s_{-2}	s_{-1}	s_{-1}	s_3	s_1
G_2	s_1	s_2	s_{-3}	s_1	s_2
G_3	s_2	s_3	s_2	s_{-1}	s_1

Table 4.2 Linguistic decision matrix R_2 (Xu, 2005)

	x_1	x_2	x_3	x_4	x_5
G_1	s_1	s_2	s_1	s_3	s_{-1}
G_2	s_{-2}	s_0	s_2	s_1	s_2
G_3	s_1	s_3	s_1	s_0	s_2

Table 4.3 Linguistic decision matrix R_3 (Xu, 2005)

	x_1	x_2	x_3	x_4	x_5
G_1	s_{-1}	s_0	s_{-2}	s_2	s_1
G_2	s_{-2}	s_2	s_1	s_0	s_3
G_3	s_3	s_1	s_2	s_2	s_{-1}

To rank the five faculty candidates, we inroduce the following decision making steps:

Step 1 Utilize the LWA$_2$ operator (4.1) to aggregate all the linguistic decision matrices $\boldsymbol{R}_k = (r_{ij}^{(k)})_{3\times5}$ $(k = 1, 2, 3)$, and get the collective linguistic decision matrix $\boldsymbol{R} = (r_{ij})_{3\times5}$ (Table 4.4):

Table 4.4 Group decision matrix R (Xu, 2005)

	x_1	x_2	x_3	x_4	x_5
G_1	$s_{-0.2}$	$s_{0.8}$	$s_{-0.3}$	$s_{2.7}$	s_0
G_2	$s_{-1.4}$	s_1	$s_{0.7}$	$s_{0.7}$	$s_{2.3}$
G_3	$s_{1.8}$	$s_{2.4}$	$s_{1.5}$	$s_{0.4}$	$s_{0.9}$

Step 2 By (4.9), we get the weight vector of the attributes G_i $(i = 1, 2, 3)$:

$$\boldsymbol{w} = (0.3096,\ 0.4219,\ 0.2685)^{\mathrm{T}}$$

Step 3 Derive the overall attribute values $z_j(\boldsymbol{w})$ $(j = 1, 2, \cdots, 5)$ of the alternatives x_j $(j = 1, 2, \cdots, 5)$ by using (4.2):

$$z_1(\boldsymbol{w}) = s_{-0.17}, \quad z_2(\boldsymbol{w}) = s_{1.31}, \quad z_3(\boldsymbol{w}) = s_{0.61}, \quad z_4(\boldsymbol{w}) = s_{1.24}, \quad z_5(\boldsymbol{w}) = s_{1.94}$$

Step 4 Rank the alternatives x_j $(j = 1, 2, \cdots, 5)$ according to $z_j(\boldsymbol{w})$ $(j = 1, 2, \cdots, 5)$:

$$x_5 \succ x_2 \succ x_4 \succ x_3 \succ x_1$$

Thus, the best faculty candidate is x_5.

If the decision makers e_k $(k = 1, 2, 3)$ provide the following uncertain linguistic decision matrices $\tilde{\boldsymbol{R}}_k$ $(k = 1, 2, 3)$ (Tables 4.5~4.7):

Table 4.5 Uncertain linguistic decision matrix $\tilde{R}_1$ (Xu, 2005)

	x_1	x_2	x_3	x_4	x_5
G_1	$[s_{-2}, s_0]$	$[s_{-1}, s_1]$	$[s_{-2}, s_{-1}]$	$[s_2, s_3]$	$[s_1, s_2]$
G_2	$[s_0, s_2]$	$[s_2, s_3]$	$[s_{-3}, s_{-2}]$	$[s_1, s_2]$	$[s_2, s_3]$
G_3	$[s_1, s_2]$	$[s_2, s_3]$	$[s_1, s_3]$	$[s_{-2}, s_{-1}]$	$[s_1, s_3]$

Table 4.6 Uncertain linguistic decision matrix $\tilde{R}_2$ (Xu, 2005)

	x_1	x_2	x_3	x_4	x_5
G_1	$[s_0, s_1]$	$[s_2, s_3]$	$[s_0, s_1]$	$[s_2, s_3]$	$[s_0, s_1]$
G_2	$[s_{-2}, s_{-1}]$	$[s_0, s_1]$	$[s_1, s_2]$	$[s_1, s_2]$	$[s_2, s_4]$
G_3	$[s_0, s_1]$	$[s_1, s_3]$	$[s_1, s_2]$	$[s_0, s_1]$	$[s_2, s_3]$

Table 4.7 Uncertain linguistic decision matrix $\tilde{R}_3$ (Xu, 2005)

	x_1	x_2	x_3	x_4	x_5
G_1	$[s_{-1}, s_1]$	$[s_0, s_1]$	$[s_{-2}, s_{-1}]$	$[s_0, s_2]$	$[s_1, s_3]$
G_2	$[s_{-3}, s_0]$	$[s_1, s_2]$	$[s_1, s_2]$	$[s_0, s_1]$	$[s_3, s_4]$
G_3	$[s_2, s_3]$	$[s_2, s_3]$	$[s_0, s_2]$	$[s_2, s_3]$	$[s_0, s_1]$

In the following, the detailed decision making steps are provided:

Step 1 Utilize the ULWA operator (4.10) to aggregate all the uncertain linguistic decision matrices $\tilde{\boldsymbol{R}}_k = (\tilde{r}_{ij}^{(k)})_{3\times 5}$ $(k = 1, 2, 3)$, and get the collective uncertain linguistic decision matrix $\tilde{\boldsymbol{R}} = (\tilde{r}_{ij})_{3\times 5}$ (Table 4.8):

Table 4.8 Group uncertain linguistic decision matrix $\tilde{R}$ (Xu, 2005)

	x_1	x_2	x_3	x_4	x_5
G_1	$[s_{-0.7}, s_{0.8}]$	$[s_{0.8}, s_2]$	$[s_{-1}, s_0]$	$[s_{1.4}, s_{2.7}]$	$[s_{0.5}, s_{1.8}]$
G_2	$[s_{-1.9}, s_{-0.1}]$	$[s_{0.7}, s_{1.7}]$	$[s_{0.2}, s_{1.2}]$	$[s_{0.7}, s_{1.7}]$	$[s_{2.3}, s_{3.8}]$
G_3	$[s_{0.8}, s_{1.8}]$	$[s_{1.5}, s_{3.0}]$	$[s_{0.7}, s_{2.2}]$	$[s_{0.2}, s_{1.2}]$	$[s_{1.2}, s_{2.4}]$

Step 2 By (4.17), we get the weight vector of the attributes G_i $(i = 1, 2, 3)$:

$$\boldsymbol{w} = (0.3449,\ 0.4599,\ 0.1952)^{\mathrm{T}}$$

Step 3 Derive the uncertain overall attribute values $\tilde{z}_j(\boldsymbol{w})$ $(j = 1, 2, \cdots, 5)$ of the alternatives x_j $(j = 1, 2, \cdots, 5)$ by using (4.10):

$$\tilde{z}_1(\boldsymbol{w}) = [s_{-0.96}, s_{0.58}], \quad \tilde{z}_2(\boldsymbol{w}) = [s_{0.89}, s_{2.06}], \quad \tilde{z}_3(\boldsymbol{w}) = [s_{-0.12}, s_{0.98}]$$

$$\tilde{z}_4(\boldsymbol{w}) = [s_{0.84}, s_{1.95}], \quad \tilde{z}_5(\boldsymbol{w}) = [s_{1.46}, s_{2.84}]$$

Step 4 Compare each pair of $\tilde{z}_j(\boldsymbol{w})$ $(j = 1, 2, \cdots, 5)$ by using (2.69), and construct a possibility degree matrix:

$$\boldsymbol{P} = \begin{pmatrix} 0.5 & 0 & 0.2652 & 0 & 0 \\ 1 & 0.5 & 0.9561 & 0.5351 & 0.2353 \\ 0.7348 & 0.0439 & 0.5 & 0.0633 & 0 \\ 1 & 0.4649 & 0.9367 & 0.5 & 0.1968 \\ 1 & 0.7647 & 1 & 0.8032 & 0.5 \end{pmatrix}$$

Then by (2.82), we get the priority vector of $\boldsymbol{P}$:

$$\boldsymbol{\zeta} = (0.1133, 0.2363, 0.1421, 0.2299, 0.2784)^{\mathrm{T}}$$

Step 5 Rank the uncertain overall attribute values $\tilde{z}_j(\boldsymbol{w})$ $(j = 1, 2, \cdots, 5)$ according to ζ_i $(i = 1, 2, \cdots, 5)$:

$$\tilde{z}_5(\boldsymbol{w}) > \tilde{z}_2(\boldsymbol{w}) > \tilde{z}_4(\boldsymbol{w}) > \tilde{z}_3(\boldsymbol{w}) > \tilde{z}_1(\boldsymbol{w})$$

by which the ranking of the alternatives x_j $(j = 1, 2, \cdots, 5)$ is

$$x_5 \succ x_2 \succ x_4 \succ x_3 \succ x_1$$

and thus, x_5 is the best faculty candidate.

4.2 Ideal Point-Based Model for Linguistic Multi-Attribute Decision Making

Now we consider a linguistic multi-attribute decision making problem where the information on attribute weights cannot be known completely. Let X and G be the sets of alternatives and attributes respectively, defined as before, and $\boldsymbol{w}=(w_1, w_2, \cdots, w_m)^{\mathrm{T}} \in H$ be the weight vector of attributes, $w_i \geqslant 0$ $(i = 1, 2, \cdots, m)$ and $\sum_{i=1}^{m} w_i = 1$, where H is a set of the known weight information, which can be constructed by the following forms (Park and Kim, 1997; Kim and Ahn, 1999; Kim et al., 1999; Xu, 2007a, 2007b; Xu and Chen, 2007):

(1) A weak ranking: $\{w_i \geqslant w_j\}$;

(2) A strict ranking: $\{w_i - w_j \geqslant \alpha_i\}$;

(3) A ranking with multiples: $\{w_i \geqslant \alpha_i w_j\}$;

(4) An interval form: $\{\alpha_i \leqslant w_i \leqslant \alpha_i + \varepsilon_i\}$;

(5) A ranking of differences: $\{w_i - w_j \geqslant w_k - w_l\}$ $(j \neq k \neq l)$, where $\{\alpha_i\}$ and $\{\varepsilon_i\}$ are the non-negative numbers.

The decision makers provide their preferences over the alternatives x_j $(j=1, 2, \cdots, n)$ with respect to the attributes $G_i \in G$ $(i = 1, 2, \cdots, m)$ by using the additive linguistic evaluation scale $S_2 = \{s_\alpha | \alpha = -\tau, \cdots, -1, 0, 1, \cdots, \tau\}$, and construct a linguistic decision matrix $\boldsymbol{R} = (r_{ij})_{m\times n}$.

Xu (2007a) defined some concepts as follows:

Definitin 4.2 (Xu, 2007a) Let $\boldsymbol{R} = (r_{ij})_{m\times n}$ be a linguistic decision matrix. Then $\boldsymbol{r}_j = (r_{1j}, r_{2j}, \cdots, r_{mj})^{\mathrm{T}}$ is called the vector of attribute values of the alternatives x_j $(j = 1, 2, \cdots, n)$.

Definition 4.3 (Xu, 2007a) Let $\boldsymbol{R} = (r_{ij})_{m\times n}$ be a linguistic decision matrix. Then $\boldsymbol{r}^+ = (r_1^+, r_2^+, \cdots, r_m^+)^{\mathrm{T}}$ is called a relative ideal point of alternatives, where $r_i^+ = \max_j\{r_{ij}\}$ $(i = 1, 2, \cdots, m)$.

Definition 4.4 (Xu, 2007a) Let $\boldsymbol{R} = (r_{ij})_{m\times n}$ be a linguistic decision matrix, and $\boldsymbol{r}^+ = (r_1^+, r_2^+, \cdots, r_m^+)^{\mathrm{T}}$ be a relative ideal point of alternatives. Then

$$z^+(\boldsymbol{w}) = \mathrm{LWA}_2\left(r_1^+, r_2^+, \cdots, r_m^+\right) = w_1 r_1^+ \oplus w_2 r_2^+ \oplus \cdots \oplus w_m r_m^+ \tag{4.18}$$

is called the overall attribute value of $\boldsymbol{r}^+$.

In real life, there always exist some differences betweem the vector $\boldsymbol{r}_j = (r_{1j}, r_{2j}, \cdots, r_{mj})^{\mathrm{T}}$ of attribute values of the alternatives x_j $(j = 1, 2, \cdots, n)$ and the relative ideal point $\boldsymbol{r}^+ = (r_1^+, r_2^+, \cdots, r_m^+)^{\mathrm{T}}$ of alternatives. Then, based on (3.17), in what follows, we define the distance between the overall attribute value $z^+(\boldsymbol{w})$ of the relative ideal point r^+ and the overall attribute value $z_j(\boldsymbol{w})$ of the alternative x_j:

$$d(z^+(\boldsymbol{w}), z_j(\boldsymbol{w})) = \sum_{i=1}^{m} w_i d(r_i^+, r_{ij}) \tag{4.19}$$

Clearly, the smaller $d(z^+(\boldsymbol{w}), z_j(\boldsymbol{w}))$, the better the alternative x_j. Therefore, the reasonable weight vector of attributes, $\boldsymbol{w} = (w_1, w_2, \cdots, w_m)^{\mathrm{T}}$, should be determined so as to make the distances $d(z^+(\boldsymbol{w}), z_j(\boldsymbol{w}))$ $(j = 1, 2, \cdots, n)$ as small as possible, which means to minimize the following distance vector:

$$d(\boldsymbol{w}) = \left(d(z^+(\boldsymbol{w}), z_1(\boldsymbol{w})),\ d(z^+(\boldsymbol{w}), z_2(\boldsymbol{w})), \cdots,\ d(z^+(\boldsymbol{w}), z_n(\boldsymbol{w}))\right) \tag{4.20}$$

under the condition $\boldsymbol{w} \in H$. In order to do that, we establish the following multi-objective optimization model:

$$\text{(M-4.3)}\quad \min \quad d(\boldsymbol{w}) = (d(z^+(\boldsymbol{w}), z_1(\boldsymbol{w})),\ d(z^+(\boldsymbol{w}), z_2(\boldsymbol{w})), \cdots, d(z^+(\boldsymbol{w}), z_n(\boldsymbol{w})))$$
$$\text{s.t.} \quad \boldsymbol{w} \in H$$

We utilize the min–max operator proposed by Zimmermann and Zysco (1980) to integrate all the differences $d(z^+(\boldsymbol{w}), z_j(\boldsymbol{w}))$ $(j = 1, 2, \cdots, n)$, i.e., we get a single-objective optimization model (Xu, 2007a):

$$\text{(M-4.4)} \qquad \begin{aligned} \min \quad & \lambda_0 \\ \text{s.t.} \quad & d(z^+(\boldsymbol{w}), z_j(\boldsymbol{w})) \leqslant \lambda_0, \quad j = 1, 2, \cdots, n \\ & \boldsymbol{w} \in H \end{aligned}$$

where

$$\lambda_0 = \max_j d(z^+(\boldsymbol{w}), z_j(\boldsymbol{w})) \tag{4.21}$$

Solving the model (M-4.4), we get the optimal solution $\boldsymbol{w}^* = (w_1^*,\ w_2^*, \cdots, w_n^*)^{\mathrm{T}}$, which can be used as the weight vector of attributes.

If the information on the attribute weights is completely unknown, then we replace (4.19) with the following form:

$$\bar{d}(z^+(\boldsymbol{w}), z_j(\boldsymbol{w})) = \sum_{i=1}^{m} w_i^2 d^2(r_i^+, r_{ij}) \tag{4.22}$$

and establish the following multi-objective optimization model (Xu, 2007a):

$$\text{(M-4.5)}\quad \min \quad \bar{d}(\boldsymbol{w}) = (\bar{d}(z^+(\boldsymbol{w}), z_1(\boldsymbol{w})),\ \bar{d}(z^+(\boldsymbol{w}), z_2(\boldsymbol{w})), \cdots, \bar{d}(z^+(\boldsymbol{w}), z_n(\boldsymbol{w})))$$
$$\text{s.t.} \quad \sum_{i=1}^{m} w_i = 1,\ w_i \geqslant 0,\quad i = 1, 2, \cdots, m$$

By linear equal weighted summation method (French et al., 1983), the model (M-4.5) can be transformed into a single-objective optimization model (Xu, 2007a):

(M-4.6)
$$\min \quad \bar{d}(\boldsymbol{w}) = \sum_{j=1}^{n} \bar{d}(z^{+}(\boldsymbol{w}), z_j(\boldsymbol{w}))$$
$$\text{s.t.} \quad \sum_{i=1}^{m} w_i = 1, \ w_i \geqslant 0, \ \ i = 1, 2, \cdots, m$$

To solve this model, we construct the Lagrange function:

$$L(\boldsymbol{w}, \lambda) = \bar{d}(\boldsymbol{w}) + 2\lambda \left(\sum_{i=1}^{m} w_i - 1 \right) \tag{4.23}$$

where λ is the Lagrange multiplier.

Differentiating $L(\boldsymbol{w}, \lambda)$ with respect to w_i $(i = 1, 2, \cdots, m)$ and λ, and setting these partial derivatives equal to zero, i.e., $\dfrac{\partial L(\boldsymbol{w}, \lambda)}{\partial w_i} = 0$ $(i = 1, 2, \cdots, m)$ and $\dfrac{\partial L(\boldsymbol{w}, \lambda)}{\partial \lambda} = 0$, the following set of equations can be obtained:

$$\begin{cases} \dfrac{\partial L(\boldsymbol{w}, \lambda)}{\partial w_i} = 2\displaystyle\sum_{i=1}^{m} d^2(r_i^{+}, r_{ij})\, w_i + 2\lambda = 0 \\ \dfrac{\partial L(\boldsymbol{w}, \lambda)}{\partial \lambda} = \displaystyle\sum_{i=1}^{m} w_i - 1 = 0 \end{cases} \tag{4.24}$$

By solving (4.24), we get a simple and exact formula for determining the attribute weights as follows:

$$w_i = \frac{1}{\displaystyle\sum_{i=1}^{m} \frac{1}{\displaystyle\sum_{j=1}^{n} d^2(r_i^{+}, r_{ij})}} \Bigg/ \sum_{j=1}^{n} d^2(r_i^{+}, r_{ij}), \quad i = 1, 2, \cdots, m \tag{4.25}$$

Obviously, $w_i \geqslant 0$ $(i = 1, 2, \cdots, m)$.

From the above analysis, we know that both the models (M-4.4) and (M-4.6) can be used to determine the attribute weights in a multi-attribute decision making problem with incomplete weight information under linguistic setting. The model (M-4.4) can be used in situations with incomplete weight information which can be constructed by various forms such as those defined previously. The model (M-4.6) can be used in situations where the information about attribute weights is completely unknown. By solving the model (M-4.6), a simple and exact formula for determining the attribute weights has also been obtained.

Based on the above models, we shall develop a practical method for solving multi-attribute decision making problems, in which the information about attribute weights is incompletely known, and the attribute values take the form of linguistic variables. The method involves the following steps (Xu, 2007a):

Step 1 Let X and G be the sets of alternatives and attributes respectively, and $\boldsymbol{w} = (w_1, w_2, \cdots, w_m)^{\mathrm{T}} \in H$ be the weight vector of attributes, where $w_i \geqslant 0$ $(i = 1, 2, \cdots, m)$, $\sum_{i=1}^{m} w_i = 1$, and H is the set of the known weight information, which can be constructed by the forms (1)~(5). Let $\boldsymbol{R} = (r_{ij})_{m\times n}$ be the linguistic decision matrix, where r_{ij} $(i = 1, 2, \cdots, m,\ j = 1, 2, \cdots, n)$ are the attribute values provided by the decision maker for the alternatives x_j $(j = 1, 2, \cdots, n)$ with respect to the attributes $G_i \in G$ $(i = 1, 2, \cdots, m)$ by using the additive linguistic evaluation sacle $S_2 = \{s_\alpha | \alpha = -\tau, \cdots, -1, 0, 1, \cdots, \tau\}$. $\boldsymbol{r}^+ = (r_1^+, r_2^+, \cdots, r_m^+)^{\mathrm{T}}$ is the relative ideal point of alternatives (see Definition 4.3).

Step 2 If the information about the attribute weights is partly known, then we solve the model (M-4.4) to obtain the attribute weights; if the information about the attribute weights is completely unknown, then we can solve the formula (4.25) to determine the attribute weights.

Step 3 Derive the overall attribute values $z_j(\boldsymbol{w})$ $(j = 1, 2, \cdots, n)$ of the alternatives x_j $(j = 1, 2, \cdots, n)$ by using the weight vector $\boldsymbol{w} = (w_1,\ w_2, \cdots, w_m)^{\mathrm{T}}$ and (4.2).

Step 4 Rank and select the alternatives x_j $(j = 1, 2, \cdots, n)$ according to the overall attribute values $z_j(\boldsymbol{w})$ $(j = 1, 2, \cdots, n)$.

If the decision maker provides his/her uncertain preferences over the alternatives x_j $(j = 1, 2, \cdots, n)$ with respect to the attributes $G_i \in G$ $(i = 1, 2, \cdots, m)$ by using the additive linguistic evaluation scale $S_2 = \{s_\alpha | \alpha = -\tau, \cdots, -1, 0, 1, \cdots, \tau\}$, and constructs an uncertain linguistic decision matrix $\tilde{\boldsymbol{R}} = (\tilde{r}_{ij})_{m\times n}$, where $\tilde{r}_{ij} = [r_{ij}^L, r_{ij}^U]$ $(r_{ij}^{\mathrm{L}}, r_{ij}^{\mathrm{U}} \in S_2,\ i = 1, 2, \cdots, m,\ j = 1, 2, \cdots, n)$. Let $\tilde{\boldsymbol{r}}_j = (\tilde{r}_{1j}, \tilde{r}_{2j}, \cdots, \tilde{r}_{mj})^{\mathrm{T}}$ be the vector of the uncertain attribute values of the alternatives x_j $(j = 1, 2, \cdots, n)$, and $\tilde{\boldsymbol{r}}^+ = (\tilde{r}_1^+, \tilde{r}_2^+, \cdots, \tilde{r}_m^+)^{\mathrm{T}}$ be the relative ideal point of alternatives, where $\tilde{r}_i^+ = [r_i^{+\mathrm{L}}, r_i^{+\mathrm{U}}]$ $(r_i^{+\mathrm{L}} = \max_j \{r_{ij}^{\mathrm{L}}\},\ r_i^{+\mathrm{U}} = \max_j \{r_{ij}^{\mathrm{U}}\},\ i = 1, 2, \cdots, m)$. Then the uncertain overall attribute value of $\tilde{\boldsymbol{r}}^+ = (\tilde{r}_1^+, \tilde{r}_2^+, \cdots, \tilde{r}_m^+)^{\mathrm{T}}$ can be expressed as

$$\tilde{z}^+(\boldsymbol{w}) = \mathrm{ULWA}(\tilde{r}_1^+, \tilde{r}_2^+, \cdots, \tilde{r}_m^+) = w_1\tilde{r}_1^+ \oplus w_2\tilde{r}_2^+ \oplus \cdots \oplus w_m\tilde{r}_m^+ \tag{4.26}$$

Based on (4.12), in what follows, we define the distance between the uncertain overall attribute value $\tilde{z}^+(\boldsymbol{w})$ of $\tilde{\boldsymbol{r}}^+$ and the uncertain overall attribute value $\tilde{z}_j(\boldsymbol{w})$ of the alternative x_j:

$$d(\tilde{z}^+(\boldsymbol{w}), \tilde{z}_j(\boldsymbol{w})) = \sum_{i=1}^{m} w_i d(\tilde{r}_i^+, \tilde{r}_{ij}) \tag{4.27}$$

Clearly, the smaller $d(\tilde{z}^+(\boldsymbol{w}), \tilde{z}_j(\boldsymbol{w}))$, the better the alternative x_j. Therefore, a reasonable weight vector $\boldsymbol{w} = (w_1, w_2, \cdots, w_m)^{\mathrm{T}}$ should be determined so as to make the distances $d(\tilde{z}^+(\boldsymbol{w}), \tilde{z}_j(\boldsymbol{w}))$ $(j = 1, 2, \cdots, n)$ as small as possible. In order to do so, similar to the model (M-4.4), we can establish the following single-objective optimization model:

(M-4.7)
$$\begin{aligned} \min\ & \lambda_1 \\ \text{s.t.}\ & d(\tilde{z}^+(\boldsymbol{w}), \tilde{z}_j(\boldsymbol{w})) \leqslant \lambda_1,\ \ j = 1, 2, \cdots, n \\ & \boldsymbol{w} \in H \end{aligned}$$

where

$$\lambda_1 = \max_j d(\tilde{z}^+(\boldsymbol{w}), \tilde{z}_j(\boldsymbol{w})) \tag{4.28}$$

Solving the model (M-4.7), we get the optimal soluation $\boldsymbol{w} = (w_1,\ w_2, \cdots, w_m)^{\mathrm{T}}$, which can be used as the weight vector of attributes.

If the information on attribute weights is unknown completely, then we replace (4.27) with the following form:

$$\bar{d}(\tilde{z}^+(\boldsymbol{w}), \tilde{z}_j(\boldsymbol{w})) = \sum_{i=1}^{m} w_i^2 d^2(\tilde{r}_i^+, \tilde{r}_{ij}) \tag{4.29}$$

based on which we establish the following single-objective optimization model:

(M-4.8)
$$\begin{aligned} \min\ & \bar{d}(\boldsymbol{w}) = \sum_{j=1}^{n} \bar{d}(\tilde{z}^+(\boldsymbol{w}), \tilde{z}_j(\boldsymbol{w})) \\ \text{s.t.}\ & \sum_{i=1}^{m} w_i = 1,\ w_i \geqslant 0,\ \ i = 1, 2, \cdots, m \end{aligned}$$

Solving the model, we get the optimal weight vector $\boldsymbol{w} = (w_1,\ w_2, \cdots, w_m)^{\mathrm{T}}$ of attributes, where

$$w_i = \frac{1}{\displaystyle\sum_{i=1}^{m} \frac{1}{\displaystyle\sum_{j=1}^{n} d^2(\tilde{r}_i^+, \tilde{r}_{ij})}} \Bigg/ \sum_{j=1}^{n} d^2(\tilde{r}_i^+, \tilde{r}_{ij}), \quad i = 1, 2, \cdots, m \tag{4.30}$$

and $w_i \geqslant 0$ $(j = 1, 2, \cdots, m)$.

Based on the above model, in what follows, we introduce a practical method for solving multi-attribute decision making problems, in which the information about attribute weights is incompletely known, and the attribute values take the form of uncertain linguistic variables. The method involves the following steps:

Step 1 Let X, G, $\boldsymbol{w}$ and H be defined as before, and $\boldsymbol{w} \in H$. The decision maker provides the uncertain linguistic decision matrix $\tilde{\boldsymbol{R}} = (\tilde{r}_{ij})_{m \times n}$.

Step 2 If the information about the attribute weights is partly known, then we solve the model (M-4.7) to obtain the attribute weights; if the information about the attribute weights is completely unknown, then we solve (4.30) to determine the attribute weights.

Step 3 Derive the uncertain overall attribute values $\tilde{z}_j(\boldsymbol{w})$ $(j = 1, 2, \cdots, n)$ of the alternatives x_j $(j = 1, 2, \cdots, n)$ by using the weight vector $\boldsymbol{w} = (w_1, w_2, \cdots, w_m)^{\mathrm{T}}$ and (4.26).

Step 4 Compare each pair of $\tilde{z}_j(\boldsymbol{w})$ $(j = 1, 2, \cdots, n)$ by using (2.69), and construct a possibility degree matrix $\boldsymbol{P} = (p_{ij})_{n\times n}$. By (2.82), we get the priority vector $\boldsymbol{\zeta} = (\zeta_1, \zeta_2, \cdots, \zeta_n)^{\mathrm{T}}$ of $\boldsymbol{P}$.

Step 5 According to ζ_i $(i = 1, 2, \cdots, n)$, we rank the uncertain overall attribute values $\tilde{z}_j(\boldsymbol{w})$ $(j = 1, 2, \cdots, n)$, from which the alternatives x_j $(j = 1, 2, \cdots, n)$ can be ranked and selected.

Example 4.2(Xu, 2007a) The propulsion/manoeuvring system selection is based on the study that has been conducted for the selection of propulsion/manoeuvring system of a double ended passenger ferry to operate across the Bosphorus in Istanbul with the aim of reducing the journey time in highly congested seaway traffic (adapted from Olcer and Odabasi (2005)). The appropriate propulsion/manoeuvring system from among three alternatives x_j $(j = 1, 2, 3)$, namely, ① x_1——conventional propeller and high lift rudder (2×1); ② x_2——Z drive (2×2); and ③ x_3——cycloidal propeller (2×1), are being tried to choose, where 2×2 means two propulsion units at both ends, and 2×1 means one propulsion unit at both ends. The selection decision is made on the basis of one objective and six subjective attributes G_i $(i = 1, 2, \cdots, 7)$. These attributes, which are critical for the selection of propulsion/manoeuvring system of a double-ended passenger ferry, are the following:

(1) G_1: investment cost analysis;

(2) G_2: operating cost analysis, including handling,repair and maintenance costs analysis;

(3) G_3: manoeuvrability analysis;

(4) G_4: vibration and noise analysis;

(5) G_5: reliability analysis, including mechanical safety, redundancy, service experience analysis;

(6) G_6: propulsive power requirement analysis, including ship geometry ship resistance, power requirement, propulsion efficiency analysis;

(7) G_7: propulsive arrangement requirement analysis, including required propulsion capacity analysis.

The decision maker utilizes the additive linguistic evaluation scale:

$S_2=\{s_{-4}=$extremely poor, $s_{-3}=$very poor, $s_{-2}=$poor, $s_{-1}=$slightly poor,

s_0= fair, s_1=slightly good, s_2=good, s_3=very good, s_4=extremely good}

to express his/her evaluation information over the alternatives x_j $(j = 1, 2, 3)$ with respect to the several attributes G_i $(i = 1, 2, \cdots, 7)$, and constructs the following decision matrix $\boldsymbol{R}$ (Table 4.9):

Table 4.9 Linguistic decision matrix R (Xu, 2007a)

	x_1	x_2	x_3
G_1	s_3	s_2	s_{-2}
G_2	s_2	s_1	s_2
G_3	s_{-1}	s_4	s_{-1}
G_4	s_0	s_2	s_3
G_5	s_2	s_3	s_0
G_6	s_1	s_1	s_3
G_7	s_3	s_2	s_1

We introduce the following decision making steps:

Step 1 Based on Table 4.9, we get the vector of attribute values of the alternatives x_j $(j = 1, 2, 3)$: $\boldsymbol{r}_j = (r_{j1}, r_{j2}, \cdots, r_{j7})^{\mathrm{T}}$ and the relative ideal point $\boldsymbol{r}^+ = (r_1^+, r_2^+, \cdots, r_7^+)^{\mathrm{T}}$ of alternatives:

$$\boldsymbol{r}_1 = (s_3, s_2, s_{-1}, s_0, s_2, s_1, s_3)^{\mathrm{T}}, \quad \boldsymbol{r}_2 = (s_2, s_1, s_4, s_2, s_3, s_1, s_2)^{\mathrm{T}}$$

$$\boldsymbol{r}_3 = (s_{-2}, s_2, s_{-1}, s_3, s_0, s_3, s_1)^{\mathrm{T}}, \quad \boldsymbol{r}^+ = (s_3, s_2, s_4, s_3, s_3, s_3, s_3)^{\mathrm{T}}$$

Step 2 If the information about the attribute weights is as follows:

$$\begin{aligned} H =\{&0.20 \leqslant w_1 \leqslant 0.30, \quad 0.05 \leqslant w_2 \leqslant 0.10, \quad 0.5 \leqslant w_3 \leqslant 0.6, \quad w_4 \leqslant 0.2w_2, \\ &w_5 - w_4 \geqslant 0.05, \quad w_6 \geqslant w_4, \quad w_5 - w_7 \geqslant w_6 - w_4, \\ &w_7 \geqslant 0.4w_6, \ w_i \geqslant 0, \ i = 1, 2, \cdots, 7, \ \sum_{i=1}^{7} w_i = 1\Big\} \end{aligned}$$

then by the model (M-4.4), we establish the following single-objective optimization model:

$$\begin{aligned} \min \quad & \lambda_0 \\ \text{s.t.} \quad & \frac{1}{8}(5w_3 + 3w_4 + w_5 + 2w_6) \leqslant \lambda_0 \\ & \frac{1}{8}(w_1 + w_2 + w_4 + 2w_6 + w_7) \leqslant \lambda_0 \\ & \frac{1}{8}(5w_1 + 5w_3 + 3w_5 + 2w_7) \leqslant \lambda_0 \end{aligned}$$

$$0.20 \leqslant w_1 \leqslant 0.30, \quad 0.05 \leqslant w_2 \leqslant 0.10$$
$$0.5 \leqslant w_3 \leqslant 0.6, \quad w_4 \leqslant 0.2w_2, \quad w_5 - w_4 \geqslant 0.05$$
$$w_6 \geqslant w_4, \quad w_5 - w_7 \geqslant w_6 - w_4, \quad w_7 \geqslant 0.4w_6$$
$$w_i \geqslant 0, \ i = 1, 2, \cdots, 7, \quad \sum_{i=1}^{7} w_i = 1$$

Solving the model, we get the attribute weight vector:

$$\boldsymbol{w} = (0.2000,\ 0.1000,\ 0.5000,\ 0.0200,\ 0.0800,\ 0.0714,\ 0.0286)^{\mathrm{T}}$$

Step 3 Derive the overall attribute values $z_j(\boldsymbol{w})$ $(j = 1, 2, 3)$ of the alternatives x_j $(j = 1, 2, 3)$:

$$z_1(\boldsymbol{w}) = s_{0.62}, \quad z_2(\boldsymbol{w}) = s_{2.91}, \quad z_3(\boldsymbol{w}) = s_{-0.40}$$

Step 4 According to the overall attribute values $z_j(\boldsymbol{w})$ $(j = 1, 2, 3)$, we rank the alternatives x_j $(j = 1, 2, 3)$ as

$$x_2 \succ x_1 \succ x_3$$

and thus, the best alternative is x_2.

If the information about the attribute weights is unknown completely, then by (4.25), we have

$$\boldsymbol{w} = (0.0243,\ 0.6315,\ 0.0126,\ 0.0632,\ 0.0632,\ 0.0789,\ 0.1263)^{\mathrm{T}}$$

and by (4.2), we derive the overall attribute values $z_j(\boldsymbol{w})$ $(j = 1, 2, 3)$ of the alternatives x_j $(j = 1, 2, 3)$:

$$z_1(\boldsymbol{w}) = s_{1.91}, \quad z_2(\boldsymbol{w}) = s_{1.38}, \quad z_3(\boldsymbol{w}) = s_{1.75}$$

by which we rank the alternatives x_j $(j = 1, 2, 3)$ as

$$x_1 \succ x_3 \succ x_2$$

and thus, the best alternative is x_1, i.e, the most desirable propulsion/manoeuvring system should be the conventional propeller and high lift rudder.

From the above results, we can see that the change of the weight information of attributes results in the difference between the final decision results.

In the example above, if the decision maker constructs the following uncertain linguistic decision matrix $\tilde{\boldsymbol{R}}$ (Table 4.10):

Table 4.10 Uncertain linguistic decision matrix $\tilde{R}$

	x_1	x_2	x_3
G_1	$[s_2, s_3]$	$[s_3, s_4]$	$[s_{-2}, s_{-1}]$
G_2	$[s_2, s_4]$	$[s_1, s_2]$	$[s_2, s_3]$
G_3	$[s_{-1}, s_1]$	$[s_2, s_4]$	$[s_{-2}, s_{-1}]$
G_4	$[s_{-1}, s_0]$	$[s_2, s_3]$	$[s_3, s_4]$
G_5	$[s_1, s_2]$	$[s_2, s_4]$	$[s_{-1}, s_0]$
G_6	$[s_0, s_2]$	$[s_1, s_2]$	$[s_1, s_2]$
G_7	$[s_3, s_4]$	$[s_0, s_1]$	$[s_0, s_2]$

Then the following decision making steps are involved:

Step 1 Based on Table 4.10, we get the vectors $\tilde{\boldsymbol{r}}_j = (\tilde{\boldsymbol{r}}_{j1}, \tilde{\boldsymbol{r}}_{j2}, \cdots, \tilde{\boldsymbol{r}}_{j7})^{\mathrm{T}}$ of attribute values of the alternatives $\boldsymbol{x}_j$ $(j = 1, 2, 3)$ and the relative ideal point $\tilde{\boldsymbol{r}}^+ = (\tilde{\boldsymbol{r}}_1^+, \tilde{\boldsymbol{r}}_2^+, \cdots, \tilde{\boldsymbol{r}}_7^+)^{\mathrm{T}}$ of alternatives:

$$\tilde{\boldsymbol{r}}_1 = ([s_2, s_3], [s_2, s_4], [s_{-1}, s_1], [s_{-1}, s_0], [s_1, s_2], [s_0, s_2], [s_3, s_4])^{\mathrm{T}}$$

$$\tilde{\boldsymbol{r}}_2 = ([s_3, s_4], [s_1, s_2], [s_2, s_4], [s_2, s_3], [s_2, s_4], [s_1, s_2], [s_0, s_1])^{\mathrm{T}}$$

$$\tilde{\boldsymbol{r}}_3 = ([s_{-2}, s_{-1}], [s_2, s_3], [s_{-2}, s_{-1}], [s_3, s_4], [s_{-1}, s_0], [s_1, s_2], [s_0, s_2])^{\mathrm{T}}$$

$$\tilde{\boldsymbol{r}}^+ = ([s_3, s_4], [s_2, s_4], [s_2, s_4], [s_3, s_4], [s_2, s_4], [s_1, s_2], [s_3, s_4])^{\mathrm{T}}$$

Step 2 If the information about attribute weights is given as

$$\begin{aligned} H = \{&0.20 \leqslant w_1 \leqslant 0.30\,, \quad 0.10 \leqslant w_2 \leqslant 0.13, \quad 0.3 \leqslant w_3 \leqslant 0.4, \\ &0 \leqslant w_4 \leqslant 0.15, \quad 0.15 \leqslant w_5 \leqslant 0.2, \quad w_4 \geqslant w_6, \quad w_7 \leqslant w_6, \quad w_7 \geqslant 0.05 \\ &w_j \geqslant 0,\ j = 1, 2, \cdots, 7, \quad \sum_{j=1}^{7} w_j = 1\Big\} \end{aligned}$$

then by the model (M-4.7), we can establish the following single-objective optimization model:

$$\begin{aligned} \min\quad & \lambda_0 \\ \text{s.t.}\quad & \frac{1}{16}(2w_1 + 6w_3 + 8w_4 + 3w_5 + w_6) \leqslant \lambda_0 \\ & \frac{1}{16}(3w_2 + 2w_4 + 6w_7) \leqslant \lambda_0 \\ & \frac{1}{16}(10w_1 + w_2 + 9w_3 + 7w_5 + 5w_7) \leqslant \lambda_0 \\ & 0.20 \leqslant w_1 \leqslant 0.30, \quad 0.10 \leqslant w_2 \leqslant 0.13 \\ & 0.3 \leqslant w_3 \leqslant 0.4, \quad 0 \leqslant w_4 \leqslant 0.15 \end{aligned}$$

$$0.15 \leqslant w_5 \leqslant 0.2, \quad w_4 \geqslant w_6, \quad w_7 \leqslant w_6$$

$$w_7 \geqslant 0.05, \quad w_j \geqslant 0,\ j=1,2,\cdots,7, \quad \sum_{j=1}^{7} w_j = 1$$

Solving the model, we get the weight vector of attributes:

$$\boldsymbol{w} = (0.20,\ 0.10,\ 0.30,\ 0.10,\ 0.15,\ 0.10,\ 0.05)^{\mathrm{T}}$$

Step 3 By (4.10), we derive the uncertain overall attribute values $\tilde{z}_j(\boldsymbol{w})$ $(j = 1,2,3)$ of the alternatives x_j $(j=1,2,3)$:

$$\tilde{z}_1(\boldsymbol{w}) = [s_{0.5}, s_2], \quad \tilde{z}_2(\boldsymbol{w}) = [s_{1.90}, s_{3.35}], \quad \tilde{z}_3(\boldsymbol{w}) = [s_{-0.55}, s_{0.5}]$$

Step 4 Compare each pair of $\tilde{z}_j(\boldsymbol{w})$ $(j = 1,2,3)$ by using (2.69), we get a possibility degree matrix:

$$\boldsymbol{P} = \begin{pmatrix} 0.5 & 0.0339 & 1 \\ 0.9661 & 0.5 & 1 \\ 0 & 0 & 0.5 \end{pmatrix}$$

and obtain the priority vector of $\boldsymbol{P}$ by using (2.82):

$$\boldsymbol{\zeta} = (0.3390, 0.4944, 0.1667)^{\mathrm{T}}$$

Then we rank the uncertain overall attribute values $\tilde{z}_j(\boldsymbol{w})$ $(j = 1,2,3)$ by using ζ_i $(i=1,2,3)$:

$$\tilde{z}_2(\boldsymbol{w}) > \tilde{z}_1(\boldsymbol{w}) > \tilde{z}_3(\boldsymbol{w})$$

from which we get the ranking of the alternatives x_j $(j=1,2,3)$ as

$$x_2 \succ x_1 \succ x_3$$

Thus, the best alternative is x_2.

If the information about the attribute weights is unknown completely, then by (4.25), we get

$$\boldsymbol{w} = (0.0082,\ 0.0857,\ 0.0073,\ 0.0126,\ 0.0148,\ 0.8573,\ 0.0141)^{\mathrm{T}}$$

Using (4.10), we derive the uncertain overall attribute values $\tilde{z}_j(\boldsymbol{w})$ $(j = 1,2,3)$ of the alternatives x_j $(j=1,2,3)$:

$$\tilde{z}_1(\boldsymbol{w}) = [s_{0.23}, s_{2.18}], \quad \tilde{z}_2(\boldsymbol{w}) = [s_{1.05}, s_{2.06}] \quad \tilde{z}_3(\boldsymbol{w}) = [s_{1.02}, s_{2.03}]$$

Step 5 Compare each pair of $\tilde{z}_j(\boldsymbol{w})$ $(j = 1,2,3)$ by using (2.69), and get a possibility degree matrix:

$$\boldsymbol{P} = \begin{pmatrix} 0.5 & 0.3818 & 0.3919 \\ 0.6182 & 0.5 & 0.5149 \\ 0.6081 & 0.4851 & 0.5 \end{pmatrix}$$

whose priority vector can be derived by (2.82):

$$\boldsymbol{\zeta} = (0.2956, 0.3555, 0.3489)^{\mathrm{T}}$$

from which the uncertain overall attribute values $\tilde{z}_j(\boldsymbol{w})$ $(j = 1, 2, 3)$ can be ranked as

$$\tilde{z}_2(\boldsymbol{w}) > \tilde{z}_3(\boldsymbol{w}) > \tilde{z}_1(\boldsymbol{w})$$

Then, the ranking of the alternatives x_j $(j = 1, 2, 3)$ is

$$x_2 \succ x_3 \succ x_1$$

i.e., the best alternative is x_2.

4.3 Approaches Based on Uncertain Aggregation Operators to Linguistic Multi-attribute Decision Making

4.3.1 WULDC-OWA Operator-Based Approach

For a linguistic multi-attribute decision making problem, in which the information about the attribute weights is unknown completely, let X, G and $\boldsymbol{w}$ be defined as in Section 4.1. The decision maker evaluates the alternatives x_j $(j = 1, 2, \cdots, n)$ with respect to the attributes G_i $(i = 1, 2, \cdots, m)$ by means of the additive linguistic evaluation scale $S_2 = \{s_\alpha | \alpha = -\tau, \cdots, -1, 0, 1, \cdots, \tau\}$, and constructs an uncertain linguistic decision matrix $\tilde{\boldsymbol{R}} = (\tilde{r}_{ij})_{m\times n}$, where $\tilde{r}_{ij}$ denotes the uncertain attribute value of the alternative x_j with respect to the attribute G_i, $\tilde{r}_{ij} = [r_{ij}^{\mathrm{L}}, r_{ij}^{\mathrm{U}}]$, and r_{ij}^{L}, $r_{ij}^{\mathrm{U}} \in S_2$ $(i = 1, 2, \cdots, m,\ j = 1, 2, \cdots, n)$. Moreover, $\tilde{\boldsymbol{r}}^+ = (\tilde{r}_1^+, \tilde{r}_2^+, \cdots, \tilde{r}_m^+)^{\mathrm{T}}$ is the relative ideal point of alternatives, where $\tilde{r}_i^+ = [r_i^{+\mathrm{L}}, r_i^{+\mathrm{U}}]$, $r_i^{+\mathrm{L}} = \max\limits_j\{r_{ij}^{\mathrm{L}}\}$, $r_i^{+\mathrm{U}} = \max\limits_j\{r_{ij}^{\mathrm{U}}\}$ $(i = 1, 2, \cdots, m)$.

Definition 4.5 (Zhang and Xu, 2005) Let $\tilde{s}_1 = [s_{\alpha_1}, s_{\beta_1}]$ and $\tilde{s}_2 = [s_{\alpha_2}, s_{\beta_2}]$ be two uncertain linguistic variables. Then

$$\mathrm{AED}(\tilde{s}_1, \tilde{s}_2) = |f_\varphi([\alpha_1, \beta_1]) - f_\varphi([\alpha_2, \beta_2])| \tag{4.31}$$

is defined as the absolute of the expected difference between $\tilde{s}_1$ and $\tilde{s}_2$, where f is the C-OWA operator (2.86).

Example 4.3 (Zhang and Xu, 2005) Let $\tilde{s}_1 = [s_{-1}, s_3]$ and $\tilde{s}_2 = [s_0, s_4]$, and let the BUM function (see Definition 2.28) be $\varphi(x) = x^2$. Then

$$f_\varphi([-1, 3]) = \int_0^1 \frac{\mathrm{d}x^2}{\mathrm{d}x}\{3 - x[3 - (-1)]\}\mathrm{d}x = \frac{1}{3}$$

$$f_\varphi([0, 4]) = \int_0^1 \frac{\mathrm{d}x^2}{\mathrm{d}x}[4 - x(4 - 0)]\mathrm{d}x = \frac{4}{3}$$

Thus, the expected difference between $\tilde{s}_1$ and $\tilde{s}_2$ is

$$\text{AED}(\tilde{s}_1, \tilde{s}_2) = |f_\varphi([-1,3]) - f_\varphi([0,4])| = 1$$

Definition 4.6 (Zhang and Xu, 2005) Let $\tilde{\boldsymbol{R}} = (\tilde{r}_{ij})_{m\times n}$ be the uncertain linguistic decision matrix, lf be the ULC-OWA operator (see Definition 2.30). Then $lf_\varphi(\tilde{\boldsymbol{R}}) = (lf_\varphi(\tilde{r}_{ij}))_{m\times n}$ is the expected linguistic decision matrix of $\tilde{\boldsymbol{R}}$, and $lf_\varphi(\tilde{\boldsymbol{r}}^+) = (lf_\varphi(\tilde{r}_1^+), lf_\varphi(\tilde{r}_2^+), \cdots, lf_\varphi(\tilde{r}_m^+))^{\mathrm{T}}$, φ is the BUM function. According to (4.31), let $d_{ij} = \text{AED}(lf_\varphi(\tilde{r}_{ij}), lf_\varphi(\tilde{r}_i^+))$, if

$$h_{\boldsymbol{w}}(d_{1j}, d_{2j}, \cdots, d_{mj}) = \sum_{i=1}^{m} w_i d_{ij}, \quad j = 1, 2, \cdots, n \tag{4.32}$$

then h is called a weighted uncertain linguistic difference C-OWA operator (WULDC-OWA, for short).

Clearly, the smaller $h_{\boldsymbol{w}}(d_{1j}, d_{2j}, \cdots, d_{mj})$, the smaller the difference between the alternative x_j and the relative ideal point of alternatives, and thus, the better the alternative x_j.

Based on the analysis above, in the following, we introduce an approach to linguistic multi-attribute decision making based on the WULC-OWA operator (Zhang and Xu, 2005):

Step 1 For the uncertain linguistic decision matrix $\tilde{\boldsymbol{R}} = (\tilde{r}_{ij})_{m\times n}$, we utilize (2.86) and (2.87) to calculate the expected linguistic decision matrix $lf_\varphi(\tilde{\boldsymbol{R}}) = (lf_\varphi(\tilde{r}_{ij}))_{m\times n}$.

Step 2 Calculate the relative ideal point $\tilde{\boldsymbol{r}}^+ = (\tilde{r}_1^+, \tilde{r}_2^+, \cdots, \tilde{r}_m^+)^{\mathrm{T}}$ of alternatives, where $\tilde{r}_i^+ = [r_i^{+\mathrm{L}}, r_i^{+\mathrm{U}}]$, $r_i^{+\mathrm{L}} = \max_j\{r_{ij}^{\mathrm{L}}\}$, $r_i^{+\mathrm{U}} = \max_j\{r_{ij}^{\mathrm{U}}\}$ $(i = 1, 2, \cdots, m)$, and calculate $lf_\varphi(\tilde{\boldsymbol{r}}^+) = (lf_\varphi(\tilde{r}_1^+), lf_\varphi(\tilde{r}_2^+), \cdots, lf_\varphi(\tilde{r}_m^+))^{\mathrm{T}}$.

Step 3 Calculate the expected differences $d_{ij} = \text{AED}(lf_\varphi(\tilde{r}_{ij}), lf_\varphi(\tilde{r}_i^+))$ $(i = 1, 2, \cdots, m,\ j = 1, 2, \cdots, n)$.

Step 4 Aggregate all the differences corresponding to each alternative by using the WULDC-OWA operator $d_j = h_{\boldsymbol{w}}(d_{1j}, d_{2j}, \cdots, d_{mj}) = \sum_{i=1}^{m} w_i d_{ij}$ $(j = 1, 2, \cdots, n)$.

Step 5 Rank the alternatives x_j $(j=1,2,\cdots,n)$ according to d_j $(j=1,2,\cdots,n)$ (the smaller the difference d_j, the better the alternative x_j), and thus, the best alternative can be obtained.

Example 4.4 (Zhang and Xu, 2005) An investment company wants to invest a sum of money in the best option adapted from Herrera et al. (2000). There is a panel with five possible companies x_j $(j = 1, 2, \cdots, 5)$ in which to invest the money: ① x_1 is a car company; ② x_2 is a food company; ③ x_3 is a computer company; ④ x_4 is an arms company; and ⑤ x_5 is a TV company. The attributes which are considered here in selection of the five possible companies are: ① G_1 is economical benefit; ② G_2 is social benefit; and ③ G_3 is the environmental pollution, whose weight vector is $\boldsymbol{w} = (0.3, 0.4, 0.2, 0.1)^{\mathrm{T}}$. If the decision maker employs the multiplicative linguistic evaluation scale:

$$S_5 = \{s_{1/5} = \text{extremely poor}, \quad s_{2/5} = \text{very poor}, \quad s_{3/5} = \text{poor}, \quad s_{4/5} = \text{slightly poor}, \\ s_1 = \text{fair}, \quad s_{5/4} = \text{slightly good}, \quad s_{5/3} = \text{good}, s_{5/2} = \text{very good}, \\ s_5 = \text{extremely good}\}$$

to evaluate the five possible companies x_j $(j = 1, 2, \cdots, 5)$ with respect to the attributes G_i $(i = 1, 2, 3, 4)$, and constructs the uncertain linguistic decision matrix (Table 4.11):

Table 4.11 Uncertain linguistic decision matrix $\tilde{R}$ (Zhang and Xu, 2005)

	x_1	x_2	x_3	x_4	x_5
G_1	$[s_1, s_2]$	$[s_2, s_3]$	$[s_{-1}, s_0]$	$[s_2, s_3]$	$[s_1, s_2]$
G_2	$[s_1, s_3]$	$[s_0, s_1]$	$[s_{-2}, s_0]$	$[s_0, s_1]$	$[s_3, s_4]$
G_3	$[s_2, s_3]$	$[s_0, s_2]$	$[s_1, s_3]$	$[s_{-1}, s_0]$	$[s_2, s_4]$
G_4	$[s_1, s_3]$	$[s_0, s_2]$	$[s_1, s_2]$	$[s_3, s_4]$	$[s_0, s_2]$

Now we utilize the above approach to find the final decision result:

Step 1 For the uncertain linguistic decision matrix $\tilde{\boldsymbol{R}} = (\tilde{r}_{ij})_{m \times n}$, we utilize (2.86) and (2.87) to calculate the expected linguistic decision matrix $lf_{\varphi}(\tilde{\boldsymbol{R}}) = (lf_{\varphi}(\tilde{r}_{ij}))_{4\times 5}$ (suppose that $\varphi(x) = x^2$), the results are listed in Table 4.12:

Table 4.12 Expected linguistic decision matrix $lf_{\varphi}(\tilde{R})$ (Zhang and Xu, 2005)

	x_1	x_2	x_3	x_4	x_5
G_1	$s_{4/3}$	$s_{7/3}$	$s_{-2/3}$	$s_{7/3}$	$s_{4/3}$
G_2	$s_{5/3}$	$s_{1/3}$	s_{-1}	$s_{1/3}$	$s_{10/3}$
G_3	$s_{7/3}$	$s_{2/3}$	$s_{5/3}$	$s_{-2/3}$	$s_{8/3}$
G_4	$s_{5/3}$	$s_{2/3}$	$s_{4/3}$	$s_{10/3}$	$s_{2/3}$

Step 2 Find the relative ideal point $\tilde{\boldsymbol{r}}^+ = (\tilde{r}_1^+, \tilde{r}_2^+, \tilde{r}_3^+, \tilde{r}_4^+)^{\mathrm{T}}$ of alternatives:

$$\tilde{r}_1^+ = [s_2, s_3], \quad \tilde{r}_2^+ = [s_3, s_4], \quad \tilde{r}_3^+ = [s_2, s_4], \quad \tilde{r}_4^+ = [s_3, s_4]$$

and calculate $lf_{\varphi}(\tilde{\boldsymbol{r}}^+) = (lf_{\varphi}(\tilde{r}_1^+), lf_{\varphi}(\tilde{r}_2^+), lf_{\varphi}(\tilde{r}_3^+), lf_{\varphi}(\tilde{r}_4^+))^{\mathrm{T}}$:

$$lf_{\varphi}(\tilde{r}_1^+) = s_{7/3}, \quad lf_{\varphi}(\tilde{r}_2^+) = s_{10/3}, \quad lf_{\varphi}(\tilde{r}_3^+) = s_{8/3}, \quad lf_{\varphi}(\tilde{r}_4^+) = s_{10/3}$$

Step 3 Calculate the expected differences $d_{ij} = \mathrm{AED}(lf_{\varphi}(\tilde{r}_{ij}), lf_{\varphi}(\tilde{r}_i^+))$ $(i = 1, 2, 3, 4,\ j = 1, 2, \cdots, 5)$:

$$d_{11}=1,\quad d_{12}=0,\quad d_{13}=3,\quad d_{14}=0,\quad d_{15}=1,\quad d_{21}=5/3,\quad d_{22}=3$$

$$d_{23}=13/3,\quad d_{24}=3,\quad d_{25}=0,\quad d_{31}=1/3,\quad d_{32}=2,\quad d_{33}=1,\quad d_{34}=10/3$$

$$d_{35}=0,\quad d_{41}=5/3,\quad d_{42}=8/3,\quad d_{43}=2,\quad d_{44}=0,\quad d_{45}=8/3$$

Step 4 Utilize the WULDC-OWA operator (4.32) to aggregate all the expected differences corresponding to each alternative, and get

$$d_1=1.2,\quad d_2=1.87,\quad d_3=3.03,\quad d_4=1.87,\quad d_5=0.57$$

Step 5 Rank the alternatives x_j $(j=1,2,\cdots,5)$ according to d_j $(j=1,2,\cdots,5)$ (the smaller the difference d_j, the better the alternative x_j):

$$x_5 \succ x_1 \succ x_2 \sim x_4 \succ x_3$$

and thus, the best alternative is x_5.

4.3.2 WULDC-OWG Operator-Based Approach

For a lingistic multi-attribute decision making problem, the decision maker employs the multiplicative linguistic evaluation scale $S_4=\{s_\alpha|\alpha=/\tau,\cdots,1/2,\ 1,\ 2,\cdots,\tau\}$ to evaluate the alternative $x_j\in X$ with respect to the attribute $G_i\in G$, and constructs an uncertain linguistic decision matrix $\tilde{\boldsymbol{R}}=(\tilde{r}_{ij})_{m\times n}$, where $\tilde{r}_{ij}$ denotes the uncertain attribute value of the alternative with respect to the attribute G_i, $\tilde{r}_{ij}=[r_{ij}^{\mathrm{L}},r_{ij}^{\mathrm{U}}]$, and $r_{ij}^{\mathrm{L}},r_{ij}^{\mathrm{U}}\in S_4$ $(i=1,2,\cdots,m,\ j=1,2,\cdots,n)$. Moreover, let $\tilde{\boldsymbol{r}}^+=(\tilde{r}_1^+,\tilde{r}_2^+,\cdots,\tilde{r}_m^+)^{\mathrm{T}}$ be the relative ideal point, where $\tilde{r}_i^+=[r_i^{+\mathrm{L}},r_i^{+\mathrm{U}}]$, $r_i^{+\mathrm{L}}=\max\limits_j\{r_{ij}^{\mathrm{L}}\}$, $r_i^{+\mathrm{U}}=\max\limits_j\{r_{ij}^{\mathrm{U}}\}$ $(i=1,2,\cdots,m)$.

Definition 4.7 (Zhang and Xu, 2005) Let $\tilde{s}_1=[s_{\alpha_1},s_{\beta_1}]$ and $\tilde{s}_2=[s_{\alpha_2},s_{\beta_2}]$ be two uncertain linguistic variables. Then

$$\mathrm{MED}(\tilde{s}_1,\tilde{s}_2)=\frac{g_\varphi([\alpha_1,\beta_1])}{g_\varphi([\alpha_2,\beta_2])} \tag{4.33}$$

is called the expected deviation degree between $\tilde{s}_1$ and $\tilde{s}_2$, where g is the C-OWG operator (2.144).

Defnition 4.8 (Zhang and Xu, 2005) Let $\tilde{\boldsymbol{R}}=(\tilde{r}_{ij})_{m\times n}$ be an uncertain linguistic decision matrix, $l\,g_\varphi(\tilde{\boldsymbol{R}})=(lg_\varphi(\tilde{r}_{ij}))_{m\times n}$, and $l\,g_\varphi(\tilde{\boldsymbol{r}}^+)=(lg_\varphi(\tilde{r}_1^+),lg_\varphi(\tilde{r}_2^+),\cdots,lg_\varphi(\tilde{r}_m^+))^{\mathrm{T}}$. Let $d_{ij}=\mathrm{MED}(l\,g_\varphi(\tilde{r}_{ij}),\,lg_\varphi(\tilde{r}_i^+))$. If

$$g_{\boldsymbol{w}}(d_{1j},d_{2j},\cdots,d_{mj})=\prod_{i=1}^m d_{ij}^{w_i},\quad j=1,2,\cdots,n \tag{4.34}$$

then g is called a weighted uncertain linguistic deviation C-OWG operator (WULDC-OWG, for short).

Obviously, the greater $g_{\boldsymbol{w}}(d_{1j}, d_{2j}, \cdots, d_{mj})$, the closer the alternative x_j to the relative ideal point of alternatives, and thus, the better the alternative x_j.

Based on the analysis above, in the following, we introduce an approach based on the WULC-OWG operator to multi-attribute decision making with uncertain linguistic information (Zhang and Xu, 2005):

Step 1 For the uncertain linguistic decision matrix $\tilde{\boldsymbol{R}} = (\tilde{r}_{ij})_{m\times n}$, we utilize (2.144) and (2.145) to calculate $lg_{\varphi}(\tilde{\boldsymbol{R}}) = (lg_{\varphi}(\tilde{r}_{ij}))_{m\times n}$.

Step 2 Find the relative ideal point $\tilde{\boldsymbol{r}}^+ = (\tilde{r}_1^+, \tilde{r}_2^+, \cdots, \tilde{r}_m^+)^{\mathrm{T}}$ of alternatives, where $\tilde{r}_i^+ = [r_i^{+\mathrm{L}}, r_i^{+\mathrm{U}}]$, $r_i^{+\mathrm{L}} = \max\limits_j\{r_{ij}^{\mathrm{L}}\}$, $r_i^{+\mathrm{U}} = \max\limits_j\{r_{ij}^{\mathrm{U}}\}$ $(i = 1, 2, \cdots, m)$, and calculate $lg_{\varphi}(\tilde{\boldsymbol{r}}^+) = (lg_{\varphi}(\tilde{r}_1^+), lg_{\varphi}(\tilde{r}_2^+), \cdots, lg_{\varphi}(\tilde{r}_m^+))^{\mathrm{T}}$.

Step 3 Calculate the expected deviation degree $d'_{ij} = \mathrm{MED}(lg_{\varphi}(\tilde{r}_{ij}), lg_{\varphi}(\tilde{r}_i^+))$ $(i = 1, 2, \cdots, m,\ j = 1, 2, \cdots, n)$.

Step 4 Aggregate all the expected deviation degrees corresponding to each alternative by using the WULDC-OWG operator $d_j = g_{\boldsymbol{w}}(d'_{1j}, d'_{2j}, \cdots, d'_{mj}) = \prod\limits_{i=1}^{m}(d'_{ij})^{w_i}$ $(j = 1, 2, \cdots, n)$.

Step 5 Rank all the alternatives x_j $(j = 1, 2, \cdots, n)$ according to d'_j $(j = 1, 2, \cdots, n)$ (the greater the deviation degree d'_j, the better the alternative x_j), and then the best alternative can be obtained.

Example 4.5 In Example 4.4, if the decision maker utilizes the multiplicative linguistic evaluation scale:

$$\begin{aligned} S_4 = \{& s_{1/5} = \text{extremely poor},\quad s_{2/5} = \text{very poor},\quad s_{3/5} = \text{poor},\quad s_{4/5} = \text{slightly poor},\\ & s_1 = \text{fair},\quad s_{5/4} = \text{slightly good},\quad s_{5/3} = \text{good},\\ & s_{5/2} = \text{very good},\quad s_5 = \text{extremely good}\} \end{aligned}$$

to evaluate the alternatives x_j $(j = 1, 2, \cdots, 5)$, and constructs an uncertain linguistic decision matrix (Table 4.13):

Table 4.13 Uncertain linguistic decision matrix $\tilde{R}$

	x_1	x_2	x_3	x_4	x_5
G_1	$[s_2, s_3]$	$[s_3, s_4]$	$[s_{1/2}, s_1]$	$[s_3, s_4]$	$[s_2, s_3]$
G_2	$[s_2, s_4]$	$[s_1, s_2]$	$[s_{1/3}, s_2]$	$[s_1, s_2]$	$[s_4, s_5]$
G_3	$[s_3, s_4]$	$[s_1, s_3]$	$[s_2, s_4]$	$[s_{1/2}, s_1]$	$[s_3, s_5]$
G_4	$[s_2, s_4]$	$[s_1, s_3]$	$[s_2, s_3]$	$[s_4, s_5]$	$[s_1, s_3]$

Below we utilize the approach introduced in this subsection to find the desirable decision result:

Step 1 For the uncertain linguistic decision matrix $\tilde{\boldsymbol{R}} = (\tilde{r}_{ij})_{4\times 5}$, we utilize

(2.143) and (2.144) to calculate $lg_{\varphi}(\tilde{\boldsymbol{R}}) = (lg_{\varphi}(\tilde{r}_{ij}))_{4\times 5}$ (suppose that $\varphi(x) = x^2$). The results are listed in Table 4.14:

Table 4.14 Expected linguistic decision matrix $lg_{\rho}(\tilde{R})$ (Zhang and Xu, 2005)

	x_1	x_2	x_3	x_4	x_5
G_1	$s_{2.289}$	$s_{3.302}$	$s_{0.630}$	$s_{3.302}$	$s_{2.289}$
G_2	$s_{2.520}$	$s_{1.260}$	$s_{0.606}$	$s_{1.260}$	$s_{4.309}$
G_3	$s_{3.302}$	$s_{1.442}$	$s_{2.520}$	$s_{0.630}$	$s_{3.557}$
G_4	$s_{2.520}$	$s_{1.442}$	$s_{2.289}$	$s_{4.309}$	$s_{1.442}$

Step 2 Find the relative ideal point $\tilde{\boldsymbol{r}}^+ = (\tilde{r}_1^+, \tilde{r}_2^+, \tilde{r}_3^+, \tilde{r}_4^+)^{\mathrm{T}}$ of alternatives:

$$\tilde{r}_1^+ = [s_3, s_4], \quad \tilde{r}_2^+ = [s_4, s_5], \quad \tilde{r}_3^+ = [s_3, s_5], \quad \tilde{r}_4^+ = [s_4, s_5]$$

and calculate $lg_{\varphi}(\tilde{\boldsymbol{r}}^+) = (lg_{\varphi}(\tilde{r}_1^+), lg_{\varphi}(\tilde{r}_2^+), lg_{\varphi}(\tilde{r}_3^+), lg_{\varphi}(\tilde{r}_4^+))^{\mathrm{T}}$:

$$lg_{\varphi}(\tilde{r}_1^+) = s_{3.302}, \quad lg_{\varphi}(\tilde{r}_2^+) = s_{4.309}, \quad lg_{\varphi}(\tilde{r}_3^+) = s_{3.557}, \quad lg_{\varphi}(\tilde{r}_4^+) = s_{4.309}$$

Step 3 Calculate the expected deviation degree $d'_{ij} = \mathrm{MED}(lg_{\varphi}(\tilde{r}_{ij}), lg_{\varphi}(\tilde{r}_i^+))$ $(i = 1, 2, 3, 4,\ j = 1, 2, 3, 4, 5)$:

$$d'_{11} = 0.693, \quad d'_{12} = 1.000, \quad d'_{13} = 0.191, \quad d'_{14} = 1.000, \quad d'_{15} = 0.693$$

$$d'_{21} = 0.585, \quad d'_{22} = 0.292, \quad d'_{23} = 0.141, \quad d'_{24} = 0.292, \quad d'_{25} = 1.000$$

$$d'_{31} = 0.928, \quad d'_{32} = 0.405, \quad d'_{33} = 0.708, \quad d'_{34} = 0.177, \quad d'_{35} = 1.000$$

$$d'_{41} = 0.585, \quad d'_{42} = 0.335, \quad d'_{43} = 0.531, \quad d'_{44} = 1.000, \quad d'_{45} = 0.335$$

Step 4 Aggregate all the exected deviation degrees corresponding to each alternative by using the WULDC-OWG operator, and get

$$d'_1 = 0.675, \quad d'_2 = 0.457, \quad d'_3 = 0.244, \quad d'_4 = 0.432, \quad d'_5 = 0.803$$

Step 5 Rank the alternatives x_j $(j = 1, 2, \cdots, 5)$ according to d'_j $(j = 1, 2, \cdots, 5)$ (the larger d_j, the better the alternative x_j):

$$x_5 \succ x_1 \succ x_2 \succ x_4 \succ x_3$$

Thus the best alternative is also x_5.

4.4 Goal Programming Model for Linguistic Multi-Attribute Decision Making

4.4.1 Goal Programming Model Based on Linguistic Labels

For a linguistic multi-attibute decision making problem, let X, G, $\boldsymbol{w}$ and H be defined as Section 4.2, $\boldsymbol{w} \in H$, and $\boldsymbol{R} = (r_{ij})_{m\times n}$ be a linguist decision matrix, where r_{ij}

is an attribute value provided by the decision maker for the alternative $x_j \in X$ with respect to the attribute $G_i \in G$ by the additive linguistic evaluation scale $S_2 = \{s_\alpha | \alpha = -\tau, \cdots, -1, 0, 1, \cdots, \tau\}$. Moreover, assume that the decision maker has preference $\pi_j \in S_2$ over the alternative $x_j \in X$.

In general, there exist some differences between the overall attribute value $z_j(\boldsymbol{w})$ of the alternative x_j derived by (4.2) and the preference value π_j provided by the decision maker for the alternative x_j. Consequently, Xu (2006a) introduced a deviation function:

$$d_j(\boldsymbol{w}) = d(z_j(\boldsymbol{w}), \pi_j) = d(s_{\varsigma_j(\boldsymbol{w})}, s_{\alpha_j}) = |\varsigma_j(\boldsymbol{w}) - \alpha_j|, \quad j = 1, 2, \cdots, n \tag{4.35}$$

where $z_j(\boldsymbol{w}) = s_{\varsigma_j(\boldsymbol{w})}$ and $\pi_j = s_{\alpha_j}$.

Clearly, a reasonable attribute weight vector $\boldsymbol{w}$ should be determined so as to make the deviation degrees of (4.35) as small as possible. As a result, we establish the following optimization model (Xu, 2006a):

$$\text{(M-4.9)} \quad \begin{aligned} \min \quad & d_j(\boldsymbol{w}) = |\varsigma_j(\boldsymbol{w}) - \alpha_j|, \quad j = 1, 2, \cdots, n \\ \text{s. t.} \quad & \boldsymbol{w} \in H \\ & w_i \geqslant 0, \quad i = 1, 2, \cdots, m, \quad \sum_{i=1}^{m} w_i = 1 \end{aligned}$$

To solve the model (M-4.9), and considering that all the objective functions are fair, the model (M-4.9) can be transformed into the following goal programming model (Xu, 2006a):

$$\text{(M-4.10)} \quad \begin{aligned} \min \quad & J = \sum_{j=1}^{n} (\delta_j^+ e_j^+ + \delta_j^- e_j^-) \\ \text{s. t.} \quad & \varsigma_j(\boldsymbol{w}) - \alpha_j - \varepsilon_j^+ + \varepsilon_j^- = 0, \quad j = 1, 2, \cdots, n \\ & \varepsilon_j^+ \geqslant 0, \ \varepsilon_j^- \geqslant 0, \quad j = 1, 2, \cdots, n \\ & \varepsilon_j^+ \varepsilon_j^- = 0, \ j = 1, 2, \cdots, n \\ & \boldsymbol{w} \in H \\ & w_i \geqslant 0, \quad i = 1, 2, \cdots, m, \quad \sum_{i=1}^{m} w_i = 1 \end{aligned}$$

where ε_j^+ is the positive deviation from the target of $\varsigma_j(\boldsymbol{w}) - \alpha_j$, ε_j^- is the negative deviation from the target of $\varsigma_j(\boldsymbol{w}) - \alpha_j$, δ_j^+ and δ_j^- are, respectively, the weighting coefficients of ε_j^+ and ε_j^-.

Utilizing the objective simplex method to solve the model (M-4.10), we can get the attribute weight vector $\boldsymbol{w}$. Then by (4.2), we get the overall attribute values $z_j(\boldsymbol{w})$ $(j = 1, 2, \cdots, n)$, by which all the alternatives can be ranked and then selected.

Example 4.6 (Xu, 2006a) Assessment and selection of cadres is a multi-attribute decision making problem. On the one hand, the decision maker should select talented people to leadership positions; on the other hand, the decision maker also hopes that in the case of the same condition, he/she would like to appoint the preferred candidate adapted from Gao (2000). The attributes which are considered by a certain unit in selection of cadre candidates are ① G_1——thought and morality; ② G_2——working attitude; ③ G_3——working; ④ G_4——literacy and knowledge structure; ⑤ G_5——leadership ability; and ⑥ G_6——develop capacity. First, the masses are asked to recommend and evaluate the initial candidates with respect to the attribute above by using the additive linguistic evaluation scale:

$$S_2=\{s_{-4}=\text{extremely poor},\ s_{-3}=\text{very poor},\ s_{-2}=\text{poor},\ s_{-1}=\text{slightly poor},\ s_0=\text{fair},\ s_1=\text{slightly good},\ s_2=\text{good},\ s_3=\text{very good},\ s_4=\text{extremely good}\}$$

Then, after the statistical processing, five candidates x_j $(j=1,2,\cdots,5)$ have been identified. The decision information on each candidate with respect to the attributes G_i $(i=1,2,\cdots,6)$ is described in the linguistic decision matrix $\boldsymbol{R}$ (Table 4.15):

Table 4.15 Linguistic decision matrix R (Xu, 2006a)

	x_1	x_2	x_3	x_4	x_5
G_1	s_3	s_4	s_3	s_4	s_2
G_2	s_1	s_2	s_4	s_2	s_4
G_3	s_{-1}	s_0	s_2	s_4	s_4
G_4	s_4	s_2	s_1	s_2	s_3
G_5	s_2	s_4	s_4	s_0	s_0
G_6	s_4	s_1	s_0	s_3	s_2

The known weight information is as follows:

$$H=\{0.18\leqslant w_1\leqslant 0.22,\ 0.11\leqslant w_2\leqslant 0.20,\ 0.15\leqslant w_3\leqslant 0.19,\ 0.10\leqslant w_4\leqslant 0.14,\ 0.16\leqslant w_5\leqslant 0.25,\ 0.20\leqslant w_6\leqslant 0.23\}$$

Suppose that the decision maker provides the following preferences over the five candidates x_j $(j=1,2,\cdots,5)$:

$$\pi_1=s_2,\quad \pi_2=s_2,\quad \pi_3=s_3,\quad \pi_4=s_4,\quad \pi_5=s_3$$

In what follows, we describe the decision making steps by using the approach introduced in this subsection:

Based on the known decision information, we first utilize the model (M-4.10) to establish a goal programming model (assume that $\delta_j^+=\delta_j^-=1$ $(j=1,2,\cdots,5)$) as follows:

$$
\begin{aligned}
\min \quad & J = \sum_{j=1}^{5} (\varepsilon_j^+ + \varepsilon_j^-) \\
\text{s. t.} \quad & 3w_1 + w_2 - w_3 + 4w_4 + 2w_5 + 4w_6 - 2 - \varepsilon_1^+ + \varepsilon_1^- = 0 \\
& 4w_1 + 2w_2 + 2w_4 + 4w_5 + w_6 - 2 - \varepsilon_2^+ + \varepsilon_2^- = 0 \\
& 3w_1 + 4w_2 + 2w_3 + w_4 + 4w_5 - 3 - \varepsilon_3^+ + \varepsilon_3^- = 0 \\
& 4w_1 + 2w_2 + 4w_3 + 2w_4 + 3w_6 - 4 - \varepsilon_4^+ + \varepsilon_4^- = 0 \\
& 2w_1 + 4w_2 + 4w_3 + 3w_4 + 2w_6 - 3 - \varepsilon_5^+ + \varepsilon_5^- = 0 \\
& \varepsilon_j^+ \geqslant 0, \ \varepsilon_j^- \geqslant 0, \ j = 1, 2, \cdots, 5 \\
& \varepsilon_j^+ \varepsilon_j^- = 0, \ j = 1, 2, \cdots, 5 \\
& 0.18 \leqslant w_1 \leqslant 0.22, \ 0.11 \leqslant w_2 \leqslant 0.20 \\
& 0.15 \leqslant w_3 \leqslant 0.19, \ 0.10 \leqslant w_4 \leqslant 0.14 \\
& 0.16 \leqslant w_5 \leqslant 0.25, \ 0.20 \leqslant w_6 \leqslant 0.23 \\
& w_1 + w_2 + w_3 + w_4 + w_5 = 1
\end{aligned}
$$

Solving the model, we get the weight vector of attributes:

$$\boldsymbol{w} = (0.22, 0.11, 0.19, 0.10, 0.16, 0.22)^{\mathrm{T}}$$

and

$$\varepsilon_1^+ = 0.18, \quad \varepsilon_1^- = 0, \quad \varepsilon_2^+ = 0.16, \quad \varepsilon_2^- = 0, \quad \varepsilon_3^+ = 0, \quad \varepsilon_3^- = 0.78$$

$$\varepsilon_4^+ = 0, \quad \varepsilon_4^- = 1.28, \quad \varepsilon_5^+ = 0, \quad \varepsilon_5^- = 0.62$$

Then we use (4.2) to derive the overall attribute values $z_j(\boldsymbol{w})$ $(j = 1, 2, \cdots, 5)$ of the alternatives x_j $(j = 1, 2, \cdots, 5)$:

$$z_1(\boldsymbol{w}) = s_{2.18}, \quad z_2(\boldsymbol{w}) = s_{2.16}, \quad z_3(\boldsymbol{w}) = s_{2.22}, \quad z_4(\boldsymbol{w}) = s_{2.72}, \quad z_5(\boldsymbol{w}) = s_{2.38}$$

by which we rank the alternatives x_j $(j = 1, 2, \cdots, 5)$:

$$x_4 \succ x_5 \succ x_3 \succ x_1 \succ x_2$$

Accordingly, the best candidate is x_4.

4.4.2 Goal Programming Model Based on Uncertain Linguistic Labels

Now we consider situations where the linguistic decision information is uncertain, i.e., the elements in the decision matrix and preferences provided by the decision maker are expressed in uncertain linguistic variables:

Let $\tilde{\boldsymbol{R}} = (\tilde{r}_{ij})_{m \times n}$ be an uncertain linguistic decision matrix, where $\tilde{r}_{ij} = [r_{ij}^{\mathrm{L}}, r_{ij}^{\mathrm{U}}]$ $(r_{ij}^{\mathrm{L}}, r_{ij}^{\mathrm{U}} \in S_2)$ is an uncertain attribute values provided by the decision makers over the alternatrives $x_j \in X$ with respect to the attribute $G_i \in G$ by using the additive linguistic evaluation scale $S_2 = \{s_\alpha | \alpha = -\tau, \cdots, -1, 0, 1, \cdots, \tau\}$. Let the uncertain preferences $\tilde{\pi}_j = [\pi_j^{\mathrm{L}}, \pi_j^{\mathrm{U}}]$ $(\pi_j^{\mathrm{L}}, \pi_j^{\mathrm{U}} \in S_2)$ be provided by the decision maker over the alternative $x_j \in X$.

Based on the uncertain linguistic decision matrix $\tilde{\boldsymbol{R}} = (\tilde{r}_{ij})_{m\times n}$, we utilize (4.11) to derive the uncertain overall attribute value $\tilde{z}_j(\boldsymbol{w}) = [z_j^{\mathrm{L}}(\boldsymbol{w}), z_j^{\mathrm{U}}(\boldsymbol{w})]$ of the alternative x_j. Considering that only the partial weight information can be given in the considered problem, we first need to determine the weight vector of attributes. It is always hoped that the uncertain overall attribute value $\tilde{z}_j(\boldsymbol{w})$ of the alternative x_j is consistent with the uncertain preference $\tilde{\pi}_j$ provided by the decision maker over the alternative x_j, i.e.,

$$\tilde{z}_j(\boldsymbol{w}) = \tilde{\pi}_j, \quad j = 1, 2, \cdots, n \tag{4.36}$$

That is to say,

$$z_j^{\mathrm{L}}(\boldsymbol{w}) = \pi_j^{\mathrm{L}}, \quad z_j^{\mathrm{U}}(\boldsymbol{w}) = \pi_j^{\mathrm{U}}, \quad j = 1, 2, \cdots, n \tag{4.37}$$

However, in general, there exist some differences between the uncertain overall attribute value $\tilde{z}_j(\boldsymbol{w})$ of the alternative x_j and the uncertain preference $\tilde{\pi}_j$ provided by the decision maker over the alternative x_j. Thus, here we introduce a deviation function:

$$d_j^{\mathrm{L}}(\boldsymbol{w}) = d(z_j^{\mathrm{L}}(\boldsymbol{w}), \pi_j^{\mathrm{L}}), \quad d_j^{\mathrm{U}}(\boldsymbol{w}) = d(z_j^{\mathrm{U}}(\boldsymbol{w}), \pi_j^{\mathrm{U}}), \quad j = 1, 2, \cdots, n \tag{4.38}$$

For convenience, let

$$\tilde{z}_j(\boldsymbol{w}) = [s_{\varsigma_j^{\mathrm{L}}(\boldsymbol{w})}, s_{\varsigma_j^{\mathrm{U}}(\boldsymbol{w})}], \quad \tilde{\pi}_j = [s_{\alpha_j}, s_{\beta_j}] \tag{4.39}$$

Then (4.39) can be transformed into

$$d_j^{\mathrm{L}}(\boldsymbol{w}) = d(s_{\varsigma_j^{\mathrm{L}}(\boldsymbol{w})}, s_{\alpha_j}) = |\varsigma_j^{\mathrm{L}}(\boldsymbol{w}) - \alpha_j|, \quad j = 1, 2, \cdots, n \tag{4.40}$$

$$d_j^{\mathrm{U}}(\boldsymbol{w}) = d(s_{\varsigma_j^{\mathrm{U}}(\boldsymbol{w})}, s_{\beta_j}) = |\varsigma_j^{\mathrm{U}}(\boldsymbol{w}) - \beta_j|, \quad j = 1, 2, \cdots, n \tag{4.41}$$

Obviously, a reasonable attribute weight vector $\boldsymbol{w}$ should be determined so as to make the above deviation function as small as possible. As a result, we establish the following optimization model (Xu, 2006a):

(M-4.11)
$$\begin{aligned} \min\quad & d_j^{\mathrm{L}}(\boldsymbol{w}) = |\varsigma_j^{\mathrm{L}}(\boldsymbol{w}) - \alpha_j|, \quad j = 1, 2, \cdots, n \\ & d_j^{\mathrm{U}}(\boldsymbol{w}) = |\varsigma_j^{\mathrm{U}}(\boldsymbol{w}) - \beta_j|, \quad j = 1, 2, \cdots, n \\ \text{s. t.}\quad & \boldsymbol{w} \in H \\ & w_i \geqslant 0, \ i = 1, 2, \cdots, m, \ \sum_{i=1}^{m} w_i = 1 \end{aligned}$$

To solve the model (M-4.11), and considering that all the obective functions are fair, we can transform the model (M-4.11) into the following goal programming model (Xu, 2006a):

(M-4.12)

$$\min \quad J' = \sum_{j=1}^{n} [(\delta_{1j}^{+}\varepsilon_{1j}^{+} + \delta_{1j}^{-}\varepsilon_{1j}^{-}) + (\delta_{2j}^{+}\varepsilon_{2j}^{+} + \delta_{2j}^{-}\varepsilon_{2j}^{-})]$$

$$\text{s. t.} \quad \varsigma_j^{\mathrm{L}}(\boldsymbol{w}) - \alpha_j - \varepsilon_{1j}^{+} + \varepsilon_{1j}^{-} = 0, \quad j = 1, 2, \cdots, n$$

$$\varsigma_j^{\mathrm{U}}(\boldsymbol{w}) - \beta_j - \varepsilon_{2j}^{+} + \varepsilon_{2j}^{-} = 0, \quad j = 1, 2, \cdots, n$$

$$\varepsilon_{1j}^{+} \geqslant 0, \ \varepsilon_{1j}^{-} \geqslant 0, \quad j = 1, 2, \cdots, n$$

$$\varepsilon_{2j}^{+} \geqslant 0, \ \varepsilon_{2j}^{-} \geqslant 0, \quad j = 1, 2, \cdots, n$$

$$\varepsilon_{1j}^{+}\varepsilon_{1j}^{-} = 0, \quad \varepsilon_{2j}^{+}\varepsilon_{2j}^{-} = 0, \quad j = 1, 2 \cdots, n$$

$$\boldsymbol{w} \in H$$

$$w_i \geqslant 0, \ i = 1, 2, \cdots, m, \ \sum_{i=1}^{m} w_i = 1$$

where ε_{1j}^{+} and ε_{2j}^{+} are the positive deviations from the targets of $\varsigma_j^{\mathrm{L}}(\boldsymbol{w}) - \alpha_j$ and $\varsigma_j^{\mathrm{U}}(\boldsymbol{w}) - \beta_j$ respectively, ε_{1j}^{-} and ε_{2j}^{-} are the negative deviations from the targets of $\varsigma_j^{\mathrm{L}}(\boldsymbol{w}) - \alpha_j$ and $\varsigma_j^{\mathrm{U}}(\boldsymbol{w}) - \beta_j$ respectively, δ_{1j}^{+}, δ_{1j}^{-}, δ_{2j}^{+} and δ_{2j}^{-} are, respectively, the weighting coefficients of ε_{1j}^{+}, ε_{1j}^{-}, ε_{2j}^{+} and ε_{2j}^{-}.

Utilizing the objective simplex method to solve the model (M-4.12), we can get the attribute weight vector $\boldsymbol{w}$, and then by (4.11), we get the uncertain overall attribute values $\tilde{z}_j(\boldsymbol{w})$ $(j = 1, 2, \cdots, n)$. After doing so, we utilize (2.69) to compare each pair of $\tilde{z}_j(\boldsymbol{w})$ $(j = 1, 2, \cdots, n)$, and construct a possibility degree matrix $\boldsymbol{P} = (p_{ij})_{n\times n}$, whose priority vector $\boldsymbol{\zeta} = (\zeta_1, \zeta_2, \cdots, \zeta_n)^{\mathrm{T}}$ can derived from (2.82). Then we use ζ_i $(i = 1, 2, \cdots, n)$ to rank the uncertain overall attribute values $\tilde{z}_j(\boldsymbol{w})$ $(j = 1, 2, \cdots, n)$, by which we rank and select the alternatives x_j $(j = 1, 2, \cdots, n)$.

Example 4.7 (Xu, 2006a) In Example 4.6, after the statistical processing, assume that the decision information on each candidate with respect to the attributes G_i $(i = 1, 2, \cdots, 6)$ is described in the uncertain linguistic decision matrix $\tilde{\boldsymbol{R}}$ (Table 4.16), and the decision maker provides his/her uncertain preferences over the five candidates x_j $(j = 1, 2, \cdots, 5)$:

$$\tilde{\pi}_1 = [s_2, s_3], \quad \tilde{\pi}_2 = [s_1, s_3], \quad \tilde{\pi}_3 = [s_2, s_3], \quad \tilde{\pi}_4 = [s_3, s_4], \quad \tilde{\pi}_5 = [s_2, s_4]$$

Table 4.16 Linguistic decision matrix $\tilde{R}$ (Xu, 2006a)

	x_1	x_2	x_3	x_4	x_5
G_1	$[s_2, s_3]$	$[s_3, s_4]$	$[s_2, s_3]$	$[s_3, s_4]$	$[s_2, s_3]$
G_2	$[s_0, s_1]$	$[s_0, s_2]$	$[s_2, s_4]$	$[s_1, s_3]$	$[s_3, s_4]$
G_3	$[s_{-1}, s_0]$	$[s_0, s_1]$	$[s_1, s_3]$	$[s_3, s_4]$	$[s_2, s_4]$
G_4	$[s_2, s_4]$	$[s_2, s_3]$	$[s_1, s_2]$	$[s_2, s_4]$	$[s_2, s_3]$
G_5	$[s_1, s_2]$	$[s_3, s_4]$	$[s_3, s_4]$	$[s_0, s_2]$	$[s_0, s_1]$
G_6	$[s_3, s_4]$	$[s_1, s_2]$	$[s_0, s_1]$	$[s_3, s_4]$	$[s_2, s_3]$

Based on the known decision information, we utilize the model (M-4.12) to establish the following goal programming model (suppose that $\delta_{1j}^+ = \delta_{1j}^- = \delta_{2j}^+ = \delta_{2j}^- = 1\ (j = 1, 2, \cdots, 5)$):

$$
\begin{aligned}
\min \quad & J' = \sum_{j=1}^{5} [(\varepsilon_{1j}^+ + \varepsilon_{1j}^-) + (\varepsilon_{2j}^+ + \varepsilon_{2j}^-)] \\
\text{s.t.} \quad & 2w_1 - w_3 + 2w_4 + w_5 + 3w_6 - 2 - e_{11}^+ + e_{11}^- = 0 \\
& 3w_1 + w_2 + 4w_4 + 2w_5 + 4w_6 - 3 - e_{21}^+ + e_{21}^- = 0 \\
& 3w_1 + 2w_4 + 3w_5 + w_6 - 1 - e_{12}^+ + e_{12}^- = 0 \\
& 4w_1 + 2w_2 + w_3 + 3w_4 + 4w_5 + 2w_6 - 3 - e_{22}^+ + e_{22}^- = 0 \\
& 2w_1 - 2w_2 + w_3 + w_4 + 3w_5 - 2 - e_{13}^+ + e_{13}^- = 0 \\
& 3w_1 + 4w_2 + 3w_3 + 2w_4 + 4w_5 + w_6 - 3 - e_{23}^+ + e_{23}^- = 0 \\
& 3w_1 + w_2 + w_3 + 2w_4 + 3w_6 - 3 - e_{14}^+ + e_{14}^- = 0 \\
& 4w_1 + 3w_2 + 4w_3 + 4w_4 + 2w_5 + 4w_6 - 4 - e_{24}^+ + e_{24}^- = 0 \\
& 2w_1 + 3w_2 + 2w_3 + 2w_4 + 2w_6 - 2 - e_{15}^+ + e_{15}^- = 0 \\
& 3w_1 + 4w_2 + 4w_3 + 3w_4 + w_5 + 3w_6 - 4 - e_{25}^+ + e_{25}^- = 0 \\
& \varepsilon_{1j}^+ \geqslant 0,\ \ \varepsilon_{1j}^- \geqslant 0,\ \ j = 1, 2, \cdots, 5 \\
& \varepsilon_{2j}^+ \geqslant 0,\ \ \varepsilon_{2j}^- \geqslant 0,\ \ j = 1, 2, \cdots, 5 \\
& \varepsilon_{1j}^+ \varepsilon_{1j}^- = 0,\ \ \varepsilon_{2j}^+ \varepsilon_{2j}^- = 0,\ \ j = 1, 2 \cdots, 5 \\
& 0.18 \leqslant w_1 \leqslant 0.22,\ \ 0.11 \leqslant w_2 \leqslant 0.20 \\
& 0.15 \leqslant w_3 \leqslant 0.19,\ \ 0.10 \leqslant w_4 \leqslant 0.14 \\
& 0.16 \leqslant w_5 \leqslant 0.25,\ \ 0.20 \leqslant w_6 \leqslant 0.23 \\
& w_1 + w_2 + w_3 + w_4 + w_5 = 1
\end{aligned}
$$

Solving the model, we get the attribute weight vector as

$$\boldsymbol{w} = (0.22, 0.11, 0.15, 0.13, 0.16, 0.23)^{\mathrm{T}}$$

and

$$\varepsilon_{11}^+ = 0, \quad \varepsilon_{11}^- = 0.6, \quad \varepsilon_{21}^+ = 0, \quad \varepsilon_{21}^- = 0.47, \quad \varepsilon_{12}^+ = 0.63$$

$$\varepsilon_{12}^- = 0, \quad \varepsilon_{22}^+ = 0, \quad \varepsilon_{22}^- = 0.26, \quad \varepsilon_{13}^+ = 0, \quad \varepsilon_{13}^- = 0.58$$

$$\varepsilon_{23}^+ = 0, \quad \varepsilon_{23}^- = 0.32, \quad \varepsilon_{14}^+ = 0, \quad \varepsilon_{14}^- = 0.83, \quad \varepsilon_{24}^+ = 0$$

$$\varepsilon_{24}^- = 0.43, \quad \varepsilon_{15}^+ = 0, \quad \varepsilon_{15}^- = 0.21, \quad \varepsilon_{25}^+ = 0, \quad \varepsilon_{25}^- = 1.06$$

Then we utilize (4.11) to get the uncertain overall attribute values $\tilde{z}_j(\boldsymbol{w})\ (j = 1, 2, \cdots, 5)$ of the alternatives $x_j\ (j = 1, 2, \cdots, 5)$:

$$\tilde{z}_1(\boldsymbol{w}) = [s_{1.40}, s_{2.53}], \quad \tilde{z}_2(\boldsymbol{w}) = [s_{1.63}, s_{2.74}], \quad \tilde{z}_3(\boldsymbol{w}) = [s_{1.42}, s_{2.67}]$$

$$\tilde{z}_4(\boldsymbol{w}) = [s_{2.17}, s_{3.57}], \quad \tilde{z}_5(\boldsymbol{w}) = [s_{1.79}, s_{2.94}]$$

According to (2.69), we compare each pair of $\tilde{z}_j(\boldsymbol{w})\ (j = 1, 2, \cdots, 5)$, and construct a possibility degree matrix:

$$\boldsymbol{P}=\begin{pmatrix} 0.5 & 0.4018 & 0.4644 & 0.1423 & 0.3246 \\ 0.5982 & 0.5 & 0.5570 & 0.2271 & 0.4204 \\ 0.5356 & 0.4430 & 0.5 & 0.1917 & 0.3693 \\ 0.8577 & 0.7729 & 0.8083 & 0.5 & 0.6980 \\ 0.6754 & 0.5796 & 0.6307 & 0.3020 & 0.5 \end{pmatrix}$$

and derive the priority vector of $\boldsymbol{P}$ by using (2.82):

$$\boldsymbol{\zeta}=(0.1667, 0.1901, 0.1770, 0.2568, 0.2094)^{\mathrm{T}}$$

from which we rank $\tilde{z}_j(\boldsymbol{w})$ $(j=1,2,\cdots,5)$ in descending order:

$$\tilde{z}_4(\boldsymbol{w}) > \tilde{z}_5(\boldsymbol{w}) > \tilde{z}_2(\boldsymbol{w}) > \tilde{z}_3(\boldsymbol{w}) > \tilde{z}_1(\boldsymbol{w})$$

and thus, the ranking of the alternatives x_j $(j=1,2,\cdots,5)$ is

$$x_4 \succ x_5 \succ x_3 \succ x_1 \succ x_2$$

4.5 Multi-Attribute Decision Making Based on Different Types of Linguistic Information

Xu (2006b) investigated the multi-attibute decision making problems where both the linguistic labels and uncertain linguistic variables are contained in a decision matrix:

Let X, G and $\boldsymbol{w}$ be defined as in Section 4.2, and $\boldsymbol{R}=(r_{ij})_{m\times n}$ be a hybrid linguistic decision matrix with different types of linguistic information, where r_{ij} is an attribute value provided by the decision maker for the alternative $x_j \in X$ with respect to the attribute $G_i \in G$ by using the additive linguistic evaluation scale $S_2=\{s_\alpha|\alpha=-\tau,\cdots,-1,0,1,\cdots,\tau\}$. Without loss of generality, suppose that $r_{ij}\in S_2$ $(i=1,2,\cdots,l,\ j=1,2,\cdots,n)$, $r_{ij}=[r_{ij}^{\mathrm{L}}, r_{ij}^{\mathrm{U}}]$, $r_{ij}^{\mathrm{L}}, r_{ij}^{\mathrm{U}}\in S_2$ $(i=l+1,\cdots,m,\ j=1,2,\cdots,n)$, and $\boldsymbol{r}_j=(r_{1j}, r_{2j},\cdots,r_{mj})^{\mathrm{T}}$ $(j=1,2,\cdots,n)$ is the vector of attribute values of the alternatives x_j $(j=1,2,\cdots,n)$.

Definition 4.9 (Xu, 2006b) Let $\boldsymbol{R}=(r_{ij})_{m\times n}$ be a hybrid linguistic decision matrix. Then $\boldsymbol{r}^+=(r_1^+, r_2^+,\cdots,r_m^+)^{\mathrm{T}}$ is called a relative ideal point, where $r_i^+=\max\limits_j\{r_{ij}\}$ $(i=1,2,\cdots,l)$, $r_i^+=[r_i^{+\mathrm{L}}, r_i^{+\mathrm{U}}]=[\max\limits_j\{r_{ij}^{\mathrm{L}}\}, \max\limits_j\{r_{ij}^{\mathrm{U}}\}]$ $(i=l+1,\cdots,m)$.

Definition 4.10(Xu, 2006b) Let $\boldsymbol{R}=(r_{ij})_{m\times n}$ be a hybrid linguistic decision matrix. Then $\boldsymbol{r}^-=(r_1^-, r_2^-,\cdots,r_m^-)^{\mathrm{T}}$ is called a relative negative ideal point, where $r_i^-=\min\limits_j\{r_{ij}\}$ $(i=1,2,\cdots,l)$, $r_i^-=[r_i^{-\mathrm{L}}, r_i^{-\mathrm{U}}]=[\min\limits_j\{r_{ij}^{\mathrm{L}}\}, \min\limits_j\{r_{ij}^{\mathrm{U}}\}]$ $(i=l+1,\cdots,m)$.

Based on the relative ideal point and the relative negative ideal point, Xu (2006b) gave a simple approach to hybrid linguistic multi-attribute decision making:

Step 1 Based on (3.17) and (4.3), we calculate the distance between the alternative x_j and the relative ideal point $\boldsymbol{r}^+$, and the distance between the alternative x_j and the relative negative ideal point $\boldsymbol{r}^-$:

$$d(\boldsymbol{r}_j, \boldsymbol{r}^+) = \sum_{i=1}^{m} w_i d(r_{ij}, r_i^+), \quad j = 1, 2, \cdots, n \tag{4.42}$$

$$d(\boldsymbol{r}_j, \boldsymbol{r}^-) = \sum_{i=1}^{m} w_i d(r_{ij}, r_i^-), \quad j = 1, 2, \cdots, n \tag{4.43}$$

Step 2 Calculate the close degrees of each of the alternative x_j $(j = 1, 2, \cdots, n)$ to the relative ideal point $\boldsymbol{r}^+$ and the relative negative ideal point $\boldsymbol{r}^+$:

$$c_j = \frac{d(\boldsymbol{r}_j, \boldsymbol{r}^-)}{d(\boldsymbol{r}_j, \boldsymbol{r}^+) + d(\boldsymbol{r}_j, \boldsymbol{r}^-)}, \quad j = 1, 2, \cdots, n \tag{4.44}$$

Clearly, the bigger the close degree c_j, the better the alternative x_j.

Step 3 Rank and select the alternatives x_j $(j = 1, 2, \cdots, n)$ according to the close degrees c_j $(j = 1, 2, \cdots, n)$.

Example 4.8(Xu, 2006b) Let us consider a customer who intends to buy a car (adapted from Herrera and Martínez (2000). Four types of cars (alternatives) x_j $(j = 1, 2, 3, 4)$ are available. The customer takes into account four attributes to decide which car to buy: ① G_1——fuel economy; ② G_2——comfort; ③ G_3——design; and ④ G_4——safety. The weight vector of the attributes is $\boldsymbol{w} = (0.2, 0.4, 0.1, 0.3)^{\mathrm{T}}$. The decision maker uses the additive linguistic evaluation scale:

$$S_2 = \{s_{-4} = \text{extremely poor},\ s_{-3} = \text{very poor},\ s_{-2} = \text{poor},\ s_{-1} = \text{slightly poor},$$
$$s_0 = \text{fair},\ s_1 = \text{slightly good},\ s_2 = \text{good},\ s_3 = \text{very good},\ s_4 = \text{extremely good}\}$$

to evaluate the alternatives x_j $(j = 1, 2, 3, 4)$ with respect to the attributes G_i $(i = 1, 2, 3, 4)$, and constructs a hybrid linguistic decision matrix $\boldsymbol{R} = (r_{ij})_{m \times n}$ (Table 4.17).

Table 4.17 Hybrid linguistic decision matrix R (Xu, 2006i)

	x_1	x_2	x_3	x_4
G_1	s_{-2}	s_1	s_4	s_3
G_2	s_1	s_3	s_3	s_2
G_3	$[s_3, s_4]$	$[s_{-1}, s_1]$	$[s_{-2}, s_0]$	$[s_0, s_1]$
G_4	$[s_2, s_3]$	$[s_1, s_2]$	$[s_0, s_1]$	$[s_2, s_4]$

In the following, we utilize the approach above to select the desirable alternative:

Step 1 Based on Table 4.17, we obtain the vectors of attribute values of the alternatives x_j $(j = 1, 2, 3, 4)$, and the relative ideal point $\boldsymbol{r}^+$ and the relative negative ideal point $\boldsymbol{r}^-$:

$$\boldsymbol{r}_1 = (s_{-2}, s_1, [s_3, s_4], [s_2, s_3])^{\mathrm{T}}, \quad \boldsymbol{r}_2 = (s_1, s_3, [s_{-1}, s_1], [s_1, s_2])^{\mathrm{T}}$$

$$\boldsymbol{r}_3 = (s_4, s_3, [s_{-2}, s_0], [s_0, s_1])^{\mathrm{T}}, \quad \boldsymbol{r}_4 = (s_3, s_2, [s_0, s_1], [s_2, s_4])^{\mathrm{T}}$$

$$\boldsymbol{r}^+ = (s_4, s_3, [s_3, s_4], [s_2, s_4])^{\mathrm{T}}, \quad \boldsymbol{r}^- = (s_{-2}, s_1, [s_{-2}, s_0], [s_0, s_1])^{\mathrm{T}}$$

Step 2 Use (4.42) to calculate the distance between each alternative and the relative ideal point:

$$d(\boldsymbol{r}_1, \boldsymbol{r}^+) = 0.269, \quad d(\boldsymbol{r}_2, \boldsymbol{r}^+) = 0.175, \quad d(\boldsymbol{r}_3, \boldsymbol{r}^+) = 0.150, \quad d(\boldsymbol{r}_4, \boldsymbol{r}^+) = 0.131$$

Step 3 Calculate the distance between each alternative and the negative ideal point by using (4.43):

$$d(\boldsymbol{r}_1, \boldsymbol{r}^-) = 0.131, \quad d(\boldsymbol{r}_2, \boldsymbol{r}^-) = 0.275, \quad d(\boldsymbol{r}_3, \boldsymbol{r}^-) = 0.250, \quad d(\boldsymbol{r}_4, \boldsymbol{r}^-) = 0.288$$

Step 4 Calculate the close degree between each alternative and the relative ideal point and the relative negative ideal point by using (4.44):

$$c_1 = 0.328, \quad c_2 = 0.611, \quad c_3 = 0.627, \quad c_4 = 0.687$$

Step 5 Rank the alternatives x_j $(j = 1, 2, 3, 4)$ according to the close degree c_j $(j = 1, 2, 3, 4)$:

$$x_4 \succ x_3 \succ x_2 \succ x_1$$

and thus, the best alternative is x_4.

4.6 Interactive Approach to Linguistic Multi-Attribute Decision Making

In many practical problems, the opinion or preferences of the decision maker sometimes cannot be provided completely one time, and thus, the decision maker needs to supplement and revise his/her opinion and preferences in the process of decision making. Xu (2007b) investigated the interactive linguistic multi-attribute decision making problem, which can be described as follows:

The decision maker evaluates each alternative $x_j \in X$ with respect to the attribute $G_i \in G$ by using the additive linguistic evaluation scale $\{s_\alpha | \alpha = -\tau, \cdots, -1, 0, 1, \cdots, \tau\}$, and constructs a linguistic decision matrix $\boldsymbol{R} = (r_{ij})_{m\times n}$, $\boldsymbol{r}_j = (r_{1j}, r_{2j}, \cdots, r_{mj})^{\mathrm{T}}$ $(j = 1, 2, \cdots, n)$ are the vector of the attribute values of the alternatives x_j $(j = 1, 2, \cdots, n)$.

Definition 4.11 (Xu, 2007b) Let $r_i^+ = s_\tau$ $(i = 1, 2, \cdots, m)$. Then $\boldsymbol{r}^+ = (r_1^+, r_2^+, \cdots, r_m^+)^{\mathrm{T}}$ is called an absolute ideal point of alternatives.

Definition 4.12 (Xu, 2007b) Let $r_i^- = s_{-\tau}$ $(i = 1, 2, \cdots, m)$. Then $\boldsymbol{r}^- = (r_1^-, r_2^-, \cdots, r_m^-)^{\mathrm{T}}$ is called an absolute negative ideal point of alternatives.

Based on the linguistic decision matrix $\boldsymbol{R} = (r_{ij})_{m\times n}$, we use (4.2) to derive the overall attribute value corresponding to the absolute ideal point:

$$z^+(\boldsymbol{w}) = w_1 s_\tau \oplus w_2 s_\tau \oplus \cdots \oplus w_m s_\tau$$

$$= (w_1 + w_2 + \cdots + w_m)\, s_\tau$$
$$= s_\tau \tag{4.45}$$

and the overall attribute value corresponding to the absolute negative ideal point:

$$\begin{aligned} z^-(\boldsymbol{w}) &= w_1 s_{-\tau} \oplus w_2 s_{-\tau} \oplus \cdots \oplus w_m s_{-\tau} \\ &= (w_1 + w_2 + \cdots + w_m)\, s_{-\tau} \\ &= s_{-\tau} \end{aligned} \tag{4.46}$$

Let $s_\alpha, s_\beta \in S_2$ be any two linguistic labels. Then

$$d(s_\alpha, s_\beta) = |\alpha - \beta| \tag{4.47}$$

is called the distance between s_α and s_β (Xu, 2007b).

Let $d(z_j(\boldsymbol{w}), z^-(\boldsymbol{w}))$ be the distance between the overall attribute value $z_j(\boldsymbol{w})$ of the alternative x_j and the overall attribute value $z^-(\boldsymbol{w})$ of the absolute negative ideal point r^-. Then the greater $d(z_j(\boldsymbol{w}), z^-(\boldsymbol{w}))$, the better the alternative x_j.

Definition 1.13 (Xu, 2007b) Let $d(z^+(\boldsymbol{w}), z^-(\boldsymbol{w}))$ be the distance between the overall attribute value $z^+(\boldsymbol{w})$ of the absolute ideal point $\boldsymbol{r}^+$ and the overall attribute value $z^-(\boldsymbol{w})$ of the absolute negative ideal point $\boldsymbol{r}^-$. Then

$$\mu(z_j(\boldsymbol{w})) = \frac{d(z_j(\boldsymbol{w}), z^-(\boldsymbol{w}))}{d(z^+(\boldsymbol{w}), z^-(\boldsymbol{w}))} \tag{4.48}$$

is called the satisfaction degree of the alternative x_j.

It can be seen from Definition 1.13 that the satisfaction degree $\mu(z_j(\boldsymbol{w}))$ of the alternative x_j is the ratio of the distance between the overall attribute value $z_j(\boldsymbol{w})$ of the alternative x_j and the overall attribute value $z^-(\boldsymbol{w})$ of the absolute negative ideal point $\boldsymbol{r}^-$ to the distance between the overall attribute value $z^+(\boldsymbol{w})$ of the absolute ideal point $\boldsymbol{r}^+$ and the overall attribute value $z^-(\boldsymbol{w})$ of the absolute negative ideal point $\boldsymbol{r}^-$.

Since

$$d(z^+(\boldsymbol{w}), z^-(\boldsymbol{w})) = d(s_\tau, s_{-\tau}) = |\tau - (-\tau)| = 2\tau \tag{4.49}$$

then (4.48) can be rewritten as

$$\mu(z_j(\boldsymbol{w})) = \frac{1}{2t} d(z_j(\boldsymbol{w}), z^-(\boldsymbol{w})) \tag{4.50}$$

Clearly, the greater the distance between the overall attribute value $z_j(\boldsymbol{w})$ of the alternative x_j and the overall attribute value $z^-(\boldsymbol{w})$ of the absolute negative ideal point $\boldsymbol{r}^-$, the higher the satisfaction degree $\mu(z_j(\boldsymbol{w}))$ of the alternative x_j is. That is, the satisfactory degree $\mu(z_j(\boldsymbol{w}))$ of the alternative x_j is a strictly monotone increasing function with respect to $d(z_j(\boldsymbol{w}), z^-(\boldsymbol{w}))$. Therefore, the higher the satisfaction degree $\mu(z_j(\boldsymbol{w}))$, the better the alternative x_j is. As a result, we establish the following optimization model (Xu, 2007b):

(M-4.13)
$$\begin{aligned} \max \quad & \mu(\boldsymbol{w}) = (\mu(z_1(\boldsymbol{w})),\ \mu(z_2(\boldsymbol{w})), \cdots, \mu(z_m(\boldsymbol{w}))) \\ \text{s.t.} \quad & \boldsymbol{w} \in H \\ & w_i \geqslant 0,\ i = 1, 2, \cdots, m,\ \sum_{i=1}^{m} w_i = 1 \end{aligned}$$

We utilize the max-min operator proposed by Zimmermann and Zysno (1980) to integrate the satisfaction degrees of all alternatives, i.e., we get the following optimization model (Xu, 2007b):

(M-4.14)
$$\begin{aligned} \max \quad & \lambda_0 \\ \text{s.t.} \quad & \mu(z_j(\boldsymbol{w})) \geqslant \lambda_0,\ \ j = 1, 2, \cdots, n \\ & \boldsymbol{w} \in H \\ & w_i \geqslant 0,\ i = 1, 2, \cdots, m,\ \sum_{i=1}^{m} w_i = 1 \end{aligned}$$

where

$$\lambda_0 = \min_j \mu(z_j(\boldsymbol{w})) \tag{4.51}$$

By solving the model (M-4.14), we can get the original optimal solution $\boldsymbol{w}^{(0)} = (w_1^{(0)},\ w_2^{(0)},\ \cdots, w_m^{(0)})^{\mathrm{T}}$, and then calculate the satisfaction degrees $\mu(z_j(\boldsymbol{w}^{(0)}))$ $(j = 1, 2, \cdots, n)$ of the alternatives x_j $(j = 1, 2, \cdots, n)$. In the process of decision making, the decision maker provides the lower bounds $\lambda_j^{(0)}$ $(j = 1, 2, \cdots, n)$ of the satisfaction degrees of the alternatives x_j $(j = 1, 2, \cdots, n)$ according to $\mu(z_j(\boldsymbol{w}^{(0)}))$ $(j = 1, 2, \cdots, n)$. Then, we establish the following optimization model (Xu, 2007b):

(M-4.15)
$$\begin{aligned} \max \quad & J = \sum_{j=1}^{n} \lambda_j \\ \text{s.t.} \quad & \mu(z_j(\boldsymbol{w})) \geqslant \lambda_j \geqslant \lambda_j^{(0)},\ \ \ j = 1, 2, \cdots, n \\ & \boldsymbol{w} \in H \\ & w_i \geqslant 0,\ i = 1, 2, \cdots, m,\ \sum_{i=1}^{m} w_i = 1 \end{aligned}$$

Solving the model (M-4.15), if there exists no optimal solution, then the decision maker needs to reconsider the lower bounds $\lambda_j^{(0)}$ $(j = 12, \cdots, n)$ of the satisfaction degrees of the alternatives x_j $(j = 1, 2, \cdots, n)$ till the optimal solution is obtained.

Theorem 4.1 (Xu, 2007b) The optimal solution of the model (M-4.15) is the Pareto solution of the model (M-4.13).

Proof Suppose that $\boldsymbol{w}^*$ is the optimal solution of the model (M-4.15), and $\boldsymbol{w}^*$ is not the Pareto solution of the model (M-4.13), then there exists $\boldsymbol{w}'$ such that $\mu(z_j(\boldsymbol{w}')) \geqslant \mu(z_j(\boldsymbol{w}^*))$ for any $x_j \in X$, and there exists $x_{j_0} \in X$, such that $\mu(z_{j_0}(\boldsymbol{w}')) > \mu(z_{j_0}(\boldsymbol{w}^*))$, then $\mu(z_j(\boldsymbol{w}')) \geqslant \lambda_j \geqslant \lambda_j^{(0)}$ for any $x_j \in X$, and there exists λ'_{j_0}, such that $\mu(z_{j_0}(\boldsymbol{w}')) \geqslant \lambda'_{j_0} > \lambda_{j_0} \geqslant \lambda_j^{(0)}$. Therefore,

$$\sum_{j=1,j\neq j_0}^{n} \lambda_j + \lambda'_{j_0} > \sum_{j=1}^{n} \lambda_j \tag{4.52}$$

which contradicts the supposition condition. This completes the proof.

Based on the above models and Theorem 4.1, we give a practical interactive procedure for linguistic multi-attribute decision making problems considered. The procedure is described as follows (Xu, 2007b):

Step 1 Use the model (M-4.14) to solve the original optimal solution $\boldsymbol{w}^{(0)} = (w_1^{(0)}, w_2^{(0)}, \cdots, w_m^{(0)})^{\mathrm{T}}$, and calculate the satisfactory degree $\mu(z_j(\boldsymbol{w}^{(0)}))$ $(j=1,2,\cdots,n)$ of the alternatives x_j $(j = 1,2,\cdots,n)$. The decision maker gives the lower bounds $\lambda_j^{(0)}$ $(j = 1,2,\cdots,n)$ of the satisfaction degrees $\mu(z_j(\boldsymbol{w}^{(0)}))$ $(j = 1,2,\cdots,n)$ of the alternatives x_j $(j = 1,2,\cdots,n)$. Let $k = 1$.

Step 2 Utilize the model (M-4.15) to derive the attribute weight vector $\boldsymbol{w}^{(k)} = (w_1^{(k)}, w_2^{(k)}, \cdots, w_m^{(k)})^{\mathrm{T}}$, and calculate the satisfaction degrees $\mu(z_j(\boldsymbol{w}^{(k)}))$ $(j = 1,2,\cdots,n)$ of the alternatives x_j $(j = 1,2,\cdots,n)$.

Step 3 If the decision maker is satisfied with the result obtained by Step 2, then we calculate the overall values $z_j(\boldsymbol{w})$ $(j = 1,2,\cdots,n)$ of the alternatives x_j $(j = 1,2,\cdots,n)$ by using (4.2), and rank all alternatives according to the values of $z_j(\boldsymbol{w})$ $(j = 1,2,\cdots,n)$; if there exists no solution for the model (M-4.15) or the result does not satisfy the decision maker, then the decision maker should increase the satisfaction degrees of some alternatives, and decrease the satisfaction degrees of some other alternatives. Let $k = k+1$, and return to Step 2.

Example 4.9 (Xu, 2007b) A multi-attribute decision making problem of evaluating university faculty for tenure and promotion (adapted from Bryson and Mobolurin (1995)) is used to illustrate the developed procedure. A practical use of the developed procedure involves the evaluation of university faculty for tenure and promotion. The criteria (attributes) used at some universities are: ① G_1——teaching; ② G_2——research; and ③ G_3——service. Five faculty candidates (alternatives) x_j $(j = 1,2,\cdots,5)$ are to be evaluated using the additive linguistic evaluation scale:

$$S_2=\{s_{-4}=\text{extremely poor},\ s_{-3}=\text{very poor},\ s_{-2}=\text{poor},\ s_{-1}=\text{slightly poor},\ s_0=\text{fair},\ s_1=\text{slightly good},\ s_2=\text{good},\ s_3=\text{very good},\ s_4=\text{extremely good}\}$$

by the decision maker under these three attributes, as listed in Table 4.18:

Table 4.18 Linguistic decision matrix R (Xu, 2007b)

	x_1	x_2	x_3	x_4	x_5
G_1	s_2	s_3	s_0	s_4	s_2
G_2	s_3	s_4	s_3	s_{-1}	s_3
G_3	s_0	s_{-2}	s_2	s_2	s_1

Suppose that the known weight information is as follows:

$$H = \{0.25 \leqslant w_1 \leqslant 0.40, 0.15 \leqslant w_2 \leqslant 0.30, w_3 > w_2\}$$

To get the most desirable alternative(s), the following steps are involved:

Step 1 Use the model (M-4.15) to establish the following optimization model:

$$\begin{aligned}
\max \quad & \lambda_0 \\
\text{s.t.} \quad & \frac{1}{8}(2w_1 + 3w_2 + 4) \geqslant \lambda_0 \\
& \frac{1}{8}(3w_1 + 4w_2 - 2w_3 + 4) \geqslant \lambda_0 \\
& \frac{1}{8}(3w_2 + 2w_3 + 4) \geqslant \lambda_0 \\
& \frac{1}{8}(4w_1 - w_2 + 2w_3 + 4) \geqslant \lambda_0 \\
& \tfrac{1}{8}(2w_1 + 3w_2 + w_3 + 4) \geqslant \lambda_0 \\
& 0.25 \leqslant w_1 \leqslant 0.40, \quad 0.15 \leqslant w_2 \leqslant 0.30, \quad w_3 > w_2 \\
& w_j \geqslant 0\,, \ \ j = 1, 2, 3, \ \ w_1 + w_2 + w_3 = 1
\end{aligned}$$

Solving the model, we get the optimal solution:

$$\boldsymbol{w}^{(0)} = (0.3751,\ 0.3000,\ 0.3429)^{\mathrm{T}}$$

and obtain the satisfaction degrees $\mu(z_j(\boldsymbol{w}^{(0)}))$ $(j = 1, 2, \cdots, 5)$ of the alternatives x_j $(j = 1, 2, \cdots, 5)$:

$$\mu(z_1(\boldsymbol{w}^{(0)})) = 0.7063, \quad \mu(z_2(\boldsymbol{w}^{(0)})) = 0.7049, \quad \mu(z_3(\boldsymbol{w}^{(0)})) = 0.6982$$

$$\mu(z_4(\boldsymbol{w}^{(0)})) = 0.7358, \quad \mu(z_5(\boldsymbol{w}^{(0)})) = 0.7491$$

The decision maker gives the lower bounds $\lambda_j^{(0)}$ $(j = 1, 2, \cdots, 5)$ of the satisfactory degrees of the alternatives x_j $(j = 1, 2, \cdots, 5)$ according to the satisfaction degrees:

$$\lambda_1^{(0)} = 0.7000, \quad \lambda_2^{(0)} = 0.7100, \quad \lambda_3^{(0)} = 0.6900$$

$$\lambda_4^{(0)} = 0.7300, \quad \lambda_5^{(0)} = 0.7300$$

Step 2 By the model (M-4.15), we establish the following optimization model:

$$\begin{aligned}
\max \quad & J = \sum_{j=1}^{5} \lambda_j \\
\text{s.t.} \quad & \frac{1}{8}(2w_1 + 3w_2 + 4) \geqslant \lambda_1 \geqslant 0.7000 \\
& \frac{1}{8}(3w_1 + 4w_2 - 2w_3 + 4) \geqslant \lambda_2 \geqslant 0.7100 \\
& \frac{1}{8}(3w_2 + 2w_3 + 4) \geqslant \lambda_3 \geqslant 0.6900 \\
& \frac{1}{8}(4w_1 - w_2 + 2w_3 + 4) \geqslant \lambda_4 \geqslant 0.7300 \\
& \frac{1}{8}(2w_1 + 3w_2 + w_3 + 4) \geqslant \lambda_5 \geqslant 0.7300 \\
& 0.25 \leqslant w_1 \leqslant 0.40, \quad 0.15 \leqslant w_2 \leqslant 0.30, \quad w_3 > w_2 \\
& w_j \geqslant 0\,, \ \ j = 1,2,3, \ \ w_1 + w_2 + w_3 = 1
\end{aligned}$$

Solving the model, we get the attribute weight vector $\boldsymbol{w}^{(1)} = (0.39,\ 0.30,\ 0.31)^{\mathrm{T}}$, and calculate the satisfaction degree $\mu(z_j(\boldsymbol{w}^{(1)}))$ $(j = 1, 2, \cdots, 5)$ of the alternatives x_j $(j = 1, 2, \cdots, 5)$:

$$\mu(z_1(\boldsymbol{w}^{(1)})) = 0.7100, \quad \mu(z_2(\boldsymbol{w}^{(1)})) = 0.7188, \quad \mu(z_3(\boldsymbol{w}^{(1)})) = 0.6900$$

$$\mu(z_4(\boldsymbol{w}^{(1)})) = 0.7350, \quad \mu(z_5(\boldsymbol{w}^{(1)})) = 0.7487$$

Step 3 The decision maker is satisfied with this result. Therefore, we can calculate the overall attribute values $z_j(\boldsymbol{w}^{(1)})$ $(j = 1, 2, \cdots, 5)$ of the alternatives x_j $(j = 1, 2, \cdots, 5)$ by using (4.2):

$$z_1(\boldsymbol{w}^{(1)}) = s_{1.68}, \quad z_2(\boldsymbol{w}^{(1)}) = s_{1.75}, \quad z_3(\boldsymbol{w}^{(1)}) = s_{1.52}$$

$$z_4(\boldsymbol{w}^{(1)}) = s_{1.88}, \quad z_5(\boldsymbol{w}^{(1)}) = s_{1.99}$$

and rank the alternatives x_j $(j = 1, 2, \cdots, 5)$ according to $z_j(\boldsymbol{w}^{(1)})$ $(j = 1, 2, \cdots, 5)$:

$$x_5 \succ x_4 \succ x_2 \succ x_1 \succ x_3$$

Hence, the most desirable alternative is x_5.

4.7 Multi-Period Multi-Attribute Decision Making under Linguistic Assessments

Multi-period (multi-stage) multi-attribute decision making occurs extensively in our daily life, such as multi-stage investment decision making, medical diagnosis, personnel dynamic examination, and military system efficiency dynamic evaluation, etc., the input arguments are usually collected from different periods (stages). In this section, we shall investigate these types of decision making problems, and apply the DLWA, LWA_2, DLWG and LWG operators to the field of multi-attribute decision making. Some approaches based on these operators to multi-period (multi-stage) multi-attribute decision making are introduced, and then extended to uncertain linguistic environments.

4.7.1 Multi-Period Multi-Attribute Decision Making Method Based on LWA_2 and DLWA Operators

Consider a multi-perid multi-attribute decision making problem, let $X = \{x_1, x_2, \cdots, x_n\}$ be a finite set of alternatives, $G = \{G_1, G_2, \cdots, G_m\}$ be the set of alternatives, and t_k $(k = 1, 2, \cdots, p)$ be p different periods, whose weight vector is $\boldsymbol{w}(t) = (w(t_1), w(t_2), \cdots, w(t_p))^{\mathrm{T}}$, where $w(t_k) \geqslant 0$ $(k = 1, 2, \cdots, p)$, $\sum_{k=1}^{p} w(t_k) = 1$, and $\boldsymbol{\omega}(t_k) = (\omega_1(t_k), \omega_2(t_k), \cdots, \omega_p(t_k))^{\mathrm{T}}$ $(k = 1, 2, \cdots, p)$ are the weight vectors of attributes at the periods t_k $(k = 1, 2, \cdots, p)$, where $\omega_i(t_k) \geqslant 0$ $(i = 1, 2, \cdots, m)$ and $\sum_{i=1}^{m} \omega_i(t_k) = 1$ $(k = 1, 2, \cdots, p)$. Let $\boldsymbol{R}(t_k) = (r_{ij}(t_k))_{m \times n}$ $(k = 1, 2, \cdots, p)$ be the linguistic decision matrices of p different periods, where $r_{ij}(t_k)$ denotes the attribute value provided by the decision maker over the alternative x_j with respect to the attribute G_i at the period t_k by using the additive linguistic evaluation scale $S_2 = \{s_\alpha | \alpha = -\tau, \cdots, -1, 0, 1, \cdots, \tau\}$.

Based on the decision information described above, in the following, we use the LWA_2 and DLWA operators to develop a simple approach to multi-period multi-attibute decision making:

Step 1 Use the LWA_2 operator:

$$\begin{aligned} z_j(t_k) &= \mathrm{LWA}_2(r_{1j}(t_k), r_{2j}(t_k), \cdots, r_{mj}(t_k)) \\ &= \omega_1(t_k) r_{1j}(t_k) \oplus \omega_2(t_k) r_{2j}(t_k) \oplus \cdots \oplus \omega_m(t_k) r_{mj}(t_k), \\ &\qquad k = 1, 2, \cdots, p, \; j = 1, 2, \cdots, n \end{aligned} \tag{4.53}$$

to derive the overall attribute values of the alternative x_j at the period t_k.

Step 2 Utilize the DLWA operator:

$$\begin{aligned} z_j &= \mathrm{DLWA}_{\boldsymbol{w}(t)}(z_j(t_1), z_j(t_2), \cdots, z_j(t_p)) \\ &= w(t_1) z_j(t_1) \oplus w(t_2) z_j(t_2) \oplus \cdots \oplus w(t_p) z_j(t_p), \quad j = 1, 2, \cdots, n \end{aligned} \tag{4.54}$$

to aggregate the overall attribute values $z_j(t_k)$ $(k = 1, 2, \cdots, p)$ collected from p different periods, and get the overall attribute value z_j of the alternative x_j.

Step 3 Rank the alternatives x_j $(j = 1, 2, \cdots, n)$ according to the overall attribute values z_j $(j = 1, 2, \cdots, n)$.

Example 4.10 The evaluation of socio-economic systems, such as the evaluation of investment environment, the effectiveness of reform measures and urban planning program, etc., involves the political, economic, technological, ecological and cultural aspects. Considering the complexity of this type of decision making problems, the decision information provided by the decision maker is usually uncertain or fuzzy. Here, we consider an investment decision making problem: There are five cities x_j $(j = 1, 2, \cdots, 5)$, the evaluation indices (attributes) used to evaluate the cities are as follows (Li et al., 2001): ① G_1—political environment; ② G_2—economic environment; ③ G_3—financial environment; ④ G_4—administrative environment; ⑤ G_5—market environment; ⑥ G_6—technical condition; ⑦ G_7—material basis; ⑧ G_8—legal environment; and ⑨ G_9—natural environment. The decision maker employs the additive linguistic evaluation scale:

$$S_2=\{s_{-4}=\text{extremely poor},\ s_{-3}=\text{very poor},\ s_{-2}=\text{poor},\ s_{-1}=\text{slightly poor},$$
$$s_0=\text{fair},\ s_1=\text{slightly good},\ s_2=\text{good},\ s_3=\text{very good},\ s_4=\text{extremely good}\}$$

to evaluate the investment environments of the cities x_j $(j = 1, 2, \cdots, 5)$ during the periods 2004~2006 according to the the evaluation indices G_i $(i = 1, 2, \cdots, 9)$, and constructs the linguistic decision matrices $R(t_k)$ $(k = 1, 2, 3)$ (Tables 4.19~4.21) respectively (here, t_1 denotes "2004", t_2 denotes "2005", and t_3 denotes "2006").

Table 4.19 Linguistic decision matrix $R(t_1)$

	x_1	x_2	x_3	x_4	x_5
G_1	s_1	s_0	s_1	s_0	s_1
G_2	s_2	s_{-1}	s_0	s_2	s_0
G_3	s_1	s_2	s_2	s_1	s_3
G_4	s_2	s_1	s_1	s_2	s_1
G_5	s_1	s_2	s_3	s_1	s_2
G_6	s_2	s_2	s_0	s_1	s_1
G_7	s_1	s_2	s_2	s_4	s_0
G_8	s_2	s_3	s_2	s_2	s_2
G_9	s_2	s_1	s_2	s_1	s_3

Table 4.20 Linguistic decision matrix $R(t_2)$

	x_1	x_2	x_3	x_4	x_5
G_1	s_2	s_1	s_2	s_{-1}	s_0
G_2	s_1	s_0	s_0	s_1	s_1
G_3	s_2	s_1	s_1	s_2	s_2
G_4	s_2	s_0	s_2	s_3	s_2
G_5	s_0	s_1	s_3	s_2	s_3
G_6	s_2	s_2	s_1	s_1	s_2
G_7	s_3	s_3	s_2	s_2	s_1
G_8	s_2	s_2	s_1	s_1	s_3
G_9	s_1	s_1	s_2	s_2	s_1

Table 4.21 Linguistic decision matrix $R(t_3)$

	x_1	x_2	x_3	x_4	x_5
G_1	s_2	s_3	s_1	s_1	s_1
G_2	s_2	s_1	s_2	s_3	s_1
G_3	s_2	s_3	s_2	s_2	s_4
G_4	s_3	s_1	s_1	s_3	s_2
G_5	s_1	s_3	s_3	s_2	s_4
G_6	s_3	s_3	s_2	s_2	s_2
G_7	s_2	s_3	s_3	s_4	s_1
G_8	s_3	s_4	s_3	s_3	s_2
G_9	s_3	s_2	s_2	s_1	s_2

The weight vector of $t_k\ (k=1,2,3)$ is $\boldsymbol{\omega}(t) = (0.2302,\ 0.3213,\ 0.4485)^{\mathrm{T}}$ (which is derived by the BUM function introduced in Definition 2.28), the weight vectors of the evaluation indices $G_i\ (i=1,2,\cdots,9)$ at the periods $t_k\ (k=1,2,3)$ are $\boldsymbol{w}(t_k)\ (k=1,2,3)$ respectively:

$$\boldsymbol{w}(t_1) = (0.12,\ 0.15,\ 0.10, 0.08, 0.10, 0.14, 0.13, 0.05, 0.13)^{\mathrm{T}}$$

$$\boldsymbol{w}(t_2) = (0.10,\ 0.16,\ 0.10, 0.07, 0.11, 0.15, 0.13, 0.06, 0.12)^{\mathrm{T}}$$

$$\boldsymbol{w}(t_3) = (0.08,\ 0.18,\ 0.11, 0.06, 0.12, 0.16, 0.12, 0.07, 0.10)^{\mathrm{T}}$$

Based on the decision information above, we introduce the following decision making steps:

Step 1 Utilize the LWA$_2$ operator (4.53) to derive the overall atribute value of the city x_j at the period t_k:

$$z_1(t_1) = s_{1.55},\quad z_2(t_1) = s_{1.15},\quad z_3(t_1) = s_{1.32},\quad z_4(t_1) = s_{1.55},\quad z_5(t_1) = s_{1.33}$$

$$z_1(t_2) = s_{1.63},\quad z_2(t_2) = s_{1.24},\quad z_3(t_2) = s_{1.48},\quad z_4(t_2) = s_{1.40},\quad z_5(t_2) = s_{1.58}$$

$$z_1(t_3) = s_{2.27},\quad z_2(t_3) = s_{2.49},\quad z_3(t_2) = s_{2.17},\quad z_4(t_1) = s_{2.37},\quad z_5(t_2) = s_{2.08}$$

Step 2 Use the DLWA operator (4.54) to aggregate the overall attribute values $z_j(t_k)\ (k=1,2,3)$ of three different periods $t_k\ (k=1,2,3)$, and get the complex overall attribute value z_j of the city x_j:

$$z_1 = s_{1.90},\quad z_2 = s_{1.78},\quad z_3 = s_{1.75},\quad z_4 = s_{1.87},\quad z_5 = s_{1.74}$$

Step 3 According to the complex overall attribute values $z_j\ (j=1,2,\cdots,5)$, we rank the cities $x_j\ (j=1,2,\cdots,5)$:

$$x_1 \succ x_4 \succ x_2 \succ x_3 \succ x_5$$

Thus, the best city is x_1.

4.7.2 Approach to Uncertain Multi-Period Multi-Attribute Decision Making Based on ULWA and UDLWA operators

In the following, we extend the decision making approach above to uncertain linguistic environments where the attribute values are expressed in uncertain linguistic variables. We introduce an approach to uncertain multi-period multi-attribute decision making based on ULWA and UDLWA operators:

Step 1 For a multi-period multi-attribute decision making problem. Let $\tilde{\boldsymbol{R}}(t_k) = (\tilde{r}_{ij}(t_k))_{n\times n}$ $(k = 1, 2, \cdots, p)$ be the uncertain linguistic decision matrices collected from p different periods, where $\tilde{r}_{ij}(t_k) = [r_{ij}^{\mathrm{L}}(t_k), r_{ij}^{\mathrm{U}}(t_k)]$ $(r_{ij}^{\mathrm{L}}(t_k), r_{ij}^{\mathrm{U}}(t_k) \in S_2)$ denotes the uncertain attribute value provided by the decision maker for the alternative x_j with respect to the attribute G_i at the period t_k by using the additive linguistic evaluation scale $S_2 = \{s_\alpha | \alpha = -\tau, \cdots, -1, 0, 1, \cdots, \tau\}$. By the ULWA operator:

$$\begin{aligned}\tilde{z}_j(t_k) =& \mathrm{ULWA}\,(\tilde{r}_{1j}(t_k), \tilde{r}_{2j}(t_k), \cdots, \tilde{r}_{mj}(t_k))\\ =&\omega_1(t_k)\tilde{r}_{1j}(t_k) \oplus \omega_2(t_k)\tilde{r}_{2j}(t_k) \oplus \cdots \oplus \omega_m(t_k)\tilde{r}_{mj}(t_k),\\ & k = 1, 2, \cdots, p,\ j = 1, 2, \cdots, n \end{aligned} \tag{4.55}$$

we get the uncertain overall attribute value $\tilde{z}_j(t_k)$ of the alternative x_j at the period t_k.

Step 2 Use the UDLWA operator:

$$\begin{aligned}\tilde{z}_j =& \mathrm{UDLWA}_{\boldsymbol{w}(t)}(\tilde{z}_j(t_1), \tilde{z}_j(t_2), \cdots, \tilde{z}_j(t_p))\\ =& w(t_1)\tilde{z}_j(t_1) \oplus w(t_2)\tilde{z}_j(t_2) \oplus \cdots \oplus w(t_p)\tilde{z}_j(t_p), \quad j = 1, 2, \cdots, n\end{aligned} \tag{4.56}$$

to aggregate the uncertain overall attribute values $\tilde{z}_j(t_k)$ $(k = 1, 2, \cdots, p)$ of the alternatives x_j at p different periods t_k $(k = 1, 2, \cdots, p)$, and get the complex overall atribute value $\tilde{z}_j$ of the alternative x_j.

Step 3 Utilize (2.69) and (2.82) to rank the uncertain overall attribute values $\tilde{z}_j (j = 1, 2, \cdots, n)$, and then rank and select the alternatives x_j $(j = 1, 2, \cdots, n)$.

Example 4.11 In Example 4.10, if the evaluation information provided by the decision maker is uncertain, i.e., the uncertain linguistic decision matrices $\tilde{\boldsymbol{R}}(t_k)$ $(k = 1, 2, 3)$ can be given (Tables 4.22~4.24).

Based on the decision information above, we introduce the following decision making steps:

Step 1 Utilize the ULWA operator (4.55) to derive the uncertain overall attribute value of the alternative x_j at the period t_k:

$$\tilde{z}_1(t_1) = [s_{1.19}, s_{2.45}], \quad \tilde{z}_2(t_1) = [s_{1.07}, s_{2.56}], \quad \tilde{z}_3(t_1) = [s_{0.69}, s_{2.08}]$$

$$\tilde{z}_4(t_1) = [s_{1.14}, s_{2.42}], \quad \tilde{z}_5(t_1) = [s_{0.68}, s_{2.43}], \quad \tilde{z}_1(t_2) = [s_{1.05}, s_{2.58}]$$

$\tilde{z}_2(t_2) = [s_{0.82}, s_{2.17}]$, $\tilde{z}_3(t_2) = [s_{0.26}, s_{1.83}]$, $\tilde{z}_4(t_2) = [s_{0.26}, s_{1.83}]$

$\tilde{z}_5(t_2) = [s_{0.45}, s_{2.03}]$, $\tilde{z}_1(t_3) = [s_{1.77}, s_{3.24}]$, $\tilde{z}_2(t_3) = [s_{1.01}, s_{2.61}]$

$\tilde{z}_3(t_3) = [s_{1.55}, s_{2.82}]$, $\tilde{z}_4(t_3) = [s_{1.56}, s_{2.06}]$, $\tilde{z}_5(t_3) = [s_{1.06}, s_{2.36}]$

Table 4.22 Uncertain linguistic decision matrix $\tilde{R}(t_1)$

	x_1	x_2	x_3	x_4	x_5
G_1	$[s_1, s_2]$	$[s_0, s_3]$	$[s_0, s_1]$	$[s_0, s_1]$	$[s_0, s_1]$
G_2	$[s_2, s_3]$	$[s_{-1}, s_1]$	$[s_0, s_2]$	$[s_2, s_3]$	$[s_0, s_3]$
G_3	$[s_1, s_2]$	$[s_2, s_3]$	$[s_1, s_2]$	$[s_1, s_2]$	$[s_1, s_4]$
G_4	$[s_2, s_3]$	$[s_0, s_1]$	$[s_0, s_1]$	$[s_2, s_3]$	$[s_1, s_2]$
G_5	$[s_0, s_1]$	$[s_2, s_4]$	$[s_1, s_3]$	$[s_0, s_2]$	$[s_1, s_4]$
G_6	$[s_2, s_3]$	$[s_2, s_3]$	$[s_0, s_2]$	$[s_1, s_2]$	$[s_1, s_2]$
G_7	$[s_{-1}, s_2]$	$[s_2, s_3]$	$[s_2, s_3]$	$[s_3, s_4]$	$[s_0, s_1]$
G_8	$[s_2, s_3]$	$[s_3, s_4]$	$[s_2, s_3]$	$[s_1, s_3]$	$[s_0, s_2]$
G_9	$[s_2, s_3]$	$[s_1, s_2]$	$[s_1, s_2]$	$[s_0, s_2]$	$[s_2, s_3]$

Table 4.23 Uncertain linguistic decision matrix $\tilde{R}(t_2)$

	x_1	x_2	x_3	x_4	x_5
G_1	$[s_1, s_3]$	$[s_1, s_2]$	$[s_2, s_3]$	$[s_{-2}, s_1]$	$[s_{-1}, s_0]$
G_2	$[s_0, s_2]$	$[s_{-1}, s_1]$	$[s_{-2}, s_{-1}]$	$[s_1, s_3]$	$[s_1, s_2]$
G_3	$[s_1, s_2]$	$[s_1, s_2]$	$[s_1, s_3]$	$[s_2, s_4]$	$[s_0, s_2]$
G_4	$[s_2, s_3]$	$[s_{-1}, s_0]$	$[s_2, s_4]$	$[s_1, s_2]$	$[s_1, s_3]$
G_5	$[s_0, s_1]$	$[s_1, s_2]$	$[s_1, s_2]$	$[s_2, s_3]$	$[s_2, s_3]$
G_6	$[s_1, s_3]$	$[s_2, s_3]$	$[s_{-1}, s_1]$	$[s_3, s_4]$	$[s_2, s_4]$
G_7	$[s_2, s_3]$	$[s_2, s_4]$	$[s_0, s_2]$	$[s_1, s_3]$	$[s_{-2}, s_1]$
G_8	$[s_1, s_2]$	$[s_1, s_3]$	$[s_1, s_2]$	$[s_{-1}, s_1]$	$[s_1, s_2]$
G_9	$[s_2, s_4]$	$[s_1, s_2]$	$[s_1, s_3]$	$[s_2, s_3]$	$[s_0, s_1]$

Table 4.24 Uncertain linguistic decision matrix $\tilde{R}(t_3)$

	x_1	x_2	x_3	x_4	x_5
G_1	$[s_2, s_3]$	$[s_2, s_4]$	$[s_1, s_2]$	$[s_1, s_2]$	$[s_0, s_1]$
G_2	$[s_2, s_4]$	$[s_{-1}, s_1]$	$[s_2, s_3]$	$[s_2, s_4]$	$[s_1, s_2]$
G_3	$[s_1, s_2]$	$[s_0, s_3]$	$[s_1, s_3]$	$[s_2, s_3]$	$[s_2, s_4]$
G_4	$[s_3, s_4]$	$[s_1, s_4]$	$[s_0, s_1]$	$[s_2, s_4]$	$[s_2, s_3]$
G_5	$[s_0, s_2]$	$[s_3, s_4]$	$[s_2, s_3]$	$[s_1, s_2]$	$[s_3, s_4]$
G_6	$[s_3, s_4]$	$[s_1, s_2]$	$[s_1, s_3]$	$[s_1, s_3]$	$[s_2, s_3]$
G_7	$[s_2, s_3]$	$[s_2, s_3]$	$[s_3, s_4]$	$[s_3, s_4]$	$[s_{-2}, s_0]$
G_8	$[s_2, s_4]$	$[s_3, s_4]$	$[s_2, s_3]$	$[s_2, s_3]$	$[s_0, s_2]$
G_9	$[s_1, s_3]$	$[s_0, s_1]$	$[s_1, s_2]$	$[s_0, s_2]$	$[s_1, s_2]$

Step 2 Utilize the UDLWA operator (4.54) to aggregate the uncertain overall attribute values $\tilde{z}_j(t_k)$ $(k = 1, 2, 3)$ of the alternatives x_j $(j = 1, 2, \cdots, 5)$ at three

different periods t_k $(k = 1, 2, 3)$, and get the complex uncertain overall attribute value $\tilde{z}_j$ of the alternative x_j:

$$\tilde{z}_1 = [s_{1.41}, s_{2.85}], \quad \tilde{z}_2 = [s_{0.96}, s_{2.46}], \quad \tilde{z}_3 = [s_{0.94}, s_{2.33}]$$

$$\tilde{z}_4 = [s_{1.35}, s_{2.40}], \quad \tilde{z}_5 = [s_{0.78}, s_{2.27}]$$

Step 3 Compare each pair of $\tilde{z}_j$ $(j = 1, 2, \cdots, 5)$ by using (2.69), and construct a possibility degree matrix:

$$\boldsymbol{P} = \begin{pmatrix} 0.5 & 0.6429 & 0.6749 & 0.6024 & 0.7065 \\ 0.3571 & 0.5 & 0.5260 & 0.4353 & 0.5619 \\ 0.3251 & 0.4740 & 0.5 & 0.4016 & 0.5382 \\ 0.3976 & 0.5984 & 0.5984 & 0.5 & 0.6378 \\ 0.2935 & 0.4381 & 0.4618 & 0.3622 & 0.5 \end{pmatrix}$$

Step 4 Derive the priority vector of $\boldsymbol{P}$ by using (2.82):

$$\zeta = (0.2313, 0.1940, 0.1869, 0.2099, 0.1778)^{\mathrm{T}}$$

from which we rank the complex uncertain overall attribute values $\tilde{z}_j$ $(j = 1, 2, \cdots, 5)$:

$$\tilde{z}_1 > \tilde{z}_4 > \tilde{z}_2 > \tilde{z}_3 > \tilde{z}_5$$

and then, the best city is x_1.

4.7.3 Approach to Multi-Period Multi-Attribute Decision Making Based on LWG and DLWG Operators

For the multi-period multi-attribute decision making problem considered in Subsection 4.7.1, if the decision maker uses the multiplicative linguistic evaluation scale $S_5 = \left\{ s_\alpha \middle| \alpha = \frac{1}{\tau}, \frac{2}{\tau}, \cdots, \frac{\tau-1}{\tau}, 1, \frac{\tau}{\tau-1}, \cdots, \frac{\tau}{2}, \tau \right\}$ to evaluate the alternatives x_j $(j = 1, 2, \cdots, n)$ with respect to the attributes G_i $(i = 1, 2, \cdots, m)$ at p different periods t_k $(k = 1, 2, \cdots, p)$, and provides the linguistic decision matrices $\boldsymbol{R}(t_k) = (r_{ij}(t_k))_{m \times n}$ $(k = 1, 2, \cdots, p)$, then we introduce an approach based on LWG and DLWG operators to multi-period multi-attribute decision making as follows:

Step 1 Utilize the LWG operator:

$$\begin{aligned} z_j(t_k) =& \mathrm{LWG}(r_{1j}(t_k), r_{2j}(t_k), \cdots, r_{mj}(t_k)) \\ =& (r_{1j}(t_k))^{\omega_1(t_k)} \otimes (r_{2j}(t_k))^{\omega_2(t_k)} \otimes \cdots \otimes (r_{mj}(t_k))^{\omega_m(t_k)}, \\ & k = 1, 2, \cdots, p, \ j = 1, 2, \cdots, n \end{aligned} \quad (4.57)$$

to derive the overall attribute value $z_j(t_k)$ of the alternative x_j at the period t_k.

Step 2 Aggregate the overall attribute values $z_j(k)$ $(k = 1, 2, \cdots, p)$ of the alternative x_j at p different periods by using the DLWG operator:

$$z_j = \text{DLWG}_{\boldsymbol{w}(t)}(z_j(t_1), z_j(t_2), \cdots, z_j(t_p))$$
$$= (z_j(t_1))^{w(t_1)} \otimes (z_j(t_2))^{w(t_2)} \otimes \cdots \otimes (z_j(t_p))^{w(t_p)}, \quad j = 1, 2, \cdots, n \quad (4.58)$$

and get the complex overall attribute value z_j of the alternative x_j.

Step 3 Rank and select the alternatives x_j $(j = 1, 2, \cdots, n)$ according to z_j $(j = 1, 2, \cdots, n)$.

Example 4.12 Consider a decision making problem that a risk investment company wants to invest a sum of money in the best option of high-tech projects, and there are four enterprises (alternatives) x_j $(j = 1, 2, 3, 4)$ to be chosen from. The risk investment company tries to evaluate the candidate enterprises from the viewpoint of enterprise capabilities, and predefines seven evaluation indices (attributes) (Song and Chen, 1999): ① G_1——marketability; ② G_2——management ability; ③ G_3——production capacity; ④ G_4——technological competence; ⑤ G_5——financial capability; ⑥ G_6——risk-taking ability; and ⑦ G_7——strategic consistency. The decision maker uses the multiplicative linguistic evaluation scale:

$$S_4 = \{s_{1/5} = \text{extremely poor}, \ s_{2/5} = \text{very poor}, \ s_{3/5} = \text{poor}, \ s_{4/5} = \text{slightly poor},$$
$$s_1 = \text{fair}, \ s_{5/4} = \text{slightly good}, \ s_{5/3} = \text{good},$$
$$s_{5/2} = \text{very good}, \ s_5 = \text{extremely good}\}$$

to evaluate the performances of the enterprises x_j $(j = 1, 2, 3, 4)$ during the periods 2001~2003 by means of the evaluation indices G_i $(i = 1, 2, \cdots, 7)$, and constructs the linguistic decision matrices $\boldsymbol{R}(t_k) = (r(t_k))_{7\times 4}$ $(k = 1, 2, 3)$ respectively (Tables 4.25~4.27). Here, t_1 denotes "2001", t_2 denotes "2002", and t_3 denotes "2003". Let $\boldsymbol{w}(t) = (1/6, 2/6, 3/6)^{\mathrm{T}}$ be the weight vector of the periods t_k $(k = 1, 2, 3)$, and

$$\boldsymbol{\omega}(t_1) = (0.15, \ 0.10, \ 0.25, \ 0.10, \ 0.15, \ 0.15, \ 0.10)^{\mathrm{T}}$$
$$\boldsymbol{\omega}(t_2) = (0.20, \ 0.10, \ 0.25, \ 0.10, \ 0.15, \ 0.10, \ 0.10)^{\mathrm{T}}$$
$$\boldsymbol{\omega}(t_3) = (0.25, \ 0.05, \ 0.25, \ 0.15, \ 0.15, \ 0.10, 0.05)^{\mathrm{T}}$$

are the weight vector of the atributes G_i $(i = 1, 2, \cdots, 7)$ at the periods t_k $(k = 1, 2, 3)$.

Table 4.25 Linguistic decision matrix $R(t_1)$

	x_1	x_2	x_3	x_4
G_1	$s_{5/4}$	$s_{5/2}$	$s_{5/3}$	$s_{5/3}$
G_2	$s_{5/2}$	$s_{5/3}$	$s_{5/3}$	$s_{5/3}$
G_3	$s_{5/2}$	s_1	$s_{5/2}$	$s_{4/5}$
G_4	s_1	$s_{5/3}$	$s_{5/4}$	$s_{5/4}$
G_5	$s_{5/4}$	$s_{5/2}$	s_5	$s_{5/2}$
G_6	$s_{5/3}$	$s_{5/4}$	$s_{5/2}$	$s_{5/4}$
G_7	$s_{5/3}$	$s_{4/5}$	$s_{5/3}$	$s_{5/4}$

Table 4.26 Linguistic decision matrix $R(t_2)$

	x_1	x_2	x_3	x_4
G_1	$s_{5/3}$	s_5	$s_{5/2}$	$s_{5/2}$
G_2	$s_{5/3}$	$s_{5/4}$	$s_{5/2}$	$s_{5/2}$
G_3	s_5	s_1	$s_{5/3}$	$s_{3/5}$
G_4	$s_{3/5}$	$s_{5/4}$	$s_{5/3}$	$s_{5/2}$
G_5	s_1	$s_{5/3}$	$s_{5/2}$	$s_{5/3}$
G_6	$s_{5/3}$	$s_{5/3}$	$s_{5/3}$	$s_{5/3}$
G_7	$s_{5/2}$	s_1	$s_{5/2}$	$s_{5/4}$

Table 4.27 Linguistic decision matrix $R(t_3)$

	x_1	x_2	x_3	x_4
G_1	$s_{5/2}$	$s_{5/3}$	$s_{5/4}$	$s_{5/2}$
G_2	$s_{5/2}$	$s_{5/2}$	$s_{5/2}$	$s_{5/3}$
G_3	$s_{5/3}$	$s_{4/5}$	$s_{5/4}$	s_1
G_4	$s_{4/5}$	$s_{5/2}$	$s_{5/3}$	$s_{5/3}$
G_5	$s_{5/3}$	$s_{5/3}$	$s_{5/3}$	$s_{5/2}$
G_6	$s_{5/2}$	$s_{5/3}$	s_5	$s_{5/3}$
G_7	$s_{5/3}$	s_1	$s_{5/3}$	$s_{5/2}$

Based on the given decision information, the following decision making steps are given:

Step 1 Derive the overall attribute value of the enterprise x_j at the period t_k by using the LWG operator (4.56):

$$z_1(t_1) = s_{1.67}, \quad z_2(t_1) = s_{1.47}, \quad z_3(t_1) = s_{2.25}, \quad z_4(t_1) = s_{1.33}$$

$$z_1(t_2) = s_{1.91}, \quad z_2(t_2) = s_{1.64}, \quad z_3(t_2) = s_{2.08}, \quad z_4(t_2) = s_{1.48}$$

$$z_1(t_3) = s_{1.76}, \quad z_2(t_3) = s_{1.47}, \quad z_3(t_2) = s_{1.64}, \quad z_4(t_1) = s_{1.76}$$

Step 2 Aggregate the overall attribute values $z_j(t_k)$ $(k = 1, 2, 3)$ of the enterprise x_j at three different periods t_k $(k = 1, 2, 3)$ by using the LWG operator (4.57), and get the complex overall attribute value z_j of the enterprise x_j:

$$z_1 = s_{1.79}, \quad z_2 = s_{1.52}, \quad z_3 = s_{1.87}, \quad z_4 = s_{1.58}$$

Step 3 Rank the alternatives z_j $(j = 1, 2, 3, 4)$ according to z_j $(j = 1, 2, 3, 4)$:

$$x_3 \succ x_1 \succ x_4 \succ x_2$$

Thus, the best enterprise is x_3.

4.7.4 Approach to Uncertain Multi-Period Multi-Attribute Decision Making Based on ULWG and UDLWG operators

In the following, we extend the decision making approach above to uncertain linguistic environments. We introduce an approach to uncertain multi-period multi-attribute decision making based on ULWA and UDLWA operators:

Step 1 For a multi-period multi-attribute decision making problem. Let $\tilde{\boldsymbol{R}}(t_k) = (\tilde{r}_{ij}(t_k))_{n\times n}$ $(k = 1, 2, \cdots, p)$ be the uncertain linguistic decision matrices at p different periods, where $\tilde{r}_{ij}(t_k) = [r_{ij}^{\mathrm{L}}(t_k), r_{ij}^{\mathrm{U}}(t_k)]$ $(r_{ij}^{\mathrm{L}}(t_k), r_{ij}^{\mathrm{U}}(t_k) \in S_4)$ denotes the uncertain attribute value provided by the decision maker over the alternative x_j with respect to the attribute G_i at the period t_k by using the multiplicative linguistic evaluation scale $S_5 = \left\{ s_\alpha | \alpha = \frac{1}{\tau}, \frac{2}{\tau}, \cdots, \frac{\tau-1}{\tau}, 1, \frac{\tau}{\tau-1}, \cdots, \frac{\tau}{2}, \tau \right\}$. Then we utilize the ULWG operator:

$$\begin{aligned}\tilde{z}_j(t_k) =& \mathrm{ULWG}(\tilde{r}_{1j}(t_k), \tilde{r}_{2j}(t_k), \cdots, \tilde{r}_{mj}(t_k)) \\ =& (\tilde{r}_{1j}(t_k))^{\omega_1(t_k)} \otimes (\tilde{r}_{2j}(t_k))^{\omega_2(t_k)} \otimes \cdots \otimes (\tilde{r}_{mj}(t_k))^{\omega_m(t_k)}, \\ & k = 1, 2, \cdots, p, \ j = 1, 2, \cdots, n \end{aligned} \tag{4.59}$$

to derive the uncertain overall attribute value $\tilde{z}_j(t_k)$ of the alternative x_j at the period t_k.

Step 2 Utilize the UDLWG operator:

$$\begin{aligned}\tilde{z}_j =& \mathrm{UDLWG}_{\boldsymbol{w}(t)}(\tilde{z}_j(t_1), \tilde{z}_j(t_2), \cdots, \tilde{z}_j(t_p)) \\ =& (\tilde{z}_j(t_1))^{w(t_1)} \otimes (\tilde{z}_j(t_2))^{w(t_2)} \otimes \cdots \otimes (\tilde{z}_j(t_p))^{w(t_p)}, \quad j = 1, 2, \cdots, n \end{aligned} \tag{4.60}$$

to aggregate the uncertain overall attribute values $\tilde{z}_j(t_k)$ $(k = 1, 2, \cdots, p)$ of the alternative x_j at p different periods t_k $(k = 1, 2, \cdots, p)$, and get the complex overall attribute value $\tilde{z}_j$ of the alternative x_j.

Step 3 Rank the complex uncertain overall attribute values $\tilde{z}_j$ $(j = 1, 2, \cdots, n)$ by using (2.69) and (2.82), and then rank and select the alternatives x_j $(j=1, 2, \cdots, n)$ according to $\tilde{z}_j$ $(j = 1, 2, \cdots, n)$.

Example 4.13 In Example 4.12, if the decision maker provides the uncertain linguistic decision matrices $\tilde{\boldsymbol{R}}(t_k)$ $(k = 1, 2, 3)$ (Tables 4.28~4.30) respectively.

Table 4.28 Uncertain linguistic decision matrix $\tilde{R}(t_1)$

	x_1	x_2	x_3	x_4
G_1	$[s_{5/4}, s_{5/3}]$	$[s_{5/3}, s_{5/2}]$	$[s_{5/3}, s_{5/2}]$	$[s_{5/3}, s_{5/2}]$
G_2	$[s_{5/2}, s_5]$	$[s_{5/4}, s_{5/3}]$	$[s_{5/4}, s_{5/2}]$	$[s_{5/4}, s_{5/3}]$
G_3	$[s_{5/3}, s_{5/2}]$	$[s_1, s_{5/4}]$	$[s_{5/2}, s_5]$	$[s_{4/5}, s_1]$
G_4	$[s_1, s_{5/4}]$	$[s_{5/4}, s_{5/2}]$	$[s_{5/4}, s_{5/3}]$	$[s_1, s_{5/4}]$
G_5	$[s_{5/4}, s_{5/3}]$	$[s_{5/4}, s_{5/3}]$	$[s_1, s_{5/4}]$	$[s_{5/3}, s_{5/2}]$
G_6	$[s_{5/3}, s_{5/2}]$	$[s_{5/4}, s_{5/3}]$	$[s_{5/3}, s_{5/2}]$	$[s_1, s_{5/4}]$
G_7	$[s_{5/4}, s_{5/3}]$	$[s_{4/5}, s_{5/4}]$	$[s_{5/3}, s_{5/2}]$	$[s_{5/4}, s_{5/3}]$

Table 4.29 Uncertain linguistic decision matrix $\tilde{R}(t_2)$

	x_1	x_2	x_3	x_4
G_1	$[s_{5/3}, s_{5/2}]$	$[s_5, s_5]$	$[s_{5/4}, s_{5/3}]$	$[s_{5/3}, s_{5/2}]$
G_2	$[s_{5/4}, s_{5/3}]$	$[s_{5/4}, s_{5/3}]$	$[s_{5/2}, s_5]$	$[s_{5/4}, s_{5/3}]$
G_3	$[s_{5/2}, s_5]$	$[s_{4/5}, s_{5/4}]$	$[s_{5/3}, s_{5/2}]$	$[s_{3/5}, s_{5/3}]$
G_4	$[s_{3/5}, s_{5/4}]$	$[s_1, s_{5/3}]$	$[s_{5/3}, s_{5/2}]$	$[s_{5/2}, s_5]$
G_5	$[s_1, s_{5/4}]$	$[s_{5/3}, s_{5/2}]$	$[s_{5/3}, s_{5/3}]$	$[s_1, s_{5/4}]$
G_6	$[s_{5/3}, s_{5/2}]$	$[s_{5/3}, s_{5/2}]$	$[s_{5/4}, s_{5/2}]$	$[s_{5/4}, s_{5/3}]$
G_7	$[s_{5/3}, s_{5/3}]$	$[s_{3/5}, s_{5/4}]$	$[s_{5/3}, s_{5/2}]$	$[s_1, s_{5/3}]$

Table 4.30 Uncertain linguistic decision matrix $\tilde{R}(t_3)$

	x_1	x_2	x_3	x_4
G_1	$[s_{5/2}, s_5]$	$[s_{5/3}, s_{5/2}]$	$[s_{5/4}, s_{5/3}]$	$[s_{5/4}, s_{5/3}]$
G_2	$[s_{5/3}, s_{5/2}]$	$[s_{5/4}, s_{5/3}]$	$[s_1, s_{5/4}]$	$[s_1, s_{5/3}]$
G_3	$[s_{5/3}, s_{5/2}]$	$[s_{4/5}, s_{5/4}]$	$[s_{5/4}, s_{5/3}]$	$[s_1, s_{5/3}]$
G_4	$[s_{4/5}, s_{5/3}]$	$[s_1, s_{5/4}]$	$[s_{4/5}, s_{5/4}]$	$[s_{5/3}, s_{5/2}]$
G_5	$[s_{5/3}, s_{5/2}]$	$[s_{5/4}, s_{5/4}]$	$[s_{5/3}, s_{5/3}]$	$[s_{5/2}, s_5]$
G_6	$[s_{5/3}, s_{5/3}]$	$[s_{5/3}, s_{5/2}]$	$[s_{5/4}, s_{5/3}]$	$[s_{5/4}, s_{5/3}]$
G_7	$[s_{5/3}, s_{5/2}]$	$[s_1, s_{5/4}]$	$[s_{5/4}, s_{5/4}]$	$[s_{5/3}, s_{5/2}]$

Then we introduce the following steps for determining the best enterprise:

Step 1 Utilize the ULWG operator (4.59) to derive the uncertain overall attribute value of the alternative x_j at the period t_k:

$$\tilde{z}_1(t_1) = [s_{1.47}, s_{2.13}], \quad \tilde{z}_2(t_1) = [s_{1.18}, s_{1.67}], \quad \tilde{z}_3(t_1) = [s_{1.61}, s_{1.72}]$$

$$\tilde{z}_4(t_1) = [s_{1.15}, s_{1.54}], \quad \tilde{z}_1(t_2) = [s_{1.50}, s_{2.31}], \quad \tilde{z}_2(t_2) = [s_{1.44}, s_{2.08}]$$

$$\tilde{z}_4(t_1) = [s_{1.15}, s_{1.54}], \quad \tilde{z}_1(t_2) = [s_{1.50}, s_{2.31}], \quad \tilde{z}_2(t_2) = [s_{1.44}, s_{2.08}]$$

$$\tilde{z}_3(t_2) = [s_{1.59}, s_{2.32}], \quad \tilde{z}_4(t_2) = [s_{1.12}, s_{1.93}], \quad \tilde{z}_1(t_3) = [s_{1.65}, s_{2.69}]$$

$$\tilde{z}_2(t_3) = [s_{1.18}, s_{1.62}], \quad \tilde{z}_3(t_3) = [s_{1.21}, s_{1.55}], \quad \tilde{z}_4(t_3) = [s_{1.37}, s_{2.13}]$$

Step 2 Aggregate the uncertain overall attribute values $\tilde{z}_j(t_k)$ $(k = 1, 2, 3)$ of the alternative x_j at three periods t_k $(k = 1, 2, 3)$ by using the UDLWG operator (4.60), and get the complex uncertain overall attribute value $\tilde{z}_j$ of the alternative x_j:

$$\tilde{z}_1 = [s_{1.57}, s_{2.46}], \quad \tilde{z}_2 = [s_{1.26}, s_{1.77}], \quad \tilde{z}_3 = [s_{1.39}, s_{1.80}], \quad \tilde{z}_4 = [s_{1.24}, s_{1.95}]$$

Step 3 Compare each pair of $\tilde{z}_j$ $(j = 1, 2, \cdots, 5)$ by using (2.69), and construct a possibility degree matrix:

$$\boldsymbol{P} = \begin{pmatrix} 0.5 & 0.8571 & 0.8231 & 0.7625 \\ 0.1429 & 0.5 & 0.4130 & 0.4344 \\ 0.1769 & 0.5870 & 0.5 & 0.5 \\ 0.2375 & 0.5656 & 0.5 & 0.5 \end{pmatrix}$$

Step 4 Utilize (2.82) to obtain the priority vector of $\boldsymbol{P}$:

$$\zeta = (0.3286, 0.2075, 0.2303, 0.2336)^{\mathrm{T}}$$

from which we rank the complex uncertain overall attribute values $\tilde{z}_j$ $(j = 1, 2, \cdots, 5)$:

$$\tilde{z}_1 > \tilde{z}_4 > \tilde{z}_3 > \tilde{z}_2$$

and then the best enterprise is x_3.

Clearly, the change of the decision information results in the difference between the rankings of the alternatives in Examples 4.12 and 4.13.

References

Bryson N, Mobolurin A. 1995. An action learning evaluation procedure for multiple criteria decision making problems. European Journal of Operational Research, 96: 379-386.

Chen S J, Hwang C L. 1992. Fuzzy Multiple Attribute Decision Making: Methods and Applications. Berlin: Springer-Verlag.

Fodor J C, Roubens M. 1994. Fuzzy Preference Modeling and Multicriteria Decision Support. Dordrecht: Kluwer.

French S, Hartley R, Thomas L C, et al. 1983. Multi-objective Decision Making. New York: Academic Press.

Gao F J. 2000. Multiple attribute decision making on plans with alternative preference under incomplete information. Systems Engineering-Theory & Practice, 20(4): 94-97.

Herrera F, Herrera-Viedma E, Martínez L. 2000. A fusion approach for managing multi-granularity linguistic term sets in decision making. Fuzzy Sets and Systems, 114: 43-58.

Herrera F, Martínez L. 2000. An approach for combining linguistic and numerical information based on the 2-tuple fuzzy linguistic representation model in decision-making. International Journal of Uncertainty, Fuzziness and Knowledge-Based Systems, 8: 539-562.

Hwang C L, Yoon K. 1981. Multiple Attribute Decision Making. Berlin: Springer-Verlag.

Keeney R L, Raiffa H. 1976. Decisions with Multiple Objectives: Preferences and Value Tradeoffs. New York: Wiley.

Kim S H, Ahn B S. 1999. Interactive group decision making procedure under incomplete information. European Journal of Operational Research, 116: 498-507.

Kim S H, Choi S H, Kim J K. 1999. An interactive procedure for multiple attribute group decision making with incomplete information: range-based approach. European Journal of Operational Research, 118: 139-152.

Li S C, Chen J D, Zhao H G. 2001. Studying on the method of appraising qualitative decision indication system. Systems Engineering-Theory & Practice, 21(9): 22-32.

Olcer A I, Odabasi A Y. 2005. A new fuzzy multiple attribute group decision making methodology and its application to propulsion/manoeuvring systems selection problem. European Journal of Operational Research, 166: 93-114.

Park K S, Kim S H. 1997. Tools for interactive multi-attribute decision making with incompletely identified information. European Journal of Operational Research, 98: 111-123.

Saaty T L. 1980. The Analytic Hierarchy Process. New York: McGraw-Hill.

Song F M, Chen T T. 1999. Research on index system of appraisal of high-tech investment projects. China Soft Science Magazine, 1: 90-93.

Wang Y M, Fu G W. 1993. Multi-criteria decision making with limited alternatives by using the method for multi-objective decision making with unlimited alternatives. Control and Decision, 8(1): 25-29.

Xu J P, Wu W. 2006. Multiple Attribute Decision Making Theory and Methods. Beijing: Tsinghua University Press.

Xu X Z. 2004. A note on the subjective and objective integrated approach to determine attribute weights. European Journal of Operational Research, 156: 530-532.

Xu Z S, Chen J. 2007. An interactive method for fuzzy multiple attribute group decision making. Information Sciences, 177: 248-263.

Xu Z S. 2004. Uncertain Multiple Attribute Decision Making: Methods and Applications. Beijing: Tsinghua University Press.

Xu Z S. 2005. Maximizing deviations procedure for multiple attribute decision making under linguistic environment. Lecture Series on Computer and Computational Sciences, 2: 150-154.

Xu Z S. 2006a. Goal programming models for multiple attribute decision making under linguistic setting. Journal of Management Sciences in China, 9: 9-17.

Xu Z S. 2006b. Multiple attribute decision making based on different types of linguistic information. Journal of Southeast University, 22(1): 134-136.

Xu Z S. 2007a. A method for multiple attribute decision making with incomplete weight information in linguistic setting. Knowledge-Based Systems, 20: 719-725.

Xu Z S. 2007b. An interactive procedure for linguistic multiple attribute decision making with incomplete weight information. Fuzzy Optimization and Decision Making, 6: 7-27.

Yager R R, Kacprzyk J. 1997. The Ordered Weighted Averaging Operators: Theory and Applications. Boston: Kluwer.

Zeeny M. 1982. Multiple Criteria Decision Making. New York: McGraw-Hill.

Zhang H M, Xu Z S. 2005. Uncertain linguistic information based C-OWA and C-OWG operators and their applications. Journal of PLA University of Science and Technology, 6(6): 604-608.

Zimmermann H J, Zysco P. 1980. Latent connectives in human decision making. Fuzzy Sets and Systems, 4: 37-51.

Index

F

G

H

I

L

M

W